Additional titles by Bernard J. Rice and Jerry D. Strange:

Algebra and Trigonometry, second edition, 1980

College Algebra, second edition, 1981

Plane Trigonometry

third edition

Plane Trigonometry

third edition

Bernard J. Rice
University of Dayton

Jerry D. Strange
University of Dayton

Prindle, Weber & Schmidt
Boston, Massachusetts

PWS PUBLISHERS

Prindle. Weber & Schmidt · ✱ · Willard Grant Press · **WG** · Duxbury Press · ♠
Statler Office Building · 20 Providence Street · Boston. Massachusetts 02116

PWS Publishers is a division of Wadsworth, Inc.

Library of Congress Cataloging in Publication Data
Rice, Bernard J
    Plane trigonometry.

    Includes index.
    1. Trigonometry, Plane.   I.   Strange, Jerry, D.,
joint author.   II.   Title.
QA533.R5   1981   516.2′4   80-26108
ISBN 0-87150-297-6

ISBN 0-87150-297-6

Cover image "Tangent" by Mark Rowland. © 1977 by Circle Fine Art Corporation. Used by permission of the publisher. The artist's work is based on uncompromisingly precise mathematical systems, with the compass and protractor as implements.

The portion on Polar Coordinates in this text has been reprinted by permission of John Wiley & Sons, Inc. and is taken from pages 359–365 and page 368 of *Analytic Geometry and Calculus with Technical Applications* by Jerry D. Strange and Bernard J. Rice, copyright 1970 by John Wiley & Sons, Inc. Publishers.

The information in Table B "Values of Trigonometric Functions — Decimal Subdivisions" is reprinted by permission of Wadsworth Publishing Company and is taken from pages 257–268 of *Analytic Trigonometry with Applications,* second edition by Raymond A. Barnett, copyright 1980 by Wadsworth, Inc.

Cover and text design by Deborah Schneider and coordinated by Susan London and the staff of Prindle, Weber & Schmidt. Chapter opening art by Julie Gecha. Composed in Helvetica and Times Roman by H. Charlesworth & Co. Ltd. Covers printed by John P. Pow Company. Text printed and bound by Rand McNally & Company.

Printed in the United States of America.
10 9 8 7 6 5 4   — 86 85 84 83

Preface

The third edition of *Plane Trigonometry* retains the essential features that made the first two editions successful. They are (1) an elementary introduction to trigonometric functions using a right triangle approach, (2) an abundance of examples and exercises and (3) a large number and variety of applications.

The material in Chapter 1, which reviews basic concepts, has been reordered to improve the flow of topics and to place more emphasis on the topic of radian measure. A section on significant digits and rounding off has been added to Chapter 1 to give the student a consistent approach to making computations with approximate numbers.

In this edition we have written the introductory definitions for the trigonometry of acute angles in terms of a right triangle imposed on the rectangular coordinate plane. In this manner the trigonometric functions of general angles are a natural extension of the definitions for acute angles. We have also completely reorganized and rewritten the material on inverse functions. It is still our belief that trigonometry is best learned, best understood, and best remembered with respect to the right triangle definitions extended to angles in standard position in the rectangular plane.

Furthermore, with this edition we have emphasized the applied nature of trigonometry by incorporating more examples and exercises showing the connection between application and theory. Besides applications to aerospace technology, navigation and the predator/prey problem, new applications to biorhythm and seasonal temperature variation have been added. Review exercise sets have been added to the end of each chapter to give the student additional practice.

With regard to the calculator and its uses, special attention has been given in this edition. Comments and expository remarks on how the calculator is used in trigonometry are included throughout the text and clearly marked with a 🖩 . At the same time, tables are included so the book can be used by those who wish to avoid calculators.

Several different types of course can be presented using this text. For a triangle-oriented course we suggest the following:

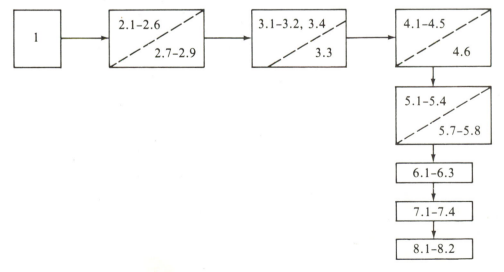

For a course emphasizing analytical trigonometry, the following would be appropriate:

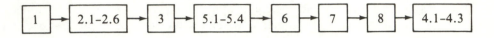

In either of these two courses optional topics such as logarithms, complex numbers and polar coordinates can be added as needed. The applications are more than plentiful and some of them must of course be omitted in any one particular course.

Once again we wish to thank Mr. Myrl H. Ahrendt and the National Aeronautics and Space Administration for their kind permission to use some of the unique applications of trigonometry to space technology from the book *Space Mathematics: A Resource for Teachers.*

This edition of *Plane Trigonometry* has benefited from critical reviews and comments from the users of the first two editions. We wish to thank the following people who assisted us with their valuable comments:

R.F. Boudria (Stephen F. Austin University)
Albert J. Froderberg (Western Washington University)
Keith S. Joseph (Metropolitan State College)
Steven Kass (University of Wisconsin-Madison)
Glenn Sands (Ferris State College)
Francesco G. Scorsone (Eastern Kentucky University)
Howard L. Wilson (Oregon State University)
Albert W. Zechmann (University of Nebraska-Lincoln)

We would also like to thank James J. Ball (Indiana State University), Leo G. Chouinard (University of Nebraska-Lincoln) and George King (Embry-Riddle Areo University) for their assistance.

Finally, it is again a pleasure to acknowledge the fine cooperation of the staff of Prindle, Weber & Schmidt, particularly our editor, Theron Shreve, our production editors, Susan London and Debbie Schneider, and the special assistance of Mary LeQuesne.

February 5, 1981   B.J.R   J.D.S.

Contents

The Solution of Oblique Triangles, *104*

Analytic Trigonometry, *130*

Identities, Equations and Inequalities, *169*

7

8

9

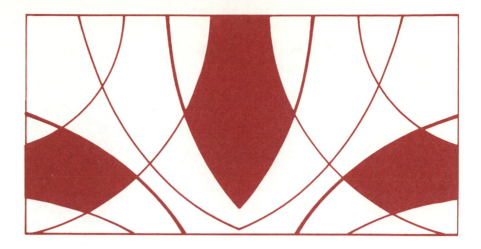

Plane Trigonometry

third edition

1

Some Fundamental Concepts

Historical Background

Trigonometry is one of the oldest branches of mathematics. An ancient scroll called the Ahmes Papyrus, written about 1550 B.C., contains problems that are solved by using similar triangles, the heart of the trigonometric idea. There is historical verification that measurements of distance and height were made by the Chinese about 1100 B.C. using what is essentially right triangle trigonometry. The subject eventually became intertwined with the study of astronomy. In fact, it is the Greek astronomer Hipparchus (180–125 B.C.) who is credited with compiling the first trigonometric tables and thus has earned the right to be known as "the father of trigonometry." The trigonometry of Hipparchus and the other astronomers was strictly a tool of measurement, and it is, therefore, difficult to refer to the early uses of the subject as either mathematics or astronomy.

In the 15th century, trigonometry was developed as a discipline within mathematics by Johann Muller (1436–1476). This development created an interest in trigonometry throughout Europe and had the effect of placing Europe in a position of prominence with respect to astronomy and trigonometry.

In the 18th century, trigonometry was systematically developed in a completely different direction, highlighted by the publication in 1748 of the now famous "Introduction to Infinite Analysis" by Leonhard Euler (1707–1783). From this new viewpoint, trigonometry did not necessarily have to be considered in relation to a right triangle. Rather, the analytic or functional properties became paramount. As this wider outlook of the subject evolved, many new applications arose, especially as a tool for describing physical phenomena that are "periodic."

In this book we proceed more or less as the subject developed historically.

1

First, we consider the trigonometry of a right triangle and only later introduce the analytic generalization that is so valuable in other areas of mathematics and physics.

To read the book profitably, you should have some ability with elementary algebra, particularly manipulative skills. Some of the specific background knowledge you will need is presented in this chapter.

1.1   The Rectangular Coordinate System

We often wish to make an association between points on a line (or in a plane) and numbers, a process called coordinatization. The number (or numbers) assigned to a point is called the **coordinate** (or coordinates) of the point.

To associate points with real numbers, choose any straight line, and choose any point on the line to be the starting point or origin. Take any unit distance and measure to the right of the origin. Then the number 0 is associated with the origin, the number 1 with the point a unit distance to the right of the origin, the number 2 with the point two units to the right of the origin, etc. In this way, the so-called **integral points** are determined. Similarly, points in between are coordinated by other real numbers. The line, illustrated in Figure 1.1, is then called a (real) number line.

FIGURE 1.1
Real number line

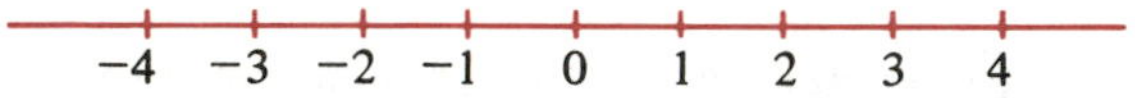

$$-4 \quad -3 \quad -2 \quad -1 \quad 0 \quad 1 \quad 2 \quad 3 \quad 4$$

For purposes of elementary trigonometry, the most important type of coordinatization is the association of each point in the plane with a pair of numbers. In this case, we choose two mutually perpendicular intersecting lines as shown in Figure 1.2. Normally, the horizontal line is called the **x-axis**, the vertical line is called the **y-axis**, and their intersection is called the **origin**. When considered together, the two axes form

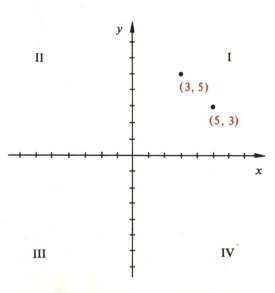

FIGURE 1.2
Cartesian coordinate system

a rectangular coordinate system.* As you can see, the coordinate axes divide the plane into four zones or **quadrants**. The upper right quadrant is called *the first quadrant*, and the others are numbered consecutively from this one in a counterclockwise direction as in Figure 1.2. The coordinate axes are not considered to be in any quadrant.

To locate points in the plane, use the origin as a reference point and lay off a suitable scale on each of the coordinate axes. The displacement of a point in the plane to the right or left of the y-axis is called the **x-coordinate**, or **abscissa**, of the point, and is denoted by x. Values of x measured to the right of the y-axis are considered to be positive and to the left, negative. The displacement of a point in the plane above or below the x-axis is called the **y-coordinate**, or **ordinate**, of the point, and is denoted by y. Values of y above the x-axis are considered to be positive and below the x-axis, negative. Considered together, the abscissa and the ordinate of a point are called the **coordinates** of the point. The coordinates of a point are conventionally written in parentheses, with the abscissa written first and separated from the ordinate by a comma, that is, (x, y).

We see that a point (x, y) lies

- in quadrant I if both coordinates are positive,
- in quadrant II if the x-coordinate is negative and the y-coordinate is positive,
- in quadrant III if both coordinates are negative,
- in quadrant IV if the x-coordinate is positive and the y-coordinate is negative.

Since the first number represents the horizontal displacement and the second the vertical displacement, we see the significance of order. For example, the ordered pair (3, 5) represents a point that is displaced 3 units to the right of the origin and 5 units above it, while the ordered pair (5, 3) represents a point that is 5 units to the right and 3 units up. The association of points in the plane with ordered pairs of real numbers is an obvious extension of a similar idea on the real line.

To be precise, we should always distinguish between the point and the ordered pair; however, it is common practice to blur the distinction and say "the point (x, y)" instead of "the point whose coordinate is (x, y)."

Each point in the plane can be described by a unique ordered pair of numbers (x, y) and each ordered pair of numbers (x, y) can be represented by a unique point in the plane called the **graph** of the ordered pair.

EXAMPLE 1   Locate the points $P(-1, 2)$, $Q(2, 3)$, $R(-3, -4)$, $S(3, -5)$, and $T(\pi, 0)$ in the plane.

SOLUTION $P(-1, 2)$ is in quadrant II because the x-coordinate is negative and the y-coordinate is positive.

$Q(2, 3)$ is in quadrant I because both coordinates are positive.

$R(-3, -4)$ is in quadrant III because both coordinates are negative.

$S(3, -5)$ is in quadrant IV because the x-coordinate is positive and the y-coordinate is negative.

$T(\pi, 0)$ is not in any quadrant, but lies on the positive x-axis.

The points are plotted in Figure 1.3.

---

*This system is also called the Cartesian coordinate system in honor of René Descartes, who invented it.

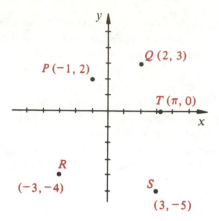

FIGURE 1.3

In plotting an entire set of ordered pairs, the corresponding set of points in the plane is called the **graph** of the set.

EXAMPLE 2   Graph the set of points whose abscissas are greater than -1 and whose ordinates are less than or equal to 4.

SOLUTION   This set is described by the two inequalities

$$x > -1, y \leq 4.$$

The shaded region in Figure 1.4 is the graph of the set. The solid line is part of the region, whereas the broken line is not.

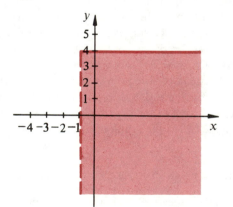

FIGURE 1.4

Sometimes we want to find the distance between two points in the plane. The following discussion explains how this is done. Consider two points P_1 and P_2 on the x-axis as shown in Figure 1.5.

FIGURE 1.5

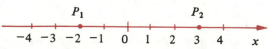

The distance between these two points can be found by counting the number of units between them, or algebraically, by subtraction. To be sure that the numerical quantity obtained in the computation of horizontal distance will be positive, we define it to be the absolute value of the difference between the coordinates of P_1 and P_2. Thus

(1.1) $d(P_1, P_2) = |P_1 - P_2|$

Computing the distance in Figure 1.5 we have

$$d(-2, 3) = |-2 - 3| = |-5| = 5$$

A similar scheme is followed if the points lie on the y-axis.

Now consider two points $P_1(x_1, y_1)$ and $P_2(x_2, y_2)$ that determine a slant line, as shown in Figure 1.6. Draw a line through P_1 parallel to the x-axis and a line through P_2 parallel to the y-axis. These two lines intersect at the point $M(x_2, y_1)$. Hence, by the Pythagorean theorem,*

(1.2) $[d(P_1, P_2)]^2 = [d(P_1, M)]^2 + [d(M, P_2)]^2$

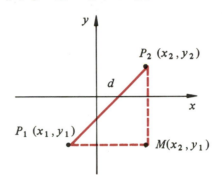

FIGURE 1.6

We see from Figure 1.6 that $d(P_1, M)$ is the horizontal distance between P_1 and P_2. Therefore, the distance $d(P_1, M)$ is given by

$$d(P_1, M) = |x_2 - x_1|$$

Likewise, the vertical distance $d(M, P_2)$ is given by

$$d(M, P_2) = |y_2 - y_1|$$

Making these substitutions into equation (1.2) and denoting $d(P_1, P_2)$ by d:

$$d^2 = (x_2 - x_1)^2 + (y_2 - y_1)^2$$

(1.3) $d = \sqrt{(x_2 - x_1)^2 + (y_2 - y_1)^2}$

---

*The Pythagorean theorem gives the relationship between the lengths of the sides of a right triangle. Specifically, in the triangle shown, $a^2 + b^2 = c^2$.

Equation (1.3) is called the **distance formula** and is used to find the distance between two points in the plane directly from the coordinates of the points. The order in which the two points are labeled is immaterial since

$$(x_2 - x_1)^2 = (x_1 - x_2)^2 \quad \text{and} \quad (y_2 - y_1)^2 = (y_1 - y_2)^2$$

EXAMPLE 3   Find the distance between $(-3, -6)$ and $(5, -2)$. (See Figure 1.7.)

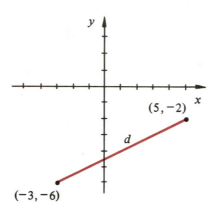

FIGURE 1.7

SOLUTION   Let $(x_1, y_1) = (-3, -6)$ and $(x_2, y_2) = (5, -2)$. Substituting these values into the distance formula,

$$d = \sqrt{(x_2 - x_1)^2 + (y_2 - y_1)^2}$$
$$= \sqrt{[5 - (-3)]^2 + [-2 - (-6)]^2}$$
$$= \sqrt{64 + 16} = \sqrt{80} = 4\sqrt{5}$$

Notice the inclusion of the numerical sign of each number in the substitution of values into the distance formula.   ■

EXAMPLE 4   Find the distance between $(2, 5)$ and $(2, -1)$.

SOLUTION   In this case the two given points lie on a vertical line since they have the same abscissa. (See Figure 1.8.) The distance between the two points, therefore, can be found directly.

$$d = |5 - (-1)| = |5 + 1| = 6 \text{ units}$$

The distance can also be found from the distance formula (equation (1.3)). Letting $(x_1, y_1) = (2, 5)$ and $(x_2, y_2) = (2, -1)$,

$$d = \sqrt{(2 - 2)^2 + (-1 - 5)^2} = \sqrt{36} = 6 \text{ units}$$

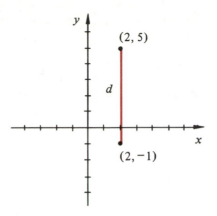

FIGURE 1.8

EXAMPLE 5   Find the distance from the origin to any point (x, y).

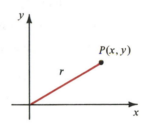

FIGURE 1.9

SOLUTION   From the distance formula, the length of r is given by

$$r = \sqrt{(x - 0)^2 + (y - 0)^2} = \sqrt{x^2 + y^2}$$

where r is always positive.

Exercises for Section 1.1

In Exercises 1–6 plot the ordered pairs.

1. $(3, 2)$ **2.** $(4, 6)$ **3.** $(-2, \frac{1}{2})$

4. $(-6, -5)$ **5.** $(\frac{1}{4}, -\frac{1}{2})$ **6.** $(-2.5, 1.7)$

7. In what two quadrants do the points have positive abscissas?

8. In what two quadrants do the points have negative ordinates?

9. In what quadrant are the abscissa and the ordinate both negative?

10. In what quadrant is the ratio y/x negative?

11. What is the ordinate of a point on the x-axis?

In Exercises 12–19 plot the pairs of points, and find the distance between the points.

12. $(1, 2), (5, 4)$ **13.** $(0, 4), (-1, 3)$

14. $(-1, 5), (-1, -6)$ **15.** $(\frac{1}{2}, \frac{1}{2}), (\frac{1}{2}, -\frac{3}{4})$

16. $(-5, 3), (2, -1)$ **17.** $(0.5, 1.6), (6.2, 7.5)$

18. $(-3, 4), (0, 4)$ **19.** $(2, -6), (-\sqrt{3}, -3)$

In Exercises 20–27 graph the set of points for which the coordinates satisfy the given condition(s).

20. $y > 0$ **21.** $x = 0$

22. $y = 2$ **23.** $x > 0$

24. $x > -1$ and $y > 0$ **25.** $x = y$

26. $x > 0$ and $y > 0$ **27.** $x > -1$ and $y < -1$

28. The point $(x, 3)$ is 4 units from $(5, 1)$. Find x.

29. Find the distance between the points $(\sqrt{x}, \sqrt{y})$ and $(-\sqrt{x}, -\sqrt{y})$.

30. Find the distance between the points (x, y) and $(-x, y)$.

31. Find the distance between the points (x, y) and $(x, -y)$.

32. Find the point on the x-axis that is equidistant from $(0, -1)$ and $(3, 2)$.

In Exercises 33 and 34 find the distance from the origin to the given point.

33. $(1, 2)$ **34.** $(-1, -5)$

1.2   Functions

There are basically two kinds of number symbols used in algebraic and trigonometric discussions: constants and variables. **Constants** have fixed values throughout a discussion, while **variables** may take on different values, called the set of permissible values of the variable, or more technically, the **domain** of the variable. If the domain is not specifically mentioned, we usually allow the variable to take on all permissible real numbers in a discussion. For example, if the expression $1/(x - 3)$ is a part of the discussion the variable, x, may not take on the value 3. (Division by zero is prohibited.) Hence, the domain consists of all real numbers except 3.

Sometimes the values of one variable determine the values of some second variable. For example, the set $\{2, 3, 4\}$ is said to determine the set $\{4, 6, 8\}$ by obeying the rule of doubling each of the elements of the first set. The notion of two variables being related to one another by some rule of correspondence is used extensively in mathematics and is basic to our understanding of the physical world.

DEFINITION 1.1

> If the rule of correspondence between two variables x and y is such that there is exactly one value of y for each value of x, then we say y **is a function of** x.

The first variable is called the **independent variable**, the second is called the **dependent variable**, and the defining relation is called a **functional relationship**. The set of values that the dependent variable can assume is called the **domain**; the corresponding set of values for the dependent variable is called the **range**. The rule of correspondence along with the domain and range make up the **function**.

EXAMPLE 1   The equation $y = x^2$ essentially defines a function with x as the independent variable and y as the dependent variable. The function has an understood domain of all real numbers and a range of nonnegative numbers.   ■

Functions are frequently defined by using some formula or expression involving x and y. The following remarks should help to clarify the important points of the definition of a function as it relates to the use of formulas.

● A formula such as $y = \pm \sqrt{x}$ does not give a functional relationship since two values of y correspond to each value of x. However, each of the formulas $y = \sqrt{x}$ and $y = -\sqrt{x}$ taken separately *does* define a function.

● The definition does not require that the value of the dependent variable change when the independent variable changes, but only that the dependent variable have a unique value corresponding to each value of the independent variable. Thus, $y = 5$ is a function, since y has the unique value 5 for any value of the independent variable. However, $x = 5$ is not a function because many (in fact, infinitely many) values of y correspond to the value $x = 5$.

● We are not saying that expressions such as $y = \pm \sqrt{x}$ and $x = 5$ do not occur or are unimportant. These formulas simply do not describe functions. Instead, they describe a broader concept called a **relation**, which is any pairing of two sets of numbers.

In some discussions, we want to indicate that y is a function of x without specifying the relationship. The notation commonly used to indicate that such a functional relationship exists between x and y is $y = f(x)$. This is read "y equals f of x." The letter f is the name of the function; it is not a variable. The notation $f(x)$ does *not* mean the multiplication of f times x. The letter x represents the domain value of the function and is sometimes called the **argument** of the function.

EXAMPLE 2   Find the value of the function $f(x) = x^2 + 3x$ at $x = 2$.

SOLUTION   We denote the value of $f(x)$ at $x = 2$ by $f(2)$. To find $f(2)$, we substitute 2 for x in $x^2 + 3x$, that is,

$$f(2) = (2)^2 + 3(2) = 10$$ ■

It is sometimes helpful to think of x as representing a blank so that the functional notation tells us what to put in the blank. The function in Example 2 could be written

$$f(\ ) = (\ )^2 + 3(\ )$$

Then, for example,

$$f(a^2) = (a^2)^2 + 3(a^2) = a^4 + 3a^2$$

Calculators are indispensible in evaluating functions for specified values of the independent variable. With a calculator, it is just as easy to evaluate $f(x) = \pi x^3 + \sqrt{x+2}$ for $x = 3.105$ as it is for $x = 2$.

A function f determines a set of ordered pairs

$\{(x, y)\}$ where $y = f(x)$;

and conversely *some* sets of ordered pairs determine a function. Can you tell which sets of ordered pairs determine a function? The key is that for any given x there must be only one value of y. Thus, if the set of ordered pairs has two pairs with the same first element, the set does not represent a function. In this case the set is said to represent a **relation**.

EXAMPLE 3 The set of ordered pairs $\{(2, 5), (3, -1), (-1, 0), (0, 2)\}$ represents a function. The domain elements are $2, 3, -1, 0$ and the range elements are $5, -1, 0, 2$. In functional notation:

$$f(2) = 5, \ f(3) = -1, \ f(-1) = 0, \ f(0) = 2. \qquad \blacksquare$$

EXAMPLE 4 The set of ordered pairs $\{(2, 3), (3, -1), (2, 5), (0, 2)\}$ does not represent a function because the first and third pairs have the same first element. This set represents a relation. $\blacksquare$

The **graph** of a function is the geometric picture of the equation defining the function. It consists of the set of points corresponding to the set of ordered pairs given by the function. Usually to graph a function we plot a "reasonable" number of the points and connect them with a smooth graph.

EXAMPLE 5 Sketch the graph of $y = \sqrt{4 - x}$.

SOLUTION A table is constructed using some reasonable values of x and determining the corresponding values of y. Both the table and graph are shown in Figure 1.10. Notice that the domain is all $x \leq 4$ and the range is all $y \geq 0$.

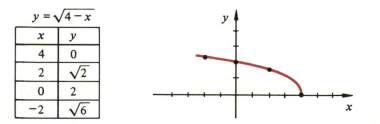

$y = \sqrt{4 - x}$

x	y
4	0
2	$\sqrt{2}$
0	2
-2	$\sqrt{6}$

FIGURE 1.10

$\blacksquare$

EXAMPLE 6   Sketch the graph of $y = 32x - 16x^2$.

SOLUTION   See Figure 1.11.

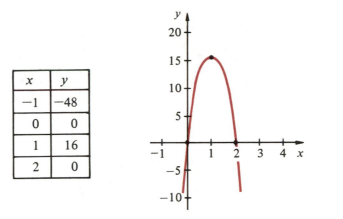

x	y
-1	-48
0	0
1	16
2	0

FIGURE 1.11

Just as certain sets of ordered pairs determine functions, so too, some sets of points of the Cartesian coordinate system determine a function. Usually, the sets of points will be in the form of continuous curves as in Figure 1.12. Not all of the graphs of Figure 1.12 determine functions; the first two do represent functions and the other two do not. Again, as in the case of sets of ordered pairs, the key lies with the fact that for any one value of x, there must be only one value of y. Geometrically, this means that if a line is drawn parallel to the y-axis, there should be, at most, one point of intersection with the graph.

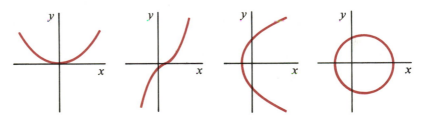

FIGURE 1.12

Exercises for Section 1.2

1. Given $f(x) = x^2 - 5x$, find $f(2)$ and $f(-1)$.

2. Given $g(x) = x^2 - 4$, find $g(2)$ and $g(-4)$.

3. Given $f(t) = (t - 2)(t + 3)$, find $f(-4)$ and $f(0)$.

4. Given $\phi(y) = y(y + 1)$, find $\phi(3)$ and $\phi(-3)$.

5. Given $h(x) = 2(x^2 - 3)$, find $h(5)$ and $h(1)$.

6. Given $g(z) = (1 - z^2)/(1 + z^2)$, find $g(-1)$ and $g(4)$.

In Exercises 7–9 using functional notation symbolize the expressions.

7. The area A of a circle as a function of the radius r.

8. The circumference C of a circle as a function of its diameter d.

9. The perimeter P of a square as a function of its side length s.

In Exercises 10–15 give the domain and range of the functions.

10. $f(x) = x^2$ **11.** $f(t) = t - 2$

12. $f(x) = x^3$ **13.** $f(x) = \sqrt{x - 25}$

14. $f(x) = \sqrt{x}$ **15.** $f(x) = 1/x$

In Exercises 16–24 graph each of the functions.

16. $y = x^3$ **17.** $f(x) = -x^2$ **18.** $z = t^2 + 4$

19. $i = r - r^2$ **20.** $y(x) = \sqrt{x}$ **21.** $\phi = w^2/2$

22. $p = z^2 - z - 6$ **23.** $v = 10 + 2t$ **24.** $y = \sqrt{16 - 4x^2}$

In Exercises 25–30 which of the sets of ordered pairs determine a function?

25. $\{(1, 1), (2, 1)\}$ **26.** $\{(1, 1), (1, 2)\}$

27. $\{(2, 0), (3, 0), (-1, 1)\}$ **28.** $\{(4, 1), (3, 2), (5, 4)\}$

29. $\{(-1, -2), (0, 0), (1, 3), (0, 5)\}$ **30.** $\{(0, 1), (1, -5), (\sqrt{2}, 3)\}$

In Exercises 31–34 which of the graphs represent a function?

31.   **32.**

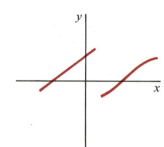

33.   **34.**

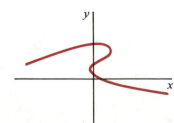

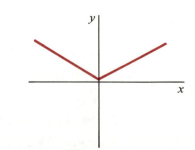

1.3   **Approximate Numbers**

Significant Digits

If you count the number of cards in a deck of 52 playing cards you can say for certain that there are 52 cards in the deck, assuming of course that none is missing. However, if using a meter stick you measure the distance between two points to be 3.5 cm you realize that 3.5 is only an approximation to the precise answer and that the accuracy of the measurement depends upon the accuracy of the measuring device being used. Most calculations in the physical sciences involve measurements that are accurate to some specified number of digits. The digits of a number that are known to be accurate are called **significant digits**. Zeros that are required to locate the decimal point are not considered to be significant digits.

EXAMPLE 1   (a)   5.793 has four significant digits.

(b)   20.781 has five significant digits.

(c)   0.000059 has two significant digits.

(d)   0.08300 has four significant digits.   ■

EXAMPLE 2   How many significant digits does 9480 have?

SOLUTION   A number like 9480 is difficult to categorize unless we know something about the number. Thus, it could represent the exact number of cards in a computer program. Or, it could represent a measurement accurate to either three or four significant figures. Sometimes we use scientific notation for numbers like this to clear up any possible confusion. Thus,

$$9.48 \times 10^3$$

is used to indicate three significant figures. While

$$9.480 \times 10^3$$

indicates four significant figures. In this text, instead of using scientific notation, we will assume that *any* zeros in whole numbers are significant. Thus 9480 would have 4 significant digits, 800 would have 3 significant digits, and so on.   ■

In passing we note that the last significant digit of an approximate number is not completely accurate. For instance, if you measure a length of wire to be 2.56 cm you realize that the length could be anywhere between 2.55 and 2.57 cm since it has been obtained by estimation.

The process of reducing a given number to a specified number of significant figures is called **rounding off**. There are several popular schemes for rounding off numbers, the one we suggest is used in many calculators and computers.

Rounding Off

(1) If the last digit is less than 5, drop the digit and use the remaining digits. For example,

8.134 rounds off as 8.13 to three significant figures.

(2) If the last digit is 5 or greater, drop the digit and increase the last remaining digit by 1. For example,

0.0225 rounds off as 0.023 to two significant figures.

Another routine used to reduce the digits in an approximate number is **truncation**. *In this scheme the unwanted digits are simply dropped or truncated. Thus, the numbers 23.157 and 23.154 both truncate to the four digit number 23.15. Likewise, under the truncation scheme $\frac{2}{3}$ to seven decimal places is carried as 0.6666666; not as 0.6666667. You should check your calculator to see if it rounds off or truncates.*

EXAMPLE 3   Round off each of the following numbers to three significant figures: (a) 18.89   (b) 0.0003725   (c) 99430   (d) 4.996.

SOLUTION

(a)   18.89 is rounded off to 18.9.

(b)   0.0003725 is rounded off to 0.000373.

(c)   99430 is rounded off to 99400.

(d)   4.996 is rounded off to 5.00.   ■

Operations With Approximate Numbers

There is a great temptation when using a calculator, to write the answer to a series of arithmetic calculations to as many digits as the calculator will display. By doing so we make the answer appear more accurate than it really is. For example, if we write

$$8.1 \times 12.137 = 98.3097$$

we would be misleading ourselves into thinking the product is accurate to seven significant figures when the numbers being multiplied are only accurate to two and five significant figures respectively. To avoid this problem when performing arithmetic operations on approximate numbers we adopt the following convention: **When an arithmetic calculation involves two or more approximate numbers the final answer can be no more accurate than the least accurate number in the calculation.** Thus the correct answer to the above operation is 98.

EXAMPLE 4

(a)   The answer to $R = 1.9(63.21) + 4.9072$ should have two significant figures.

(b)   The answer to $3.005\sqrt{2}$ should have four significant figures.

(c)   The answer to $y = 3 - \sqrt{29}$ where 3 is an exact number can be written to as many significant figures as desired since there are no approximate numbers being used.   ■

EXAMPLE 5   The length of a rectangle is measured with a meter stick to be 95.7 cm and the width is measured with a vernier caliper to be 8.433 cm. What is the area of the rectangle?

SOLUTION   The area is given by

$$A = 8.433 \times 95.7 = 807.0381 \text{ sq cm}$$

Since the length has only three significant figures, the area must be rounded off to three significant figures. Therefore, $A = 807$ sq cm is the proper answer to the question. ■

NOTE: The answers appearing in the answer section have been rounded off to the appropriate number of significant digits.

Exercises for Section 1.3

In Exercises 1–10 indicate the number of significant figures in the given numbers.

1. 3.37	**2.** 2.002	**3.** 812.0	**4.** 6161	**5.** 0.03
6. 0.000215	**7.** 0.40	**8.** 57.001	**9.** 500.0	**10.** 0.06180

In Exercises 11–20 round off the given number to three significant figures.

11. 9818	**12.** 72267	**13.** 54.745	**14.** 1.002	**15.** 0.06583
16. 2435	**17.** 39.75	**18.** 0.4896	**19.** 0.9997	**20.** 900,498

In Exercises 21–32 perform the indicated operations and round off the answer to the appropriate number of significant figures.

21. $23.45(0.91669)$

22. $4.7(54.75)$

23. $0.5782 + 1.34 + 0.0057$

24. $50.68 + 9.666 - 24.059$

25. $2.9(3.57 + 10.28) + 25.0$

26. $0.20 + 3.86(0.127 - 0.097)$

27. $\sqrt{2.4^2 + 1.93^2}$

28. $2.176\sqrt{3}$

29. $\dfrac{25(0.9297)}{0.0102}$

30. $\dfrac{5.0887(2.20)}{8813}$

31. $\dfrac{0.9917(771.33)}{\sqrt{30.04}}$

32. $\sqrt{2.14^2 + 3.9^2}$

1.4   Angles

When two line segments meet, they form an **angle**. We ordinarily think of an angle as formed by two half-lines OA and OB that extend from some common point O, called the **vertex**. The half-lines are called the **sides** of the angle. (See Figure 1.13.)

We refer to an angle by mentioning a point on each of its sides and the vertex. Thus the angle in Figure 1.13 is called "the angle *AOB*," and is written ∠ *AOB*. If there is only one angle under discussion whose vertex is at *O*, we sometimes simply say, "the angle at *O*," or more simply, "angle *O*." It is also customary to use Greek letters to designate angles. For example, ∠ *AOB* might also be called the angle θ (read "theta").

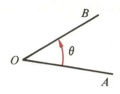

FIGURE 1.13

Often, in trigonometry, we must conceive of an angle as being "formed" by rotating one of the sides about its vertex while keeping the other side fixed as shown in Figure 1.14. If we think of *OA* as being fixed and *OB* as rotating about the vertex, *OA* is called the **initial side** and *OB* the **terminal side** of the generated angle. Other terminal sides such as *OB′* and *OB″* result in different angles. The *size* of the angle depends on the amount of rotation of the terminal side. Thus, ∠ *AOB* is considered smaller than ∠ *AOB′* which, in turn, is smaller than ∠ *AOB″*. Two angles are equal (in size) if they are formed by the same amount of rotation of the terminal side.

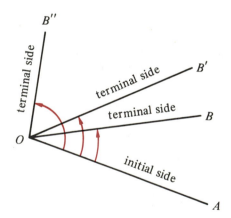

FIGURE 1.14

The Degree

The most commonly used unit of angular measure is the **degree**. We will take as definition that the measure of an angle formed by one complete revolution of the terminal side about its vertex is 360 degrees, also written 360°. One half of this angle, 180°, is called a **straight angle** and one fourth of it, 90°, is called a **right angle**. (See Figure 1.15.)

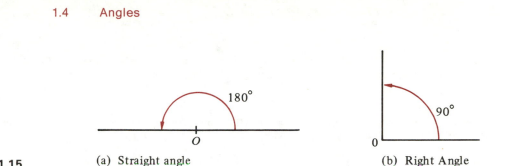

FIGURE 1.15   (a) Straight angle   (b) Right Angle

An angle is **acute** if it is less in size than a right angle and is **obtuse** if it is larger than a right angle but smaller than a straight angle (see Figure 1.16). Figure 1.17 shows two angles that are larger than a straight angle.

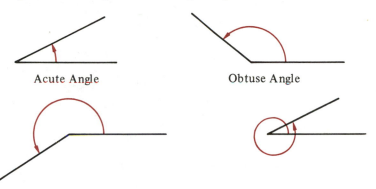

FIGURE 1.16   Acute Angle   Obtuse Angle

FIGURE 1.17
Angles larger than
a straight angle

Angles with the same initial and terminal sides are said to be **coterminal**. The two angles shown in Figure 1.18 are coterminal, but they are obviously not equal. There are many important considerations, both practical and theoretical, which require that we know how the angle was formed.

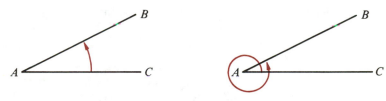

FIGURE 1.18
Coterminal angles

Sometimes it is necessary to consider the angle to be "directed," that is, to determine the direction of rotation of the terminal side in forming the angle. The almost universal convention is to consider those angles obtained by a counterclockwise rotation of the terminal side as *positive* and those obtained by a clockwise rotation as *negative* angles as shown in Figure 1.19.

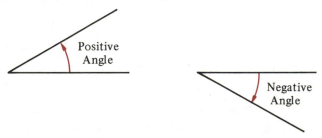

FIGURE 1.19

The measure of an angle has no numerical limit since a terminal side may be rotated either clockwise or counterclockwise as much as desired.

EXAMPLE 1 Draw the following angles: (a) θ (theta) of measurement $42°$, (b) ϕ (phi) of $-450°$, (c) β (beta) of $1470°$ and (d) α (alpha) of $-675°$.

SOLUTION See Figure 1.20.

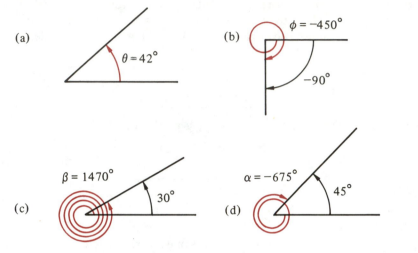

FIGURE 1.20

EXAMPLE 2 Determine angles whose measurements are between $-180°$ and $180°$ and which are coterminal with angles whose measure is the same as those of Example 1.

SOLUTION Using Figure 1.20, we can see that (a) $42°$ is the desired angle, (b) $-90°$ is coterminal with $-450°$, (c) $30°$ is coterminal with $1470°$ and (d) $45°$ is coterminal with $-675°$.

In passing, we note that if the sum of the measures of two angles is $90°$, the two angles are said to be **complementary**. If the sum of the measures is $180°$, they are **supplementary** (see Figure 1.21).

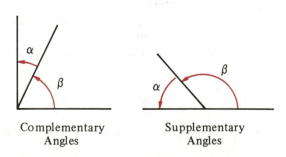

Complementary
Angles

Supplementary
Angles

FIGURE 1.21

The basic angular unit of the degree is subdivided into 60 parts, each of which is called a **minute** and denoted by the symbol ('). The minute is further subdivided into

60 parts, each of which is called a **second** and is denoted by the symbol ("). As the next example shows, arithmetic calculations are sometimes a bit more cumbersome with these subdivisions than with the decimal system.

EXAMPLE 3   Find the sum and difference of the two angles whose measurements are $45°41'09''$ and $32°52'12''$.

SOLUTION   The sum of the two angles is found by adding the corresponding units; that is, degrees to degrees, minutes to minutes, and seconds to seconds. Thus,

$$45°41'09'' + 32°52'12'' = (45 + 32)° \, (41 + 52)' \, (09 + 12)'' = 77°93'21''$$

Here, we see that $93' = 1°33'$ so we write our answer as $78°33'21''$.
To find the difference in the two angles, we must write $45°41'09''$ in the following form:

$$45°41'09'' = 45°40''69'' = 44°100'69''$$

Thus,

$$45°41'09'' - 32°52'12'' = (44 - 32)° \, (100 - 52)' \, (69 - 12)''$$
$$= 12°48'57''$$ ■

The use of minutes and seconds for angular subdivisions (and, indeed, the degree measurement itself) is based upon an ancient Babylonian numeral system. While the decimalization of angular measurement has been accelerated with the widespread use of the hand calculator, we will continue to use both systems for the foreseeable future. Therefore, you should know how to make conversions between the two systems.

To convert an angle measured in degrees, minutes and seconds to a decimal representation in degrees simply divide the minutes by 60 (since $60' = 1$ degree) and the seconds by 3600 (since 3600 sec. $= 1$ deg.) and then add the results. A comment on conversion accuracy is appropriate at this point. So that the converted decimal does not suggest more accuracy than the given angle, we adopt the following convention:

●   An angle measured to the nearest minute should contain two decimal places in the converted form.

●   An angle measured to the nearest second should contain four decimal places in the converted decimal form.

EXAMPLE 4   Convert $15°35'$ to decimal degrees.

SOLUTION $$15°35' = 15° + (35/60)°$$
$$= 15° + (0.583...)°$$
$$= 15.58°$$ ■

EXAMPLE 5   Convert $37°47'23''$ to decimal degrees.

SOLUTION

$$37°47'23'' = 37° + (47/60)° + (23/3600)°$$
$$= 37° + (0.78333...)° + (0.006388...)°$$
$$= 37.7897°$$ ■

To convert decimal notation to degrees, minutes and seconds multiply the fractional part of a degree by 60 to obtain minutes; multiply the fractional part of this result by 60 to obtain seconds.

EXAMPLE 6   Convert 67.8235° to degrees, minutes and seconds.

SOLUTION

$$67.8235° = 67° + (0.08235 \times 60)'$$
$$= 67° + (49.41)'$$
$$= 67°49' + (0.41 \times 60)''$$
$$= 67°49'25''$$ To the nearest second ■

You may be lucky enough to have a calculator with a built in conversion button, usually designated by DMS *. For such a calculator enter the angle in the order degrees/ minutes/seconds followed by the* DMS *key to obtain the angle in degree decimal. If the angle is in degree decimal push* inv DMS *to convert to degree/minutes/seconds. As always you should check your manual for specific instructions.*

In trigonometry, we often locate angles in the Cartesian plane. An angle is said to be in **standard position** in the plane if its vertex is at the origin and its initial side is along the positive x-axis as shown in Figure 1.22. The magnitude of an angle in standard position is measured from the positive x-axis to the terminal side.

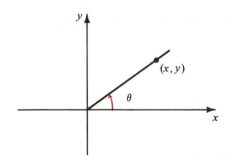

FIGURE 1.22
An Angle In
Standard Position

EXAMPLE 7   Draw an angle in standard position whose terminal side passes through $(-2, -2)$. What is the measure of this angle in degrees?

SOLUTION   The terminal side of θ obviously bisects the third quadrant and therefore $\theta = 180° + 45° = 225°$. See Figure 1.23. You should also observe that the measure of this angle is not unique, since there are many angles with the indicated side as the terminal side. Each of these angles differs by 360°. Thus, $\theta = 225° + m \cdot 360°$ where m is any integer.

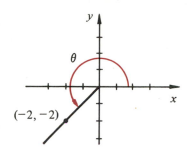

FIGURE 1.23

Any angle in standard position is coterminal with one whose measure is between 0 and 360°. For example, −45° is coterminal with an angle of 315°.

An angle is called a **first quadrant angle** if the terminal side is in the first quadrant, **a second quadrant angle** if the terminal side is in the second quadrant, and so on for the other quadrants.

Measuring Instruments

Figure 1.24 shows a simple form of a **protractor**, the simplest instrument used to measure angles. It is marked in degrees around its rim.

A more accurate device used by engineers and surveyors is called a **transit**. A transit measures an angle by locating two different line of sight objects. See Figure 1.25.

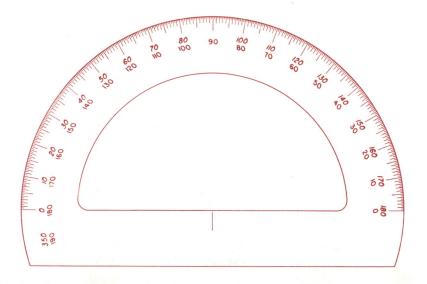

FIGURE 1.24

FIGURE 1.25

Exercises for Section 1.4

1. What is the sum of two complementary angles?

2. What is the difference in degree measure of two coterminal angles?

In Exercises 3–12, find the sum $A + B$ and the difference $A - B$ of the two given angles.

3. $A = 45°10'$, $B = 30°5'$

4. $A = 72°12'$, $B = 30°38'$

5. $A = 58°35'40''$, $B = 50°34'20''$

6. $A = 42°40'10''$, $B = 65°50'50''$

7. $A = 60°10'15''$, $B = 70°45'$

8. $A = 138°40'20''$, $B = 23°52'30''$

9. $A = 240°45'40''$, $B = 333°25'14''$

10. $A = 320°50'20''$, $B = -30°55'10''$

11. $A = -40°42'57''$, $B = -80°18'13''$

12. $A = -90°0'49''$, $B = 269°57'1''$

In Exercises 13–20 convert the given angle to degree/decimal representation. Express angles to four decimal places.

13. $18°25'36''$ **14.** $54°50'16''$ **15.** $94°17'08''$ **16.** $-90°5'48''$

17. $283°36'30''$ **18.** $480°45'45''$ **19.** $183°14'40''$ **20.** $71°12'20''$

In Exercises 21–28 convert the given angle to degree/minute/second representation.

21. $48.2572°$ **22.** $-34.5618°$ **23.** $-235.4500°$ **24.** $30.5052°$

25. $45.7575°$ **26.** $234.5831°$ **27.** $15.2575°$ **28.** $68.3040°$

In Exercises 29–44 draw the angle and name the initial and terminal sides. Indicate an angle between $-180°$ and $180°$ that is coterminal with the given one.

29. $300°$ **30.** $-300°$ **31.** $-317.5°$ **32.** $500°$

33. $-225°$ **34.** $-270°$ **35.** $290°$ **36.** $189.1°$

37. $720°$ **38.** $780°$ **39.** $840°$ **40.** $765°$

41. $1485°$ **42.** $2000°$ **43.** $-290°5'$ **44.** $-205°16'$

In Exercises 45–50 draw an angle in standard position whose terminal side passes through the given point. Give a degree measure of the angle.

45. $(-1, 1)$ **46.** $(5, 0)$ **47.** $(0, -3)$

48. $(4, -4)$ **49.** $(-4, 0)$ **50.** $(1000, -1000)$

51. In surveying a building site the angle 79.473° is recorded in the log. Convert this angle to degrees, minutes, and seconds.

52. A road that makes an angle of 30°40′ North of East intersects a road that makes an angle of 76°45′ South of East. What is the angle between the two roads?

53. During a lab experiment a student measures an angle as 16°50′. Another member of the group measures the same angle as 16.75°. What is the difference to the nearest one hundredth of a degree in the two measurements?

54. A pine tree grows vertically on a hill side that makes an angle of 25.7° with the horizontal. What angle θ does the tree make with the hill side above it?

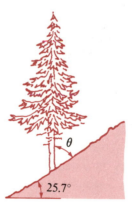

1.5   **Radian Measure**

In the previous section we reviewed the degree as a popular unit for measuring an angle. Another common unit of angular measure is the **radian**. Although less familiar the radian is in some ways a more natural choice as a unit of measure for an angle. The radian is used extensively in more advanced applications of trigonometry and is the standard unit of angular measurement in the International System.

DEFINITION 1.2

One **radian** is the measure of an angle whose vertex is at the center of a circle and whose sides subtend an arc on the circle equal in length to the radius of the circle (see Figure 1.26).

FIGURE 1.26

One radian

Hence, the radian is a measure of the ratio of arc length to radius. Since the circumference, C, of a circle of radius r is known to be $2\pi r$, it follows that $C/r = 2\pi$. Thus there are 2π radians in an angle of $360°$. In equation form,

$$2\pi \text{ radians} = 360°$$

from which we get the following important conversion formulas.

1 degree $= \pi/180$ **radians** $\approx$ **0.0175 radians**

(1.4)   and

1 radian $= 180/\pi$ **degrees** $\approx$ **57.3 degrees**

Sometimes the formulas in (1.4) are combined into one formula such as

$$\frac{\text{Degree measure of an angle}}{180°} = \frac{\text{Radian Measure}}{\pi}$$

The next two examples show you how to use these formulas to convert from degrees to radians and conversely.

EXAMPLE 1   Express in radian measure,

(a) $60°$ (b) $225°$ (c) $24.8°$

SOLUTION
(a)   60 degrees $= 60(\pi/180)$ radians $= \pi/3$ radians
(b)   225 degrees $= 225(\pi/180)$ radians $= 5\pi/4$ radians
(c)   24.8 degrees $= 24.8(0.0175)$ radians $= 0.434$ radians

When the radian measure is a convenient multiple of π, you will usually find it beneficial *not* to convert it to a decimal fraction. When such a conversion is necessary, 3.1416 is often used as a decimal approximation to π.   ■

EXAMPLE 2   Express $\pi/6$ radians, $3\pi/4$ radians, and 1.13 radians in degrees.

SOLUTION
$$\pi/6 \text{ radians} = (\pi/6)(180/\pi) \text{ degrees} = 30 \text{ degrees}$$
$$3\pi/4 \text{ radians} = (3\pi/4)(180/\pi) \text{ degrees} = 135 \text{ degrees}$$
$$1.13 \text{ radians} = 1.13(57.3) = 64.7 \text{ degrees}$$

      ■

Some calculators have a [d↔r] *key that allows for direct conversion from degrees to radians and from radians to degrees. Thus, the entry 180 followed by* [d↔r] *will yield 3.1416 radians. If you do not have a* [d↔r] *key you will have to multiply by $\pi/180$ to convert from degrees to radians and by $180/\pi$ to convert from radians to degrees.*

Most calculators have a special button for the number π. If yours does not have a [π] *button you can use 3.1416 as an approximation, but your answers may be slightly different from those given in the answer section.*

Table 1.1 is a conversion table showing frequently occurring angles with both their degree and radian measure. **You should get to know the entries in this table without making the conversion calculation**.

TABLE 1.1

Table of Degree
and Radian
Measure for
Commonly
Occurring Angles

Angle in Degrees	Angle in Radians
0	0
30	$\pi/6$
45	$\pi/4$
60	$\pi/3$
90	$\pi/2$
120	$2\pi/3$
135	$3\pi/4$
150	$5\pi/6$
180	π
270	$3\pi/2$
360	2π

The word "radian" is understood without being written. This is not the case with degree measurement; its units must always be included. The next example emphasizes the difference between these two angular measures.

EXAMPLE 3   Compare the angle of 60 degrees with that of 60 radians. (Figure 1.27).

SOLUTION   Note from Figure 1.27 that the angle of 60 radians is obtained by 9 repeated revolutions of the terminal side (each revolution being 2π radians) plus an additional 3.45 radians. To obtain the value 3.45, divide 60 by 2π; the remainder is 3.45. Thus the angle of 60 radians is coterminal with an angle of 3.45 radians.

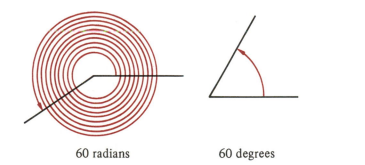

FIGURE 1.27   60 radians   60 degrees   ■

The question of whether to measure an angle in degrees or radians is sometimes a matter of personal preference, but more often than not the unit is dictated by the particular problem under discussion. For example, the length, s, of the arc intercepted by the central angle, θ, in Figure 1.28, can be found by the formula

(1.5) $s = r\theta$

if the angle θ is given in radians. This formula follows immediately from the definition of a radian as the ratio of arc length to radius, that is, $\theta = s/r$. **Formula 1.5 is not valid if θ is stated in degrees**.

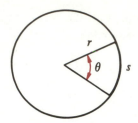

FIGURE 1.28

EXAMPLE 4   Find the length of arc on a circle of radius 5.0 cm which subtends a central angle of 38°.

SOLUTION   To use the above formula, the degree measure must first be converted to radian measure. Thus

$$38 \text{ degrees} \times \pi/180 \text{ radians} = 19\pi/90 \text{ radians}$$

Therefore, $s = 5.0 \times 19\pi/90$ cm ≈ 3.3 cm.   ■

EXAMPLE 5   As the drum in Figure 1.29 rotates counter clockwise, the cord is wound around the drum. How far will the weight on the end of the cord be moved when the drum is rotated through an angle of 53.8°, if $r = 4.5$ ft?

SOLUTION   As the drum rotates, the cord will be wound around the drum, so the distance that the weight will move is equal to the arc length along the edge of the drum formed by a rotation of 53.8°. To use Formula (1.2) we must specify the angle of rotation in radians. Thus,

$$\theta = 53.8(\pi/180) = 0.939$$

The distance s is then given by

$$s = r\theta = 4.5(0.939) = 4.2 \text{ ft}$$

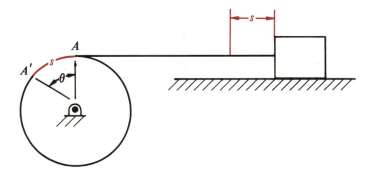

FIGURE 1.29   ■

Exercises for Section 1.5

1. What is the difference in radian measure of two coterminal angles?

2. What is the radian measure of a right angle? Of a straight angle?

Express the following angles in radian measure. Express the measure in multiples of π when convenient.

3. 75°   **4.** −30°   **5.** 480°   **6.** 210°

7. −240°   **8.** 42°   **9.** 95°   **10.** 1485°

11. 750°   **12.** −300°   **13.** 92.1°   **14.** 105.7°

15. 0.092°   **16.** 34°39′   **17.** 253°36′   **18.** 311°48′

19. 0°27′   **20.** 400°40′

In Exercises 21–28 give the radian measure of an angle between $-\pi$ and π whose terminal side passes through the given point.

21. (2, 2)   **22.** (0, 3)   **23.** (−π, 0)   **24.** (5, −5)

25. (−1, −1)   **26.** (−25, 0)   **27.** (0, −1)   **28.** (−3, −3)

In Exercises 29–38 draw the angle, name the initial and terminal sides, indicate an angle between $-\pi$ and π coterminal with the one given. Express each of the given angles in degrees.

29. 4   **30.** 2π   **31.** π   **32.** $-\dfrac{7\pi}{6}$ **33.** −3π

34. −100   **35.** 100   **36.** 30   **37.** 100π   **38.** −100π

39. A pendulum 10.0 ft long swings through an arc of 30°. How long is the arc described by its midpoint?

40. A racing car travels a circular course about the judges' stand. If the angle subtended by the line of sight is 120° while the car travels 1.0 mi, how large is the entire track?

41. How high will the weight in the following figure be lifted if the drum is rotated through an angle of 81.5°?

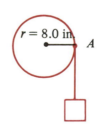

$r = 8.0$ in. A

42. Referring to the figure in Exercise 41, through what angle, in degrees, must the drum be rotated to raise the weight 10 in.?

43. Referring to the figure in Exercise 41, compute the radius of the drum if the weight is raised 5.2 in. by a rotation of 70.7°.

44. The scale on an ammeter is 8.0 cm long. If the scale is an arc of a circle having a radius of 3.2 cm, what angle in degrees will the needle make between the zero reading and a full scale reading? (See figure on next page.)

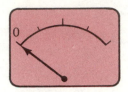

45. A voltage of 6.2 volts causes the needle on a voltmeter to deflect through an angle of 48°. If the needle is 1.75 inches long, how far did the tip of the needle move in indicating the applied voltage?

46. The diameter of the earth is approximately 8000 miles. Find the distance between two points on the Equator whose longitude differs by 3°.

47. The two wheels shown in Figure 1.30 make contact so that the rotation of one wheel causes the other to rotate. How much rotation occurs in the larger wheel when the smaller one rotates through an angle of 100°?

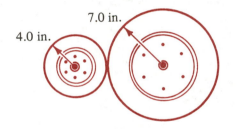

FIGURE 1.30

48. Referring to Figure 1.30, how much rotation must occur in the smaller wheel to cause the larger one to rotate a quarter of a revolution?

49. A schematic drawing of a typical bicycle drive chain is shown in Figure 1.31. How far will the bicycle move forward for each complete revolution of the drive sprocket?

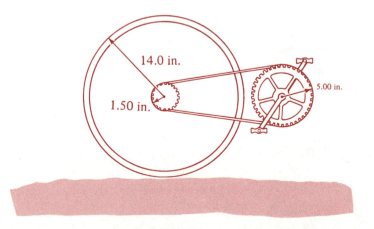

FIGURE 1.31

50. Referring to Figure 1.31 how much rotation of the drive sprocket must occur to move the bicycle 5.00 feet forward?

1.6   Some Facts About Triangles

Much of Chapters 2 and 3 is devoted to a discussion of triangles and how the subject of trigonometry can be used to compute unknown parts of a triangle. Therefore, in this section, some geometrical facts about triangles are summarized.

A triangle is said to be *equiangular* if the measures of each of its three angles are exactly the same; it is said to be *equilateral* if all three sides have the same length. A theorem of geometry tells us that *a* **triangle** *is* **equiangular if and only if it is equilateral**. A triangle is said to be **isosceles** if two of its sides are equal and in such a triangle, the angles opposite the two equal sides are also equal (see Figure 1.32).

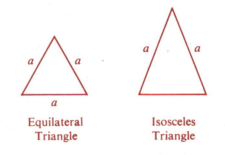

Equilateral   Isosceles
Triangle   Triangle

FIGURE 1.32

A **right** triangle is one in which one of the angles is a right angle. An **oblique** triangle is one without a right angle. In any triangle, the sum of the measures of the angles is 180°. Thus, in an equilateral triangle, each of the angles measures 60°. In a right triangle, each of the two remaining angles is acute and the sum of their measures is 90°.

In any triangle, for example, a triangle with vertices A, B, and C, there is a relatively standard method of referencing the sides and the angles. The sides AB and AC are called the sides **adjacent** to the angle at vertex A. The side BC is called the side **opposite** angle A. There are similar statements concerning the sides opposite and adjacent to angle B and those opposite and adjacent to angle C. In the special case of a right triangle the side opposite the right angle is called the **hypotenuse**.

Referring to the right triangle in Figure 1.33, we see that side AC is called the adjacent side to angle A, side BC is called the side opposite angle A, and side AB is called the hypotenuse.

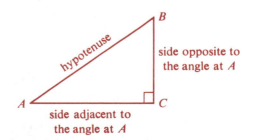

FIGURE 1.33

In trigonometry, as in practically every other branch of mathematics, we use the famous theorem of Pythagoras relating the squares of the lengths of the sides of a right triangle.

THEOREM 1.1

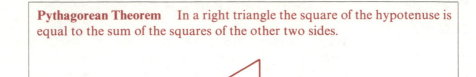

> **Pythagorean Theorem**   In a right triangle the square of the hypotenuse is equal to the sum of the squares of the other two sides.
>
> $$c^2 = a^2 + b^2$$

With the use of the Pythagorean theorem, a third side of a right triangle may be found if any two of the sides are known. For instance, if a and b are given, the hypotenuse c can be computed by

$$c = \sqrt{a^2 + b^2}$$

EXAMPLE 1   The line of sight distance to the top of an antenna attached to the chimney of a house is known to be 15 m. If the sighting is taken 6.0 m from the house, how high is the top of the antenna above the ground?

SOLUTION   A diagram of the situation is shown in Figure 1.34. As you can see, the unknown measurement is the third side of a right triangle in which two of the sides are known. Hence from the Pythagorean theorem,

$$h^2 + 6^2 = 15^2$$

so $\qquad\qquad h^2 = 225 - 36$

and $\qquad\qquad h = \sqrt{189} = 13.75 = 14\ m$ (To two significant figures.)

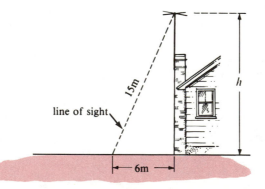

FIGURE 1.34

Another basic concept of trigonometry is that of similar triangles. Generally, two triangles are **similar** if they have the same shape (not necessarily the same size). Thus, similar triangles have equal angles but not necessarily equal sides. The relationship between the sides of similar triangles has been known for centuries and is given in the following theorem of Euclid.

THEOREM 1.2

> If two triangles are similar, their sides are proportional.

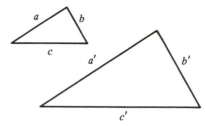

FIGURE 1.35

Two similar
triangles

Referring to Figure 1.35, the content of Theorem 1.2 can be written

$$\frac{a}{a'} = \frac{b}{b'} = \frac{c}{c'}$$

Combinations of any two of the three ratios will yield an equation with four parts. If we know three of these parts, we can find the fourth. Similar triangles are commonly used to compute distances that are difficult to measure by direct means. The next two examples show how this is done.

EXAMPLE 2   High divers tend to dive from some unusually large heights. A spectator who knows he is 200 yd from the diving site notes that his pencil of length 6.0 in. is just large enough to cover the diving height when he holds the pencil about 3.00 ft from his eye. How high is the dive? (See Figure 1.36.)

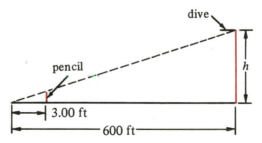

FIGURE 1.36

SOLUTION   Because of the relatively large distances involved we may, for the sake of approximation, ignore the fact that the sighting is taken somewhere about 5 ft or so above ground level. Then, since the small triangle and the larger triangle are obviously similar, we have that

$$\frac{3 \text{ ft}}{600 \text{ ft}} = \frac{1/2 \text{ ft}}{h \text{ ft}}$$

from which

$$h = \tfrac{1}{6}(600) = 100 \text{ ft} \qquad \blacksquare$$

EXAMPLE 3   A group of physics students was provided with a pencil and a 12 inch ruler and told to determine the height of the steeple on the campus chapel. This is how they did it.

SOLUTION   First they measured the length of the shadow of the steeple on the ground using the ruler and found it to be 28 ft. Then, holding the ruler vertically with one end on the sidewalk, they marked the end of its shadow. Finding the length of the shadow of the ruler to be 5.3 in., they used the pencil to write

$$\frac{h \text{ ft}}{28 \text{ ft}} = \frac{12 \text{ in}}{5.3 \text{ in}}$$

$$h = \frac{12}{5.3}(28) = 63 \text{ ft} \qquad \text{(Two significant figures)}$$

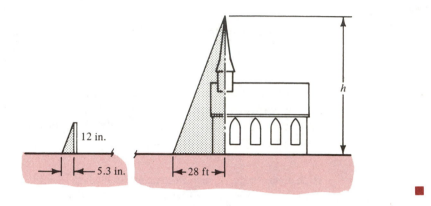

FIGURE 1.37

HISTORICAL NOTE

The early Greek mathematician Erathostehenes (240 B.C.) closely approximated the circumference of the Earth by simple arc length and shadow measurements. He noticed that at Syene (now the site of the Aswan dam) the rays of the noon sun on the summer solstice shone straight down a deep well. Simultaneously at Alexandria, approximately 500 miles due North, the sun rays were measured to be 7.5° from the vertical. Thus the circumference of the earth must be 360/7.5 times the distance between Syene and Alexandria, or approximately 24,000 miles. (The actual value is 24,875 miles.)

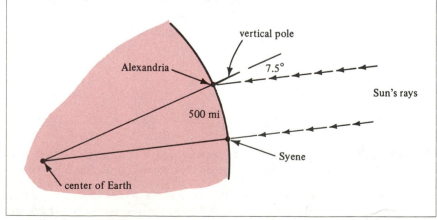

Exercises for Section 1.6

1. If the angle between the equal sides of an isosceles triangle is 32°, how large is each of the other angles?

2. How many degrees are there in each angle of an equilateral triangle?

3. A baseball diamond is a square 90.0 feet on a side. What is the distance across the diamond from first to third base?

4. What is the line of sight distance to an airplane known to be directly over the center of a city, which is 3 mi away, if the plane is flying at 5000 ft?

5. The solar collector of a water heating system is held in a right angle bracket as shown in the following figure. Calculate the length of the solar collector.

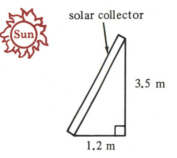

6. One side of a rectangle is half as long as the diagonal. The diagonal is 4 m long. How long is the other side of the rectangle?

7. What is the length of the diagonal of a cube that is 5.0 cm on an edge?

8. If a room is 21 ft long, 15 ft wide, and 10 ft high, what is the length of the diagonal of the floor; of an end wall; of a side wall of the room?

In Exercises 9–15 show that the triangles with sides having the given measures are right triangles.

9. 6, 8, 10   **10.** 5, 12, 13   **11.** 7, 24, 25   **12.** 9, 40, 41

13. 11, 60, 61   **14.** 10, 24, 26   **15.** 28, 21, 35

The two triangles shown in Figure 1.38 are similar. Find the unknown sides for the following given conditions.

FIGURE 1.38

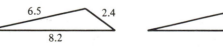
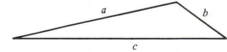

16. Given $a = 9.2$, find b and c.   **17.** Given $b = 3.0$, find a and c.

18. Given $b = 2.5$, find a and c.   **19.** Given $c = 12$, find a and b.

20. Given $c = 8.7$, find a and b.

21. If the sides of a triangle are 2, 4, and 5 cm, what is the perimeter of a similar triangle in which the longest side is 15 cm?

22. A snapshot is 7.60 cm wide and 10.1 cm long. It is enlarged so that it is 25.3 cm wide. How long is the enlarged picture? What is its area? its perimeter?

23. Is every equilateral triangle similar to every other equilateral triangle? Is every isosceles triangle similar to every other isosceles triangle? Give reasons.

24. At the same time that a yardstick held vertically casts a 5.0-foot shadow, a vertical flagpole casts a 30-foot shadow. How high is the flagpole?

25. At a certain time of day, a television relay tower casts a shadow 100 m long, and a nearby pole 12 m tall casts a shadow 15 m long. How tall is the tower?

26. Assume that the three triangles in Figure 1.39 are similar. Find the measure of the unknown sides.

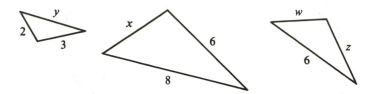

FIGURE 1.39

27. If the measure of the hypotenuse of the right triangle in the following figure is $(m/2)+1$, and one leg has measure $(m/2)-1$, find the measure of the other leg.

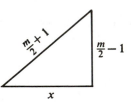

28. Find angles α and ϕ in the figure if $\theta = 215°$ and $\beta = 26.6°$.

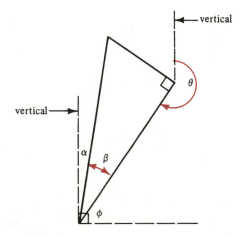

29. Find angles α and ϕ if $\theta = 211°14'$ and $\beta = 24°12'$.

Key Topics for Chapter 1

Define and/or discuss each of the following.

Cartesian Coordinate System

Function

Significant Figures

Rounding Off

Angles

Coterminal Angles

Degree Measure

Complementary Angles

Supplementary Angles

Standard Position of an Angle

Radian Measure

Length of Arc Formula

Pythagorean Theorem

Similar Triangles

Review Exercises for Chapter 1

1. If $f(x) = 2x^3 - x^2 + 1$, find $f(1), f(0)$ and $f(-1)$.

2. If $f(x) = 3x - 2$, find $f(r + s)$.

3. Graph $y = f(x)$ if $f(x) = 2x - 3$.

4. Graph $y = f(x)$ if $f(x) = x^2 - 5$.

5. Find the distance from $(1, f(1))$ to $(2, f(2))$ if $f(x) = x^2$.

6. Find the distance from $(0, 5)$ to $(3, 2)$.

7. Determine if the points $(0, 1), (4, 2)$ and $(3, 5)$ are vertices of a right triangle.

8. Let $y = \dfrac{3 - x}{x + 1}$. Find the distance from $(1, f(1))$ to $(0, f(0))$.

9. Determine an angle between $0°$ and $360°$ which is coterminal with an angle whose measure is $3400°$.

10. Determine an angle whose radian measure is between 0 and 2π and which is coterminal with the angle whose measure is 80 radians.

11. Change $38°43'23''$ to degree/decimal form.

12. Change $67.5428°$ to degree/minute/second form.

13. Subtract: $45°23'14'' - 35°35'54''$.

14. Express $3000°$ in radian measure.

15. Express $-85.43°$ in radian measure.

16. Express 435 radians in degree measure.

17. Express 180π radians in degree measure.

18. The hypotenuse and leg of a right triangle are 54.6 ft and 34.9 ft respectively. Find the other leg.

19. Find the length of arc subtended by a $32°$ angle on a circle of radius 4.6 ft.

20. Find the line of sight distance to the top of a 100 ft tower if you are standing 50 ft from its base.

In Exercises 21–26 round off to four significant figures.

21. 44.653	**22.** 131,450	**23.** 0.0021245
24. 0.871351	**25.** 352.97	**26.** 12999

In Exercises 27–30 indicate how many significant figures the answer should contain.

27. 22.6(0.9087)   **28.** 0.023(67.445)

29. $\dfrac{1.83\sqrt{15}}{2.005}$ **30.** $3.7^2 + 0.936^2 + \pi^2$

Test 1 for Chapter 1

In Exercises 1–10, answer *true* or *false*.

1. Coterminal angles have the same degree measure but a different radian measure.

2. An angle larger than a straight angle is said to be obtuse.

3. Conventionally, positive angles are obtained by a clockwise rotation of the terminal side.

4. One radian is equal to π degrees.

5. An angle in standard position whose measure is 60 radians and one whose measure is 60 degrees are coterminal.

6. Similar triangles have the same shape.

7. In order for a formula to represent a function there must be one and only one real value of $f(x)$ for every real x.

8. The distance from a to b is the same as that from b to a.

9. The number 0.0217 has four significant figures.

10. The answer to 0.029(17.93) should have two significant figures.

11. How many degrees are there in 45 radians?
How many radians are there in 45 degrees?

12. The Gateway Arch in St. Louis is known to be approximately 670 ft high. If you are 500 ft away from the foot of the arch, what is your line of sight distance to the top?

13. Make a careful sketch of the function

$$f(x) = x^2 + 4x - 6$$

14. Find the distance from $(2, \sqrt{2})$ to $(-2, 0)$.

15. Graph the set of points $x - 1$. Is this a function?

Test 2 for Chapter 1

1. Find $\alpha + \beta$. if $\alpha = 145°14'56''$ and $\beta = 19°47'23''$.

2. Find the angle between 0° and 360° that is coterminal with 825°.

3. Find the angle between 0 and 2π that is coterminal with 15 radians.

4. Express 125° in radian measure.

5. Express $2\pi/3$ radians in degrees.

6. A woman walks 1.8 km due East and then turns and walks 2.4 km due North. How far is she from her initial starting point? Round off the answer to the correct number of significant figures.

7. Find x and y if the two triangles are similar triangles.

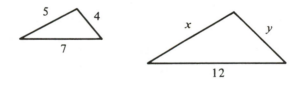

8. Plot the points $(-1, 5)$ and $(4, -7)$ and determine the distance between the two points.

9. Draw the graph of $y = \sqrt{3 - x}$. What is the domain and range of this function?

10. Under certain conditions the path of a thrown ball can be represented mathematically by the equation $h = a + bx + cx^2$, where x is the horizontal displacement of the ball, h is the corresponding height, and a, b, and c are constants. Draw the path of a ball for the interval $x = 0$ to $x = 7$, if $h = 7x - x^2$.

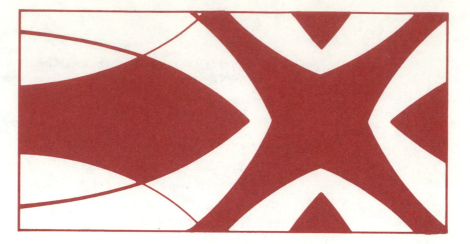

Right Triangle Trigonometry

2.1   Definitions of the Trigonometric Functions

Trigonometry was invented as a means of indirectly measuring the parts of a right triangle; in fact, the word **trigonometry** means "three angle measure." Today, trigonometry has many applications that have nothing to do with triangles, but the basic concepts are still best understood relative to the right triangle. We begin our study of trigonometry with a brief discussion of generated angles in standard position and then move immediately to right triangle trigonometry.

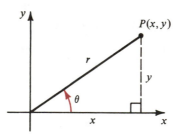

FIGURE 2.1

In Figure 2.1 an acute angle θ is shown in standard position. There are three important numbers relative to any point on the terminal side of angle θ; namely, the x- and y-coordinates of the point and the distance r from the origin to the point. For a given angle θ, we are interested in the ratios of the numbers x, y and r. By inspection you can see that the following six ratios can be formed from the three numbers.

$$\frac{y}{r}, \frac{x}{r}, \frac{y}{x}, \frac{x}{y}, \frac{r}{x}, \frac{r}{y}$$

For a given angle θ these six ratios are independent of the point chosen on the terminal side. Proof that this is true may be seen from the following argument. Choose two distinct points on the terminal side of angle θ as shown in Figure 2.2 and then draw a line through each point perpendicular to the x-axis. In this way we visualize two similar triangles with a common vertex at the origin. From our knowledge of similar triangles, we know that corresponding sides of the two triangles are proportional. Therefore,

$$\frac{y}{y_1} = \frac{r}{r_1} \text{ and } \frac{x}{x_1} = \frac{r}{r_1} \text{ and } \frac{y}{y_1} = \frac{x}{x_1}$$

or, rearranging terms,

$$\frac{y}{r} = \frac{y_1}{r_1} \text{ and } \frac{x}{r} = \frac{x_1}{r_1} \text{ and } \frac{y}{x} = \frac{y_1}{x_1}$$

Each proportion says that the ratio of the two numbers associated with the smaller triangle is equal to the ratio of the corresponding numbers in the larger triangle. Recognizing that these are the first three ratios of the six ratios mentioned earlier and that the other three could be handled in the same manner, we conclude that the six ratios are independent of the point chosen on the terminal side of θ.

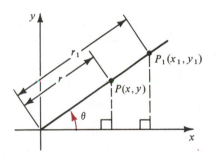

FIGURE 2.2

    Although the six ratios are independent of the point on the terminal side, they are dependent upon the magnitude of the generated angle θ. For instance, if in Figure 2.2 angle θ increases, the x-coordinate of P decreases and the y-coordinate increases. Consequently, the ratio x/r would decrease as θ increases and the ratio y/r would increase as θ increases. The six ratios are functions of the angle θ and have come to be called the **trigonometric functions**. To facilitate discussion each of the ratios is given a name as indicated in Definition 2.1.

    In this chapter we focus our attention on acute angles (called first quadrant angles). For such angles the trigonometric functions are always positive. However, Definition 2.1 is valid for angles of any magnitude and as you will see in the next chapter, the trigonometric functions may be negative.

DEFINITION 2.1

With reference to Figure 2.1, the six trigonometric functions of angle θ are as follows.

$$\textbf{sine } \theta = \frac{y}{r} \qquad \text{(abbreviated sin } \theta)$$

$$\textbf{cosine } \theta = \frac{x}{r} \qquad \text{(abbreviated cos } \theta)$$

$$\textbf{tangent } \theta = \frac{y}{x} \qquad \text{(abbreviated tan } \theta)$$

$$\textbf{cotangent } \theta = \frac{x}{y} \qquad \text{(abbreviated cot } \theta)$$

$$\textbf{secant } \theta = \frac{r}{x} \qquad \text{(abbreviated sec } \theta)$$

$$\textbf{cosecant } \theta = \frac{r}{y} \qquad \text{(abbreviated csc } \theta)$$

You should take time to learn the definition of each of the trigonometric functions since they are the building blocks of trigonometry. You should know them so well that when someone mentions sin θ you automatically think "y to r".

HISTORICAL NOTE

The word "sine" has an interesting origin. The first trigonometry tables were of chords of a circle corresponding to an angle θ as shown in the figure below. As you can see if the radius of the circle is one, then sin θ is just one half the chord length. The Hindus gave the name "jiva" to the half chord and the Arabs used the word "jiba". In the Arabic language there is also a word "jaib" meaning "bay" whose Latin translation is "sinus". A medieval translator inadvertently confused the words jiba and jaib. Thus the word "sine" is used instead of "half chord".

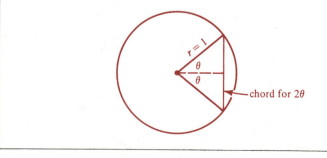

EXAMPLE 1   Determine the trigonometric functions of the angle θ whose terminal side passes through the point (3, 5) as shown in Figure 2.3.

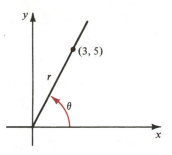

FIGURE 2.3

SOLUTION   Recall from Example 5, Section 1.1 that r is given by $r = \sqrt{x^2 + y^2}$. Therefore, in this case $r = \sqrt{3^2 + 5^2} = \sqrt{34}$. Using $x = 3, y = 5, r = \sqrt{34}$ in Definition 2.1, we find that

$$\sin \theta = \frac{5}{\sqrt{34}} \qquad\qquad \csc \theta = \frac{\sqrt{34}}{5}$$

$$\cos \theta = \frac{3}{\sqrt{34}} \qquad\qquad \sec \theta = \frac{\sqrt{34}}{3}$$

$$\tan \theta = \frac{5}{3} \qquad\qquad \cot \theta = \frac{3}{5}$$

Of course, $\sin \theta$ and $\cos \theta$ can be written in rationalized form as $\sin \theta = \dfrac{5\sqrt{34}}{34}$ and $\cos \theta = \dfrac{3\sqrt{34}}{34}$ respectively.   ■

EXAMPLE 2   A support line from the top of a 150 m antenna is anchored 50 m from the base of the antenna. Find the tangent of the angle of elevation of the cable. (The angle of elevation is the angle between the horizontal and the line of sight when looking up at the object.)

SOLUTION   Here, we draw the support line and the antenna in the Cartesian plane with the anchor point at the origin. The situation is shown in Figure 2.4.

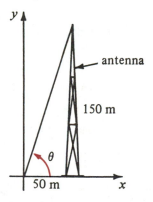

FIGURE 2.4

From the figure we see that $x = 50$ and $y = 150$, so the tangent of the angle of elevation of θ is

$$\tan \theta = \frac{y}{x} = \frac{150}{50} = 3.0$$ ■

As the following two examples are intended to show, a knowledge of one of the trigonometric functions is sufficient to determine the other five functions.

EXAMPLE 3   Given that $\sin \theta = \frac{1}{2}$, find the other five trigonometric functions of the acute angle θ.

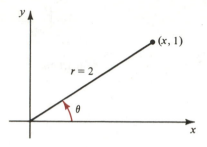

FIGURE 2.5

SOLUTION   Since $\sin \theta = 1/2$, we know that $y/r = 1/2$ and so we assume $y = 1$ and $r = 2$. From the formula $r = \sqrt{x^2 + y^2}$, we have $x = \sqrt{r^2 - y^2} = \sqrt{2^2 - 1^2} = \sqrt{3}$. Therefore, $(\sqrt{3}, 1)$ is a point on the terminal side of angle θ. Now using $x = \sqrt{3}$, $y = 1$, and $r = 2$ in Definition 2.1, we get

$$\cos \theta = \frac{\sqrt{3}}{2}, \tan \theta = \frac{1}{\sqrt{3}}, \cot \theta = \frac{\sqrt{3}}{1} = \sqrt{3}$$

$$\sec \theta = \frac{2}{\sqrt{3}}, \csc \theta = \frac{2}{1} = 2$$

■

In the previous example, the values of the trigonometric functions are left in a form involving a radical. If the ratios are actually used for computational purposes, you will need to convert to approximate decimal values. For instance, in Example 3 we would write

$$\cos \theta = \frac{\sqrt{3}}{2} \approx \frac{1.732}{2} = 0.866$$

EXAMPLE 4   Given $\cot \beta = 3$, find $\cos \beta$ and $\csc \beta$. (See Figure 2.6.)

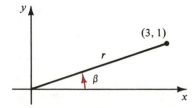

FIGURE 2.6

SOLUTION   Since $\cot \beta = x/y$ we interpret $\cot \beta = 3$ to mean $x = 3$ and $y = 1$, so that the point $(3, 1)$ is a point on the terminal side of angle β. Then $r = \sqrt{3^2 + 1^2} = \sqrt{10}$. It follows that

$$\cos \beta = \frac{x}{r} = \frac{3}{\sqrt{10}}$$

$$\csc \beta = \frac{r}{y} = \frac{\sqrt{10}}{1} = \sqrt{10}$$

We note that any other point on the terminal side of β could have been used to obtain the other trigonometric values. For example, the point $x = 6$, $y = 2$ or $x = 9/2$, $y = 3/2$ would also work just fine. ∎

Exercises for Section 2.1

In Exercises 1–12 compute the six trigonometric functions of the angle in standard position whose terminal side passes through the given point. In each case draw the angle. The numbers are to be interpreted as exact.

1. $(3, 4)$ **2.** $(4, 3)$ **3.** $(5, 12)$ **4.** $(24, 7)$

5. $(5, 1)$ **6.** $(2, 7)$ **7.** $(2, 2)$ **8.** $(9, 9)$

9. $(2, \sqrt{3})$ **10.** $(\sqrt{5}, 1)$ **11.** $(\frac{5}{2}, \frac{6}{5})$ **12.** $(\frac{1}{2}, \frac{3}{10})$

In Exercises 13–26 find the exact values of the other five trigonometric functions of the acute angle θ. In each case draw the angle θ.

13. $\cos \theta = 3/5$ **14.** $\tan \theta = 2$ **15.** $\sin \theta = 1/3$

16. $\sin \theta = 1/6$ **17.** $\cos \theta = 1/2$ **18.** $\sec \theta = \sqrt{2}$

19. $\tan \theta = \sqrt{5}$ **20.** $\cot \theta = \sqrt{3}$ **21.** $\csc \theta = 11/10$

22. $\sin \theta = 3/10$ **23.** $\sin \theta = u/v$ **24.** $\tan \theta = u/v$

25. $\cos \theta = u$ **26.** $\sin \theta = 1/v$

27. A 6.0-foot man casts a shadow of 4.0 ft. Find the tangent of the angle that the rays of the sun make with the horizontal.

28. A wire 30 ft long is used to brace a flagpole. If the wire is attached to the pole 25 ft above the level ground, what is the cosine of the angle made by the wire with the ground?

29. The line of sight distance to the top of a 128-foot high building is 456 ft. What is the tangent of the angle of elevation?

30. Suppose that a boy is flying a kite at the end of a 100-m string that makes an angle of 45° with the ground. Find the cosine of the angle that the string makes with the ground.

31. A man on a 255-meter cliff looks down on a rowboat known to be 75.0 m from the base of the cliff. What is the sine of the angle of depression? (The angle of depression is defined as the angle between the horizontal and the line of sight when looking down on an object.)

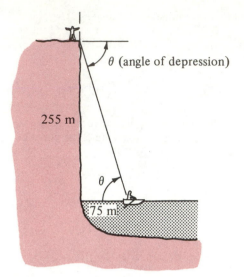

θ (angle of depression)

255 m

θ

75 m

2.2   **Fundamental Relations**

The values of the trigonometric functions are interrelated by some interesting and useful formulas. Recall from Definition 2.1 that $\sin \theta = y/r$ and $\csc \theta = r/y$, consequently,

$$\sin \theta \csc \theta = \frac{y}{r} \cdot \frac{r}{y} = 1$$

Similarly, we note that

$$\cos \theta \sec \theta = \frac{x}{r} \cdot \frac{r}{x} = 1$$

and

$$\tan \theta \cot \theta = \frac{y}{x} \cdot \frac{x}{y} = 1$$

Rearranging, we have

$$\sin \theta = \frac{1}{\csc \theta}$$

(2.1) $$\cos \theta = \frac{1}{\sec \theta}$$

$$\tan \theta = \frac{1}{\cot \theta}$$

The three relations in (2.1) are called the **reciprocal relations** for the trigonometric functions. Of course, they can also be written $\csc \theta = \dfrac{1}{\sin \theta}$, $\sec \theta = \dfrac{1}{\cos \theta}$, and

$\cot \theta = \dfrac{1}{\tan \theta}$.

EXAMPLE 1   Given $\sin \theta = \dfrac{2}{3}$, use (2.1) to find $\csc \theta$.

SOLUTION $\csc \theta = \dfrac{1}{\sin \theta} = \dfrac{1}{2/3} = \dfrac{3}{2}$ ■

EXAMPLE 2   Given $\sec \alpha = 2$, use (2.1) to find $\cos \alpha$.

SOLUTION $\cos \alpha = \dfrac{1}{\sec \alpha} = \dfrac{1}{2}$ ■

Another relation that is of considerable importance is

(2.2) $\tan \theta = \dfrac{\sin \theta}{\cos \theta}$

Relation (2.2) is verified by the following sequence of operations.

$\tan \theta = \dfrac{y}{x}$ (definition of $\tan \theta$)

$\quad\;\; = \dfrac{y/r}{x/r}$ (divide numerator and denominator by r)

$\quad\;\; = \dfrac{\sin \theta}{\cos \theta}$ (definition of $\sin \theta$ and $\cos \theta$)

EXAMPLE 3   Given $\sin \phi = 1/\sqrt{5}$ and $\cos \phi = 2/\sqrt{5}$, use (2.2) to find $\tan \phi$.

SOLUTION $\tan \phi = \dfrac{\sin \phi}{\cos \phi} = \dfrac{1/\sqrt{5}}{2/\sqrt{5}} = \dfrac{1}{2}$ ■

Finally, with the aid of the Pythagorean theorem we derive an important relation between the sine and cosine functions. For any angle θ whose terminal side passes through the point $P(x, y)$, the Pythagorean theorem requires that

$y^2 + x^2 = r^2$

Dividing both sides of this equation by r^2, we obtain

$$(y/r)^2 + (x/r)^2 = 1$$

or, in terms of the trigonometric functions,

$$(\sin \theta)^2 + (\cos \theta)^2 = 1$$

It is customary to write $(\sin \theta)^2$ as $\sin^2 \theta$ and $(\cos \theta)^2$ as $\cos^2 \theta$. (A similar convention holds for expressing powers of the other trigonometric functions.) Thus the equation reads:

(2.3) $\sin^2 \theta + \cos^2 \theta = 1$

a formula that is often called the *Pythagorean relation* of trigonometry.

EXAMPLE 4   Given that $\sin \theta = \frac{1}{4}$, use the Pythagorean relation to find $\cos \theta$. Find $\tan \theta$.

SOLUTION   Solving the Pythagorean relation for $\cos \theta$, we have $\cos \theta = \sqrt{1 - \sin^2 \theta}$, where the positive square root is chosen because $\cos \theta > 0$ (when θ is acute). Therefore,

$$\cos \theta = \sqrt{1 - \left(\frac{1}{4}\right)^2} = \sqrt{1 - \left(\frac{1}{16}\right)} = \frac{\sqrt{15}}{4}$$

Finally, using $\sin \theta = 1/4$ and $\cos \theta = \sqrt{15}/4$ in (2.2) we get

$$\tan \theta = \frac{\sin \theta}{\cos \theta} = \frac{1/4}{\sqrt{15}/4} = \frac{1}{\sqrt{15}}$$

$\blacksquare$

Exercises for Section 2.2

In Exercises 1–25 use the fundamental relations (2.1)–(2.3) to find the exact value of the indicated trigonometric function. Assume the indicated angle is acute.

1. $\sin \theta = \frac{1}{2}$, find $\csc \theta$

2. $\cos \phi = \frac{2}{3}$, find $\sec \phi$

3. $\sec \beta = 3$, find $\cos \beta$

4. $\tan \theta = \frac{10}{7}$, find $\cot \theta$

5. $\sin A = \frac{\sqrt{3}}{2}$, find $\cos A$

6. $\sin \alpha = \frac{\sqrt{2}}{2}$, find $\cos \alpha$

7. $\cot \theta = \sqrt{2}$, find $\tan \theta$

8. $\csc \theta = \frac{2}{\sqrt{3}}$, find $\sin \theta$

9. $\csc \alpha = 2$, find $\cos \alpha$

10. $\cos x = \frac{1}{2}$, find $\sin x$

11. $\sin \phi = \frac{5}{13}$, $\cos \phi = \frac{12}{13}$, find $\tan \phi$

12. $\sin \beta = \frac{2}{\sqrt{7}}$, $\cos \beta = \sqrt{\frac{3}{7}}$, find $\tan \beta$

13. $\tan \theta = \frac{1}{2}$, $\cos \theta = \frac{2}{\sqrt{5}}$, find $\sin \theta$

14. $\tan \theta = \frac{2}{3}$, $\sin \theta = \frac{2}{\sqrt{13}}$, find $\cos \theta$

15. $\sin x = \dfrac{1}{\sqrt{10}}$, $\cos x = \dfrac{3}{\sqrt{10}}$, find $\tan x$ **16.** $\sec \theta = \frac{13}{12}$, $\tan \theta = \frac{5}{12}$, find $\sin \theta$

17. $\csc B = \dfrac{\sqrt{5}}{2}$, find $\tan B$ **18.** $\sin \gamma = \frac{2}{3}$, find $\cot \gamma$

19. $\cos \phi = \dfrac{\sqrt{2}}{3}$, find $\cot \phi$ **20.** $\sec \alpha = 2$, find the other five functions

21. $\csc \theta = 3$, find the other five functions

22. $\sin A = \frac{2}{3}$, find the other five functions

23. $\cos \theta = \dfrac{2}{\sqrt{5}}$, find the other five functions

24. $\tan \alpha = 1$, find the other five functions

25. $\tan \beta = \sqrt{2}$, find the other five functions

Use the relations in (2.1) and a calculator to find the indicated trigonometric functions.

26. $\sin \theta = 0.4313$, find $\csc \theta$ **27.** $\cos x = 0.1155$, find $\sec x$

28. $\tan \phi = 2.397$, find $\cot \phi$ **29.** $\csc A = 1.902$, find $\sin A$

30. $\sec t = 2.030$, find $\cos t$

2.3   The Values of the Trigonometric Functions

In the previous sections we discussed and computed values of the trigonometric functions from known points on the terminal side of an angle or by using the fundamental relations. No attempt was made to relate the measure of the angle to the values of its trigonometric functions. In practice it is important to know how to obtain the trigonometric functions for a specified angle measured either in degrees or radians. The values of the trigonometric functions of certain angles can be found from purely geometric considerations as illustrated in the next two examples.

EXAMPLE 1   Find the values of the trigonometric functions for a 45° angle.

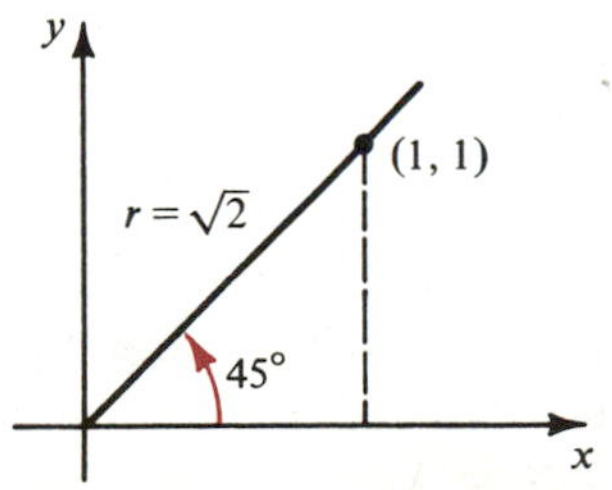

FIGURE 2.7

SOLUTION   Drawing a 45°Angle in standard position, we observe that the terminal side will bisect the first quadrant. Consequently, the x-coordinate of any point on the terminal side of a 45° angle will equal the y-coordinate. For convenience we choose the point (1, 1) as shown in Figure 2.7. Then $r = \sqrt{1^2 + 1^2} = \sqrt{2}$. Using $x = 1$, $y = 1$, $r = \sqrt{2}$ in the definitions, we get

$$\sin 45° = \frac{y}{r} = \frac{1}{\sqrt{2}} \approx 0.707 \qquad \csc 45° = \frac{r}{y} = \frac{\sqrt{2}}{1} = \sqrt{2} \approx 1.414$$

$$\cos 45° = \frac{x}{r} = \frac{1}{\sqrt{2}} \approx 0.707 \qquad \sec 45° = \frac{r}{x} = \frac{\sqrt{2}}{1} = \sqrt{2} \approx 1.414$$

$$\tan 45° = \frac{y}{x} = \frac{1}{1} = 1 \qquad \cot 45° = \frac{x}{y} = \frac{1}{1} = 1$$

EXAMPLE 2   Find the trigonometric functions for a 60° angle.

FIGURE 2.8–2.9

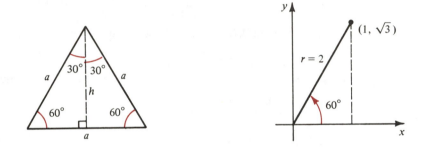

SOLUTION   Consider the equilateral triangle shown in Figure 2.8. The bisector of any one of the angles divides the equilateral triangle into two congruent right triangles. Since a line that bisects an angle of an equilateral triangle also bisects the side opposite that angle, the length of the side opposite the 30° angle is one-half the length of the hypotenuse. By the Pythagorean theorem, the length of h is $h = \sqrt{a^2 - (\frac{1}{2}a)^2} = \frac{a\sqrt{3}}{2}$. If $a = 2$, then $h = \sqrt{3}$ and $\frac{1}{2}a = 1$. From this we can conclude that the terminal side of a 60° angle in standard position will pass through the point $(1, \sqrt{3})$ if r is 2. See Figure 2.9. Hence, by definition

$$\sin 60° = \frac{y}{r} = \frac{\sqrt{3}}{2} \approx 0.866 \qquad \csc 60° = \frac{r}{y} = \frac{2}{\sqrt{3}} \approx 1.155$$

$$\cos 60° = \frac{x}{r} = \frac{1}{2} = 0.5 \qquad \sec 60° = \frac{r}{x} = \frac{2}{1} = 2$$

$$\tan 60° = \frac{y}{x} = \frac{\sqrt{3}}{1} = \sqrt{3} \approx 1.732 \qquad \cot 60° = \frac{x}{y} = \frac{1}{\sqrt{3}} \approx 0.577$$

The values of the trigonometric functions for a 30° angle are found by the same right triangle relationship used for a 60° angle. Table 2.1 summarizes the trigonometric functions for 30°, 45°, and 60°. Study it carefully. You should know how to *derive* each of the values in the table. (Keep in mind Figures 2.7 and 2.9.)

	θ (degrees)	θ (radians)	$\sin \theta$	$\cos \theta$	$\tan \theta$	$\cot \theta$	$\sec \theta$	$\csc \theta$
TABLE 2.1 Values for Some Important Angles	30	$\pi/6$	$1/2$	$\sqrt{3}/2$	$1/\sqrt{3}$	$\sqrt{3}$	$2/\sqrt{3}$	2
	45	$\pi/4$	$\sqrt{2}/2$	$\sqrt{2}/2$	1	1	$\sqrt{2}$	$\sqrt{2}$
	60	$\pi/3$	$\sqrt{3}/2$	$1/2$	$\sqrt{3}$	$1/\sqrt{3}$	2	$2/\sqrt{3}$

A REMARK ON
NOTATION

As mentioned in Section 1.5, the word "radian" may be omitted when radian measure is used. Thus $\sin (\frac{1}{6}\pi$ radians$) = \sin \frac{1}{6}\pi$. When degrees are used, the degree symbol must always be included: $\sin 30° \neq \sin 30$.

An interesting and useful observation that can be made about the values in Table 2.1 is that for the complementary angles 30° and 60°, we have

$$\sin 30° = 1/2 = \cos 60°$$

$$\tan 30° = 1/\sqrt{3} = \cot 60°$$

$$\sec 30° = 2/\sqrt{3} = \csc 60°$$

Two trigonometric functions that have equal values for complementary angles are called **cofunctions**. The various cofunction pairs are apparent from the names of the trigonometric functions; for example, the names "sine" and "cosine" reflect the cofunction relationship. Similarly, the tangent and the cotangent are cofunctions as are the secant and cosecant. Making use of the fact that θ and $90° - \theta$ are complementary angles, we have

(2.4) $\left\{ \begin{matrix} \textbf{trigonometric function of} \\ \textbf{an acute angle } \boldsymbol{\theta} \end{matrix} \right\} = \left\{ \textbf{Cofunction of } \boldsymbol{(90° - \theta)} \right\}$

EXAMPLE 3

(a) $\sin 40° = \cos 50°$

(b) $\tan 5.6° = \cot 84.4°$

(c) $\cos 13°15' = \sin 76°45'$

(d) $\sec \pi/3 = \csc (\pi/2 - \pi/3) = \csc \pi/6$ ■

By the nature of the definitions, we can determine how the values of the trigonometric functions change as an angle varies from 0° to 90°. As shown in Figure 2.10, for a fixed value of r, the ordinate of a point on the terminal side increases from 0 to a length equal to r as the angle increases from 0° to 90°. On the other hand, the abscissa of the point decreases from r to 0 as the angle increases from 0° to 90°.

Thus, we can construct the table of variation of the values (Table 2.2). While not giving the exact values, this table tells approximately how the functions vary. (Again, study it carefully and be able to explain the table.)

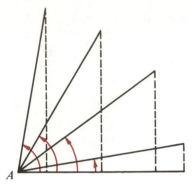

FIGURE 2.10

Variation of
ordinates as
angle *A* increases
(fixed radius)

TABLE 2.2

Variation in the
Trigonometric
Function As *A*
increases from 0°
to 90°

sin *A*	increases from 0 to 1
cos *A*	decreases from 1 to 0
tan *A*	increases from 0 and runs through all real numbers
cot *A*	decreases to 0, running through all real numbers
sec *A*	increases from 1, running through all real numbers ≥ 1
csc *A*	decreases to 1, running through all real numbers ≥ 1

Tables 2.1 and 2.2 are obviously incomplete; the first one is incomplete because it tabulates the functional values corresponding to only three angles; the second, because it gives only a general idea of the nature of the variation. Specific values of the trigonometric ratios, which are computed by methods beyond the scope of this book, are listed in table form for our use. Table 2.3 is a table of trigonometric ratios for angles measured to the nearest degree. Three additional tables of trigonometric ratios are included in the appendix. Table A is tabulated in degrees to the nearest 10′ increment; Table B is tabulated in degrees to the nearest 0.1° increment; and Table C is tabulated in radians to the nearest 0.01 radian increment.

Table 2.3 is representative of most trigonometry tables that are tabulated in degrees in that it apparently includes only those angles between 0° and 45°. This is so because the values of the functions for angles between 45° and 90° are the same as the values of the cofunction between 0° and 45°. Thus sin 57° = cos (90° − 57°) = cos 33°. Further, most tables that take advantage of this relation between the cofunctions of complementary angles place the complementary angle to the right of the table. The names of the function to be read for that particular angle are then located at the bottom; thus the table does "double duty."

To summarize the use of Table 2.3 in this chapter and Tables A and B in the Appendix:

(1)   To find the values of the trigonometric ratios for angles between 0° and 45°, locate the angle at the left hand side of the table and the name of the function at the top of the column.

(2)   To find the values of the trigonometric ratios for angles between 45° and 90°, locate the angle at the right hand side of the table and the name of the function at the bottom.

(3)   Opposite the angle, in the appropriate column, is found the value of the trigonometric function.

TABLE 2.3	θ	$\sin \theta$	$\cos \theta$	$\tan \theta$	$\cot \theta$	$\sec \theta$	$\csc \theta$	
Three Place Table of Trigonometric Functions of Degrees	0°	.000	1.00	.000	—	1.00	—	90°
	1°	.018	1.00	.018	57.3	1.00	57.3	89°
	2°	.035	.999	.035	28.6	1.00	28.7	88°
	3°	.052	.999	.052	19.1	1.00	19.1	87°
	4°	.070	.998	.070	14.3	1.00	14.3	86°
	5°	.087	.996	.088	11.4	1.00	11.5	85°
	6°	.105	.995	.105	9.51	1.01	9.57	84°
	7°	.122	.993	.123	8.14	1.01	8.21	83°
	8°	.139	.990	.141	7.12	1.01	7.19	82°
	9°	.156	.988	.158	6.31	1.01	6.39	81°
	10°	.174	.985	.176	5.67	1.02	5.76	80°
	11°	.191	.982	.194	5.14	1.02	5.24	79°
	12°	.208	.978	.213	4.70	1.02	4.81	78°
	13°	.225	.974	.231	4.33	1.03	4.45	77°
	14°	.242	.970	.249	4.01	1.03	4.13	76°
	15°	.259	.966	.268	3.73	1.04	3.86	75°
	16°	.276	.961	.287	3.49	1.04	3.63	74°
	17°	.292	.956	.306	3.27	1.05	3.42	73°
	18°	.309	.951	.325	3.08	1.05	3.24	72°
	19°	.326	.946	.344	2.90	1.06	3.07	71°
	20°	.342	.940	.364	2.75	1.06	2.92	70°
	21°	.358	.934	.384	2.61	1.07	2.79	69°
	22°	.375	.927	.404	2.48	1.08	2.68	68°
	23°	.391	.921	.425	2.36	1.09	2.56	67°
	24°	.407	.914	.445	2.25	1.10	2.46	66°
	25°	.423	.906	.466	2.14	1.10	2.37	65°
	26°	.438	.899	.488	2.05	1.11	2.28	64°
	27°	.454	.891	.510	1.96	1.12	2.20	63°
	28°	.470	.883	.554	1.88	1.13	2.13	62°
	29°	.485	.875	.532	1.80	1.14	2.06	61°
	30°	.500	.866	.577	1.73	1.16	2.00	60°
	31°	.515	.857	.601	1.66	1.17	1.94	59°
	32°	.530	.848	.625	1.60	1.18	1.89	58°
	33°	.545	.839	.649	1.54	1.19	1.84	57°
	34°	.559	.829	.675	1.48	1.21	1.79	56°
	35°	.574	.819	.700	1.43	1.22	1.74	55°
	36°	.588	.809	.727	1.38	1.24	1.70	54°
	37°	.602	.799	.754	1.33	1.25	1.66	53°
	38°	.616	.788	.781	1.28	1.27	1.62	52°
	39°	.629	.777	.810	1.23	1.29	1.59	51°
	40°	.643	.766	.839	1.19	1.31	1.56	50°
	41°	.656	.755	.869	1.15	1.33	1.52	49°
	42°	.669	.743	.900	1.11	1.35	1.50	48°
	43°	.682	.731	.933	1.07	1.37	1.47	47°
	44°	.695	.719	.966	1.04	1.39	1.44	46°
	45°	.707	.707	1.00	1.00	1.41	1.41	45°
		$\cos \theta$	$\sin \theta$	$\cot \theta$	$\tan \theta$	$\csc \theta$	$\sec \theta$	θ

EXAMPLE 4   Use Table 2.3 to verify that

(a) $\sin 34° = 0.559$

(b) $\cos 7° = 0.993$

(c) $\tan 76° = 4.01$

(d) $\sec 21° = 1.07$

(e) $\sin 81° = 0.988$ ■

EXAMPLE 5   Use Tables A and B to verify that

Table A   (a) $\sin 34°10' = 0.5616$

(b) $\cos 54°30' = 0.5807$

(c) $\tan 65°40' = 2.2113$

Table B   (a) $\cos 17.3° = 0.9548$

(b) $\tan 39.9° = 0.8361$

(c) $\csc 70.4° = 1.602$ (Find sin 70.4° and use csc 70.4° = 1/sin 70.4°)   ■

EXAMPLE 6   Use Table C to verify that

(a) $\sin 0.72 = 0.6594$

(b) $\cot 1.11 = 0.4964$

(c) $\sec 0.23 = 1.027$ ■

In closing this section, we remark that trigonometric tables are used in two ways. The first is to find the value of a trigonometric function if an angle is given. The other is to find the value of the angle when the value of the trigonometric function is known. For example, if you are given that $\cos \theta = \frac{1}{2}$, you know that $\theta = 60°$.

EXAMPLE 7   Use Table 2.3 to find the angle θ if $\sin \theta = 0.358$.

SOLUTION   We examine Table 2.3 running down the column headed "sin θ" until we come to 0.358. Then we read across and find that this corresponds to an angle of 21°. This is written

$$\sin \theta = 0.358$$

$$\theta = 21°$$ ■

EXAMPLE 8   Use Table 2.3 to find angle β if $\tan \beta = 1.60$.

SOLUTION   Running down the column headed "tan θ" at the top of the page we fail to reach 1.60 before coming to 45°; therefore, we continue looking in the column headed "tan θ" at the bottom until we come to 1.60. Reading "angle" from the right hand column, we find that $\beta = 58°$, that is,

$$\tan \beta = 1.60$$

$$\beta = 58°$$ ■

EXAMPLE 9   Use Table A to find ϕ if tan $\phi = 1.5900$.

SOLUTION   Running down the column headed "tan θ" at the top of the pages of Table A we fail to reach 1.5900 before coming to 45°; therefore, we continue looking in the column headed "tan θ" at the bottom until we come to 1.5900. Reading "angle" from the right hand column, we find that $\phi = 57°50'$, that is

$$\tan \phi = 1.5900; \quad \phi = 57°50'$$

EXAMPLE 10   Use Table C to find x if sec $x = 1.229$.

SOLUTION   Scanning the column headed "sec t", until we come to 1.229, we conclude that $x = 0.62$ radians. This is written

$$\sec x = 1.229$$
$$x = 0.62 \text{ radians}$$

Exercises for Section 2.3

Use Table A, B or C to find the values of the trigonometric functions indicated in Exercises 1–24.

1. sin 13°	**2.** sin 46°10′	**3.** tan 17°30′
4. cos 61°40′	**5.** sec 5°10′	**6.** tan 44°
7. cot 17°50′	**8.** csc 38°20′	**9.** sin 75.5°
10. cot 56.5°	**11.** cos 80°	**12.** sec 49°10′
13. sin 14°20′	**14.** cos 47°40′	**15.** sec 64°50′
16. tan 25°30′	**17.** cot 57°20′	**18.** csc 70°40′
19. tan 1.23	**20.** sin 0.54	**21.** cot 1.00
22. sec 0.02	**23.** cot 0.78	**24.** csc 0.50

Using Table A, find the degree measure of the angle ϕ for Exercises 25–33.

25. sin $\phi = 0.4617$	**26.** cot $\phi = 0.5354$	**27.** tan $\phi = 3.7321$
28. sec $\phi = 2.0957$	**29.** cos $\phi = 0.5568$	**30.** csc $\phi = 1.2335$
31. cot $\phi = 0.3089$	**32.** sin $\phi = 0.7451$	**33.** cos $\phi = 0.9983$

Using Table C, find the radian measure of the angle β for Exercises 34–39.

34. sin $\beta = 0.8415$	**35.** cos $\beta = 0.8253$	**36.** tan $\beta = 4.072$
37. cot $\beta = 1.462$	**38.** sec $\beta = 1.180$	**39.** csc $\beta = 5.295$

2.4   Trigonometric Ratios from a Calculator

The trigonometric functions can also be evaluated using a calculator. Most scientific calculators have keys for the sine, the cosine and the tangent. The other three trigonometric values are obtained from these three by using the reciprocal button. Many scientific calculators operate with both degrees and radians. For such calculators a separate button or switch enables the user to operate in either mode, thus making conversion unnecessary. However, degree and radian "modes" are not available on every calculator that has trigonometric function capability. Some have an automatic conversion key from degrees to radians that must be pushed each time; while still others require that the radian measure be multiplied (each time) by $180/x$ to obtain degrees. Be sure to determine which units you are using before activating your calculator.

EXAMPLE 1   Use a calculator to find (a) sin 34°18′; (b) sec 1.25.

SOLUTION   (a) Put the calculator in the degree mode and express 18′ in degrees. Thus, dividing 18′ by 60′/degree we have

$$34°18′ = 34.3°$$

Enter 34.3 and push the $\boxed{\text{sin}}$ button to obtain 0.5635. Thus,

$$\sin 34°18′ = 0.5635$$

(b) To find sec 1.25 we note that $\sec \theta = \dfrac{1}{\cos \theta}$. Put the calculator in the radian mode. Enter 1.25 and push the $\boxed{\text{cos}}$ button to obtain 0.3153. Now push the $\boxed{1/x}$ button to obtain 3.171. Thus,

$$\sec 1.25 = \frac{1}{\cos 1.25} = 3.171. \qquad \blacksquare$$

The procedure for finding the angle corresponding to a given functional value varies from brand to brand, but most calculators require that you enter the given number and then push an $\boxed{\text{inv}}$ or $\boxed{\text{arc}}$ button prior to the trigonometric function button. Other models have single buttons labeled $\boxed{\sin^{-1}}$, $\boxed{\cos^{-1}}$, or $\boxed{\tan^{-1}}$ for this purpose. The mathematical nature of this operation is explained in the chapter on inverse functions. The procedure for finding θ when either sin θ, cos θ, or tan θ is known is straightforward and is explained in the following example.

EXAMPLE 2   Use a calculator to find θ in degrees and minutes, if tan $\theta = 1.0455$.

SOLUTION   To find θ, put the calculator in the degree mode and enter 1.0455. Now push the $\boxed{\text{inv}}$ button followed by the $\boxed{\text{tan}}$ button to obtain 46.2743°. To convert the decimal part of the angle to minutes, subtract 46 from the number in the register to obtain 0.2743°. Now multiply this by 60 to get 16.458′. Rounding off to the nearest minute we have

$$\theta = 46°16′ \qquad \blacksquare$$

The procedure for finding θ when either csc θ, sec θ, or cot θ is known makes use of the reciprocal relationships (2.1) and is outlined in the next example. The reasoning behind the procedure must wait until inverse trigonometric functions are introduced in Chapter 8.

EXAMPLE 3   Find the measure of θ to the nearest tenth of a degree if sec $\theta = 2.178$.

SOLUTION   First, put the calculator in the degree mode. Enter 2.178 and push the ⎡1/x⎤ button. Now push the ⎡ inv ⎤ button followed by the ⎡ cos ⎤ button to obtain 62.7. Thus, if sec $\theta = 2.178$, $\theta = 62.7°$. ■

EXAMPLE 4   Find the measure of θ to the nearest minute if cot $\theta = 0.6247$.

SOLUTION   With the calculator in the degree mode, enter 0.6247 and push the ⎡1/x⎤ button. (Remember tan $\theta = \dfrac{1}{\cot \theta}$.) Now push the ⎡ inv ⎤ button followed by the ⎡ tan ⎤ button to obtain 58.007. To convert 0.007 to minutes multiply by 60 to obtain 0.42'. Thus, $\theta = 58°0'$. ■

EXAMPLE 5   Find the measure of ϕ to the nearest hundredth of a radian if sin $\phi = 0.6843$.

SOLUTION   With your calculator in the radian mode, enter 0.6843 and push ⎡ inv ⎤ ⎡ sin ⎤ to obtain 0.7536. Rounding-off to the nearest hundredth yields 0.75 radians. ■

Exercises for Section 2.4

In Exercises 1–24 use a calculator to find the values of the trigonometric functions indicated.

1. cos 39.5°   **2.** sin 42.6°   **3.** csc 60.7°   **4.** tan 17.5°

5. sec 87.3°   **6.** sin 77.9°   **7.** sin 1.29   **8.** cot 1.08

9. sec 0.33   **10.** tan 0.53   **11.** cos 0.08   **12.** csc 1.27

13. sin 5°16'   **14.** csc 17°29'   **15.** tan 72°55'   **16.** cot 63°36'

17. sin 80°33'   **18.** cos 22°22'   **19.** sec 48°3'   **20.** cot 61°10'

21. csc 20°9'   **22.** $\sin^2 35° + \cos^2 35°$

23. $\sin^2 52.4° + \cos^2 52.4°$ **24.** $\sin^2 0.84 + \cos^2 0.84$

In Exercises 25–33 use a calculator to find the value of θ to the nearest tenth of a degree.

25. sin $\theta = 0.5546$ **26.** cos $\theta = 0.0442$ **27.** sec $\theta = 1.225$ **28.** tan $\theta = 0.8439$

29. cot $\theta = 1.053$ **30.** cot $\theta = 0.8091$ **31.** cos $\theta = 0.1138$ **32.** sin $\theta = 0.0016$

33. csc $\theta = 1.401$

In Exercises 34–42 use a calculator to find the value of x to the nearest minute.

34. tan $x = 2.232$ **35.** sec $x = 3.007$ **36.** sin $x = 0.3035$ **37.** csc $x = 1.599$

38. $\cot x = 0.7892$ **39.** $\tan x = 0.0874$ **40.** $\cos x = 0.5729$ **41.** $\sin x = 0.6666$

42. $\csc x = 4.526$

In Exercises 43–50, use a calculator to show that the left member is equal to the right member.

43. $\sin 42.7° = \dfrac{\tan 42.7°}{\sec 42.7°}$ **44.** $\tan 33.2° = \dfrac{\sin 33.2°}{\cos 33.2°}$ **45.** $\cot 2.1 = \dfrac{\cos 2.1}{\sin 2.1}$

46. $\tan 1.3 = \dfrac{\sin 1.3}{\cos 1.3}$ **47.** $\tan 0.6 = \dfrac{\sin 0.6}{\cos 0.6}$ **48.** $\cot 62.2° = \dfrac{\cos 62.2°}{\sin 62.2°}$

49. $\sin^2 15° + \cos^2 15° = 1$ **50.** $\sin^2 0.75 + \cos^2 0.75 = 1$

51. The computation of the displacement of the end of a vibrating spring from its equilibrium position uses $\sin 1.316$. Find this value.

52. In solving the equations for the trajectory of a projectile the expression $\cos \theta = 0.6578$ arises. Find angle θ to the nearest minute.

2.5

SKIP

Interpolation

Interpolation is a method of estimating a value between two given values. All mathematical tables, and in particular trigonometric values, are of necessity tabulated in discrete steps. Thus, when using such tables you will often find it necessary to use interpolated values between tabulated values. For example, if Table A is being used for the trigonometric values, then the values for $41°15'$ are not immediately available but must be estimated by using those in the table for $41°10'$ and $41°20'$.

   Obviously, a calculator with trigonometric function capability makes the use of interpolation unnecessary since the table itself in that case is unnecessary. However, many other tables of science and engineering are not available on calculators and hence the technique of interpolation will often prove useful. We explain the interpolation process in the context of the trigonometric tables, since that is an immediate application.

   The type of interpolation that is the easiest to use and understand is called **interpolation by proportional parts** or **linear interpolation**. Although it is only approximately true, we assume that a change in the angular measurement is proportional to a linear change in the value of the trigonometric function.

   A graphical display of a typical error introduced by the assumption of linearity is shown in Figure 2.11 for a trigonometric function whose values are increasing from θ_1 to θ_2. The value for the given trigonometric ratio at θ_1 and θ_2 are assumed to be known from a table, and we assume no such tabulated value is known for the angle θ. Hence the necessity to use an interpolated value.

   The interpolated value of the function at any θ between θ_1 and θ_2 is the distance from the x-axis to the point P, and the actual value is the distance from the x-axis to the point Q. Thus, the error in using linear interpolation is length of the line segment PQ. As long as the interval from θ_1 to θ_2 is relatively small and the difference in the known tabulated values is not too large, the error introduced will usually be acceptable.

Using Figure 2.11, we can derive an equation for determining the interpolated value in terms of tabulated values. Since triangles OPA and OCB are similar, the corresponding sides are proportional, and hence,

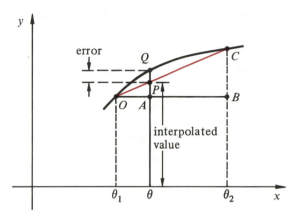

FIGURE 2.11

$$\frac{\overline{PA}}{\overline{CB}} = \frac{\theta - \theta_1}{\theta_2 - \theta_1}$$

where $\overline{PA}$ and $\overline{CB}$ are the distances between the points P and A and C and B, respectively. In words,

(2.5) $\dfrac{\textbf{interpolated value at } \theta - \textbf{tabulated value at } \theta_1}{\textbf{tabulated value at } \theta_2 - \textbf{tabulated value at } \theta_1} = \dfrac{\theta - \theta_1}{\theta_2 - \theta_1}$

The next few examples will help you to learn the method of linear interpolation.

EXAMPLE 1   Use Table 2.3 and the method of linear interpolation to approximate sin 31°35′.

SOLUTION   From Table 2.3, we have sin 31° = 0.515 and sin 32° = 0.530. For an increase of 1° in the angle, the sine increases by 0.530 − 0.515 = 0.015. Then for an angular increase of 35/60 degree, the sine is assumed to increase by (35/60)(0.015) = 0.009. Because sin 31°35′ > sin 31°00′, this correction is *added* to 0.515, giving sin 31°35′ = 0.524. The above discussion is summarized in the table below.

		Angle	Sine		
60	35	31°00′	0.515	c	0.015
		31°35′	....		
		32°00′	0.530		

$$\frac{c}{0.015} = \frac{35}{60}$$

$$c = \frac{35}{60}(0.015)$$

$$= 0.009$$

We round off c to three decimal places since this is a three place table. Thus,

$$\sin 31°35' = \sin 31° + 0.009 = 0.515 + 0.009 = 0.524$$

■

EXAMPLE 2   Use Table 2.3 to approximate $\cos 52°15'$.

SOLUTION   The interpolation is shown below.

		Angle	Cosine		
60	15	52°00' 52°15' 53°00'	0.616 . . . . 0.602	c	0.014

$$\frac{c}{0.014} = \frac{15}{60}$$

$$c = \frac{15}{60}(0.014)$$

$$= 0.004$$

Since $\cos 52°15' < \cos 52°$, the correction c is *subtracted* from 0.616. Thus,

$$\cos 52°15' = 0.616 - 0.004 = 0.612.$$

■

Naturally, an interpolated value will depend on the table used to perform the interpolation. We usually consider an interpolated value to be more accurate if a table with finer divisions is used.

EXAMPLE 3   Repeat the previous example using Table A in the Appendix.

SOLUTION   From Table A, $\cos 52°10' = 0.6134$ and $\cos 52°20' = 0.6111$.

		Angle	Cosine		
10	5	52°10' 52°15' 52°20'	0.6134 . . . . 0.6111	c	0.0023

$$\frac{c}{0.0023} = \frac{5}{10}$$

$$c = \frac{5}{10}(0.0023)$$

$$= 0.0012$$

Since $\cos 52°15' < \cos 52°10'$, we subtract the correction from 0.6134. Thus,

$$\cos 52°15' = 0.6134 - 0.0012 = 0.6122$$

■

EXAMPLE 4   Use Table C to find $\tan \sqrt{2}$.

SOLUTION   Table C lists angles in radians to two decimal places. We can therefore use interpolation to find trigonometric functions of angles to three decimal places. Using $\sqrt{2} = 1.414$, we seek $\tan 1.414$.

		Angle	Tangent		
.010	.004	1.410 1.414 1.420	6.165 . . . . 6.581	c	.416

$$\frac{c}{.416} = \frac{.004}{.010}$$

$$c = 0.166$$

Therefore, $\tan \sqrt{2} = 6.165 + 0.166 = 6.331$ ∎

Interpolation is also used to approximate the measure of an unknown angle whose trigonometric ratio is given but does not correspond exactly to any of the tabulated values.

EXAMPLE 5   Use Table A to find angle θ to the nearest minute if $\tan \theta = 0.3$.

SOLUTION   From Table A, we find $\tan 16°40' = 0.2994$ and $\tan 16°50' = 0.3026$.

	Angle	Tangent	
10 $\begin{bmatrix} c \begin{bmatrix} \end{bmatrix} \end{bmatrix}$	16°40' .... 16°50'	$\begin{bmatrix} 0.2994 \\ 0.3000 \end{bmatrix}$ 0.0006 0.3026	0.0032

$$\frac{c}{10} = \frac{0.0006}{0.0032}$$

$$c = \frac{0.0006}{0.0032}(10) = 2'$$

Adding this 2′ correction to 16°40′, we have $\theta = 16°42'$.   ∎

EXAMPLE 6   Use Table C to estimate the value of angle θ if $\cos \theta = 0.8145$.

SOLUTION   From the table, we have $\cos 0.61 = 0.8196$ and $\cos 0.62 = 0.8139$.

	Angle	Cosine	
0.010 $\begin{bmatrix} c \begin{bmatrix} \end{bmatrix} \end{bmatrix}$	0.610 .... 0.620	$\begin{bmatrix} 0.8196 \\ 0.8145 \end{bmatrix}$ 0.0051 0.8139	0.0057

$$\frac{c}{0.010} = \frac{0.0051}{0.0057}$$

$$c = 0.009$$

The correction c is rounded off to the nearest thousandth to be consistent with the accuracy of the table. Therefore, $\theta = 0.619$ radian.   ∎

Exercises for Section 2.5

By interpolation, find the value of the trigonometric functions in Exercises 1–9. Use Table A in the Appendix.

1. $\sin 38°38'$ **2.** $\tan 15°32'$ **3.** $\cos 26°55'$

4. $\sec 42°15'$ **5.** $\cot 38°45'$ **6.** $\csc 70°5'$

7. $\sin 65°26'$ **8.** $\cos 15°9'$ **9.** $\tan 50°43'$

By interpolation, find the value of the trigonometric functions in Exercises 10–18. Use **Table C** in the Appendix.

10. $\tan 1.425$ **11.** $\sin \pi/8$ **12.** $\sin 1/8$

13. $\cos(\sqrt{3}/2)$

14. $\sec 1.083$

15. $\tan(\sqrt{2}/2)$

16. $\sin 0.531$

17. $\cos 0.012$

18. $\cot(\sqrt{3}/2)$

Use Table A to find the value of θ in degrees and minutes to the nearest minute by interpolation in Exercises 19–30.

19. $\sin\theta = 0.776$

20. $\cos\theta = 0.3000$

21. $\tan\theta = 1.4$

22. $\cot\theta = 2$

23. $\cos\theta = 0.5108$

24. $\sin\theta = 0.4804$

25. $\tan\theta = 0.5$

26. $\cot\theta = 0.5$

27. $\sec\theta = 1.8$

28. $\tan\theta = 0.3692$

29. $\csc\theta = 4$

30. $\sin\theta = 0.1234$

Use Table C to find the radian measure of θ to the nearest thousandth, where θ is a positive acute angle (Exercises 31–42). Use interpolation where necessary.

31. $\sin\theta = 0.1365$

32. $\cos\theta = 0.7976$

33. $\tan\theta = 0.4040$

34. $\sin\theta = \pi/6$

35. $\cos\theta = \pi/8$

36. $\tan\theta = \pi/3$

37. $\sin\theta = 0.6541$

38. $\tan\theta = 2$

39. $\sin\theta = 0.105$

40. $\cot\theta = 0.9602$

41. $\cos\theta = 0.7$

42. $\sec\theta = 2$

2.6 Solution of Right Triangles

One of the principal uses of the trigonometric functions is to compute dimensions of right triangles. This may seem to be largely of academic interest, but, as you will see in this section, the types of applications can be quite modern.

A right triangle is composed basically of six parts, the three sides and the three angles. To *solve a triangle* means that we must find the values of each of these six parts. Of course, they are not all independent. For example, if two of the angles are known, so is the other one. If two of the sides of a right triangle are known, the remaining side is not of arbitrary length but must be such that all three sides, together, satisfy the Pythagorean theorem.

In discussing right triangles it is customary to designate the vertices and corresponding angles by the capital letters A, B, and C as shown in Figure 2.12(a). In this convention the right angle is usually denoted by the letter C and the lower case letters a, b, and c are used to designate the sides opposite angles A, B, and C respectively. Side c is then the hypotenuse of the triangle.

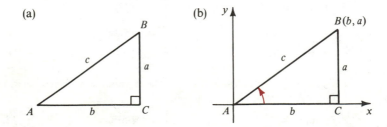

FIGURE 2.12

To see how the trigonometric functions are related to the parts of a right triangle, consider the right triangle shown in Figure 2.12(b). By locating angle A in standard position the coordinates of the vertex at B are (b, a) and the distance from A to B of a (A, B) is c. Thus, the six trigonometric functions of angle A in terms of the sides of the standard right triangles are

(2.6)

$$\sin A = a/c \qquad \csc A = c/a$$
$$\cos A = b/c \qquad \sec A = c/b$$
$$\tan A = a/b \qquad \cot A = b/a$$

The sides of a right triangle are often referenced to one of the two acute angles. For example, the side of length a is called the **side opposite** angle A, the side of length b is called the **side adjacent** to angle A, and the side of length c is called the **hypotenuse**. Using this terminology, the six trigonometric functions in Definition 2.1 become

(2.7)

$$\sin A = \frac{a}{c} = \frac{\text{opposite side}}{\text{hypotenuse}} \qquad \csc A = \frac{c}{a} = \frac{\text{hypotenuse}}{\text{opposite side}}$$

$$\cos A = \frac{b}{c} = \frac{\text{adjacent side}}{\text{hypotenuse}} \qquad \sec A = \frac{c}{b} = \frac{\text{hypotenuse}}{\text{adjacent side}}$$

$$\tan A = \frac{a}{b} = \frac{\text{opposite side}}{\text{adjacent side}} \qquad \cot A = \frac{b}{a} = \frac{\text{adjacent side}}{\text{opposite side}}$$

The definitions in (2.7) are convenient for solving right triangles.

When a right triangle is given, the six parts may be completely determined if you know two parts, at least one of which is a side.

- If an angle and one of the sides is given, then the third angle is simply the complement of the one given. The other two sides are obtained from the values of the known trigonometric functions.

- If two sides are given, the value of the third side is obtained from the Pythagorean theorem. The angles may then be determined by taking ratios of the sides which will uniquely determine the value of some trigonometric function.

Thus, in solving right triangles, we make use of the trigonometric functions, the Pythagorean theorem, and the fact that the two acute angles are complementary. You will usually find it to your advantage to make a rough sketch of the triangle. This will help you to determine what is given and which trigonometric functions must be used to find the unknown parts.

The relationship between the accuracy of the sides and the angles is given in the following table. We will use this convention in the examples and the answers to the exercises.

Accuracy of Sides	Accuracy of Angles
2 significant figures	Nearest degree
3 significant figures	Nearest 0.1° or 10′
4 significant figures	Nearest 0.01° or 1′

EXAMPLE 1   Solve the triangle in Figure 2.13.

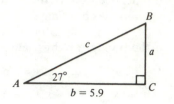

FIGURE 2.13

SOLUTION   Since A and B are complementary angles, $B = 90° - 27° = 63°$. Also, $\tan A$ = opposite side/adjacent side, so

$$\tan 27° = \frac{a}{5.9}$$

Solving for a, we get

$$a = 5.9 \tan 27° = 5.9(0.5095) = 3.0$$

Similarly,

$$\cos 27° = \frac{5.9}{c}$$

so

$$c = \frac{5.9}{\cos 27°} = \frac{5.9}{0.8910} = 6.6$$ ■

EXAMPLE 2   A ladder of 20.4 feet is placed against a building so that its lower end is 4.75 feet from the base of the building. What angle does the ladder make with the ground?

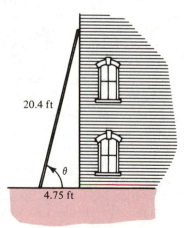

20.4 ft

θ

4.75 ft

FIGURE 2.14

SOLUTION   The desired angle is designated θ in Figure 2.14. From the figure, we see that

$$\cos \theta = \frac{\text{adjacent side}}{\text{hypotenuse}} = \frac{4.75}{20.4} = 0.233$$

From Table A, we find $\theta = 76°30'$ to the nearest 10′.   ■

EXAMPLE 3   An engineer wishing to know the width of a river walks 100 yd downstream from a point directly across from a tree on the opposite bank. If the angle between the river bank and the line of sight to the tree at this point is 55.1°; what is the width of the river? (See Figure 2.15.)

SOLUTION   From Figure 2.15, you can see that this is a typical problem in right triangle trigonometry where the "side opposite" the 55.1° angle represents the unknown distance across the river. Thus,

$$\frac{d}{100} = \tan 55.1°$$

from which

$$d = 100(1.4335) = 143 \text{ yd}$$

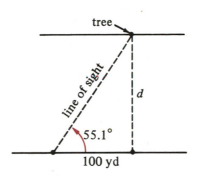

FIGURE 2.15

EXAMPLE 4   In Figure 2.16, a radar station tracking a missile indicates the angle of elevation to be 20.7° and the line of sight distance (called the **slant range**) to be 38.2 km. Determine the altitude and horizontal range of the missile.

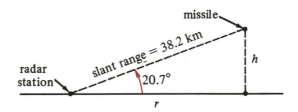

FIGURE 2.16

SOLUTION   The altitude is

$$h = 38.2 \sin 20.7° = 38.2(0.3535) = 13.5 \text{ km}$$

and the horizontal range is

$$r = 38.2 \cos 20.7° = 38.2(0.9354) = 35.7 \text{ km}$$

EXAMPLE 5   An airplane flying at a speed of 185 ft/sec starts to descend to the runway on a straight line glide path that is 7.5° below the horizontal. If the plane is at a 2980 ft altitude at the start of the glide path, how long does it take for the plane to touch down?

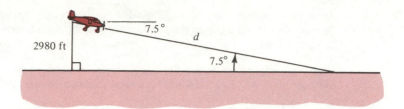

FIGURE 2.17

SOLUTION Referring to Figure 2.17 you can see that the length of the glide path is the hypotenuse of a right triangle. Therefore, we can write

$$\sin 7.5° = \frac{2980}{d}$$

$$d = \frac{2980}{\sin 7.5°} = \frac{2980}{0.1305} = 22{,}800 \text{ ft}$$

To find the time required for descent recall that

$$\text{velocity} = \frac{\text{distance}}{\text{time}}$$

Or, solving for time

$$\text{time} = \frac{\text{distance}}{\text{velocity}} = \frac{22{,}800 \text{ ft}}{185 \text{ ft/sec}} = 123 \text{ sec}$$

Thus, the descent takes 123 sec = 2.05 min. ■

Example 6 shows how right triangle trigonometry can be used to find the height of an object when a side of a right triangle cannot be obtained. The procedure is to measure the angle of elevation to the top of the object at two different locations and the distance between the two locations.

EXAMPLE 6 Two observers that are 4250 feet apart measure the angle of elevation to the top of a mountain to be 18.7° and 25.3° respectively. See Figure 2.18. Find the height of the mountain.

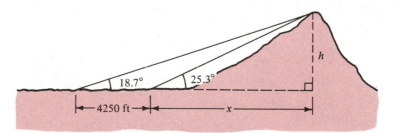

FIGURE 2.18

SOLUTION From Figure 2.18 we can write the two equations

(1) $$\tan 18.7° = \frac{h}{x + 4250}$$

(2) $\tan 25.3° = \dfrac{h}{x}$

The only two unknowns in these two equations are x and h. Solving (2) for x, we get

$$x = \dfrac{h}{\tan 25.3°} = h \cot 25.3°$$

Substituting this expression into equation (1), we have

$$\tan 18.7° = \dfrac{h}{h \cot 25.3° + 4250}$$

To solve this equation for h we proceed as follows.

$$(h \cot 25.3° + 4250) \tan 18.7° = h$$

$$h \cot 25.3° \tan 18.7° + 4250 \tan 18.7° = h$$

$$(1 - \cot 25.3° \tan 18.7°)h = 4250 \tan 18.7°$$

$$h = \dfrac{4250 \tan 18.7°}{1 - \cot 25.3° \tan 18.7°}$$

$$= \dfrac{4250(0.3385)}{1 - (2.116)(0.3385)}$$

$$= 5070$$

Thus, the top of the mountain is 5070 feet above the observers.   ■

Exercises for Section 2.6

Solve the right triangles in Exercises 1–5.

1.

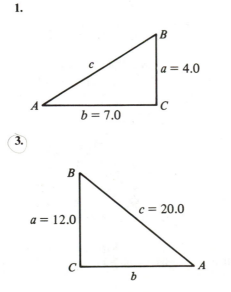

2.

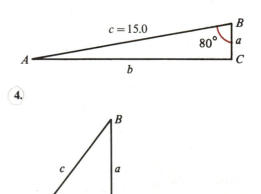

3.

4.

5.

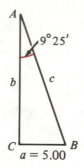

6. One side of a rectangle is half as long as the diagonal. The diagonal is 5.0 m long. How long are the sides of the rectangle? Solve without using the Pythagorean theorem.

7. A woman is walking along the prairie and stops to measure the angle of inclination to a high mountain. It measures 30°. The woman then walks a kilometer toward the mountain and measures again. This time the angle of inclination is 45°. How high is the mountain?

8. In Japan, it is common to brace trees by means of poles. Suppose that a pole 14 ft long is used to brace a tree standing on level ground. Suppose that the end of the pole touching the ground is 10 ft away from the base of the tree. What is the size of the angle that the pole makes with the ground?

9. One morning, a man 6.0 ft tall wanted to know the time, but no one in the vicinity had a timepiece of any kind. The man knew that on that day the sun rose at 6.00 A.M. and would be directly overhead at noon. A friend measured the man's shadow which was 10 ft long. Approximately what time was it?

10. A solar collector is placed on the roof of a house as shown in Figure 2.19. What angle does the collector make with the vertical?

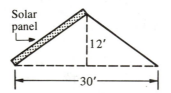

FIGURE 2.19

11. A solar panel (as shown in Figure 2.20) is to be tilted so that angle $\phi = 100°$ when the elevation angle of the sun is 27°. Find h, if the length of the panel is 6.4 m.

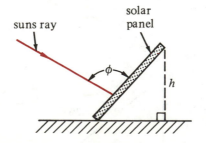

FIGURE 2.20

12. A television antenna stands on top of a house which is 20 ft tall. The angle subtended by the antenna from a point 30 ft from the base of the building is 15°. Find the height of the antenna.

13. In designing a steel truss for a bridge as shown in Figure 2.21, it is desired that BC shall be 10.0 m and that AC shall be 7.00 m. What angle will AB make with AC? with BC?

FIGURE 2.21

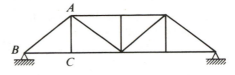

14. At noon in the tropics when the sun is directly overhead, a fisherman holds his 5.0-meter pole inclined 30° to the horizontal. How long is the shadow of the pole? How high is the tip of the pole above the level of the other end?

15. A boy walking southward along a straight road turns to the left at a point P and continues along a path that runs straight for a distance of 500 m to a spring S. He then takes another path running at right angles to the road and returns to the road at point Q, which is 190 m from P. Find the angle at P that the left hand path makes with the road and the angle at S between the two paths.

16. A flagpole broken over by the wind forms a right triangle with the ground. The angle that the broken part makes with the ground is 60°, and the distance from the tip of the pole to the foot is 40 ft. How tall was the pole?

17. An observer stands 550 m from the base of a building and measures the angles of elevation of the top and bottom of a radio tower on top of the building to be 33.6° and 32.1°, respectively. Find the height of the tower.

18. The length of each blade of a pair of shears from the pivot to the point is 6.0 in. When the points of the open shears are 4.0 in. apart, what angle do the blades make with each other?

19. From the top of a building that is 220.0 ft high, an observer looks down on a parking lot. If the lines of sight of the observer to two different cars in the lot are 28°15′ and 36°20′ below the horizontal, respectively, what is the distance between the two cars?

20. Lock-on for a certain automatic landing system used when visibility is poor occurs when the airplane is 4.0 miles (slant range) from the runway and at an altitude of 3800 ft. If the glide path is a straight line to the runway, what angle does it make with the horizontal?

21. An airplane takes off with an airspeed of 265 ft/sec and climbs at 8.7° with the horizontal until it reaches an altitude of 5800 ft. How long does it take the plane to reach this altitude?

22. A civil engineering student is given the following sketch of a survey and asked to find the distance x. Show how this can be done using right triangles and then compute x.

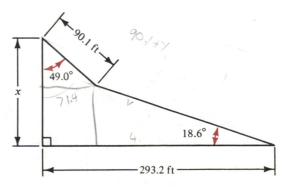

23. An observer at the base of a hill knows that the television antenna on top of the hill is 550 ft high. If the angle of elevation from the observer to the base of the antenna is 16.4° and to the top of the antenna is 29.1°, how high is the hill?

24. A fishing boat sailing due north sites a light house as 16.2° east of north. If 12.7 miles later the light house is sited as 43.7° east of north, how close will the boat come to the light house?

2.7   Components of a Vector

The trigonometric ratios have many applications in physics, some of which you will learn in this chapter. In this section, we are interested in applications to vectors. **Vectors** are entities, such as velocity and force, which require both a magnitude and direction for their description. Graphically, we may think of a vector as an arrow whose length represents the magnitude of the vector and whose angle with a reference line represents its direction. For example, an automobile traveling 60 mph at an angle of 30° north of east can be represented by the vector diagram in Figure 2.22.

FIGURE 2.22

Velocity
vector

By convention, we usually place vectors with their initial point at the origin of a rectangular coordinate system and use the positive x-axis as the reference line for the direction angle. We will designate vectors in bold type. The magnitude of a vector $\mathbf{F}$ is denoted by $|\mathbf{F}|$.

The projections of the vector onto the x- and y-axes are called the **components** of the vector. We say that a vector is resolved into its x and y components, called the **horizontal** and **vertical components** of $\mathbf{F}$ respectively. Resolving a vector into its x and y components is a simple problem of trigonometry. From Figure 2.23,

$$F_x = x \text{ component of } \mathbf{F} = |\mathbf{F}| \cos \theta$$

$$F_y = y \text{ component of } \mathbf{F} = |\mathbf{F}| \sin \theta$$

Obviously,

$$F_x^2 + F_y^2 = |\mathbf{F}|^2$$

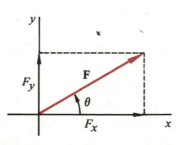

FIGURE 2.23

Components of a
vector

EXAMPLE 1   Find the horizontal and vertical components of a force vector of magnitude 15 lb acting at an angle of 30° to the horizontal.

SOLUTION

$$F_x = |\mathbf{F}| \cos \theta = (15)(0.866) = 13$$

$$F_y = |\mathbf{F}| \sin \theta = (15)(0.5) = 7.5$$

EXAMPLE 2   Find the magnitude and direction of the vector whose components are shown in Figure 2.24.

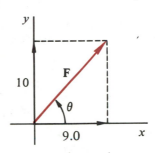

FIGURE 2.24

SOLUTION   The magnitude is

$$|\mathbf{F}| = \sqrt{10^2 + 9^2} = \sqrt{181} \approx 13 \text{ (To two significant figures)}$$

The angle that the vector makes with the horizontal is determined from

$$\tan \theta = \frac{10}{9.0} = 1.111$$

$$\theta \approx 48°$$

EXAMPLE 3   A boat that can travel at the rate of 3.0 km/hr in still water is pointed directly across a stream having a current of 4.0 km/hr. What will be the actual speed of the boat, and in which direction will the boat go?

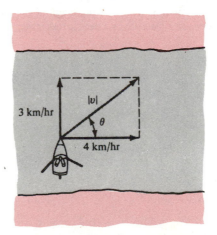

FIGURE 2.25

SOLUTION   In still water, the boat would go at right angles to the bank at the rate of 3 km/hr. But the current carries it downstream 4 units for every 3 units that it goes across. Thus, 4 is the X component of velocity and 3 is the Y component. Hence, by the Pythagorean Theorem the magnitude of the velocity is $|v| = \sqrt{3^2 + 4^2} = 5.0$. Let θ be the angle that the velocity vector makes with the bank. Then $\tan \theta = \dfrac{3}{4} = 0.75$ and hence,

$$\theta \approx 37°$$

In summary, the boat will travel at 5.0 km/hr at an angle of 37° with the bank. ■

EXAMPLE 4   A physics lab experiment on components of forces is shown schematically in Figure 2.26. Calculate the force parallel to the plane.

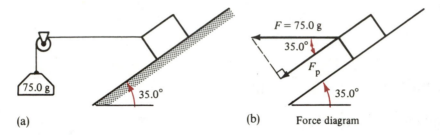

FIGURE 2.26   (a)   (b)   Force diagram

SOLUTION   The vector representation of the 75.0 g force is shown in Figure 2.26(b). Since we want the component of force parallel to the plane we must use

$$\cos 35.0° = F_p/75.0$$

Solving this for F_p we get

$$F_p = 75.0 \cos 35.0° = 75.0(0.8192) = 61.4 \text{ g}$$

as the parallel component of the applied force. ■

EXAMPLE 5   In Figure 2.27, find the force in each of the two supporting cables if the system is in equilibrium. (A force system is in static equilibrium if the algebraic sum of the horizontal forces is zero and the algebraic sum of the vertical forces is zero.)

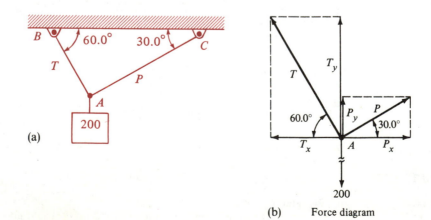

FIGURE 2.27   (b)   Force diagram

SOLUTION   The force in cable AB is represented by T and that in cable AC by P. The horizontal and vertical components of these forces at point A are shown in Figure 2.27(b). The horizontal components are

$$T_x = -T \cos 60.0° = -\frac{1}{2} T$$

$$P_x = P \cos 30.0° = \frac{\sqrt{3}}{2} P$$

and the vertical components are

$$T_y = T \sin 60.0° = \frac{\sqrt{3}}{2} T$$

$$P_y = P \sin 30.0° = \frac{1}{2} P$$

The minus sign in T_x indicates that T_x is directed to the left. Since the forces are in equilibrium, we have the following system of equations.

(1) $-\dfrac{1}{2} T + \dfrac{\sqrt{3}}{2} P = 0$ [Horizontal forces must add up to zero]

(2) $\dfrac{\sqrt{3}}{2} T + \dfrac{1}{2} P - 200 = 0$ [Vertical forces must add up to zero]

Or, multiplying each equation by 2,

(1) $-T + \sqrt{3} P = 0$

(2) $\sqrt{3} T + P = 400$

Multiplying (1) by $\sqrt{3}$ and adding to (2), we get

(1) $-\sqrt{3T} + 3P = 0$

(2) $\dfrac{\sqrt{3}T + P = 400}{4P = 400}$

$$P = 100 \text{ lb}$$

Finally, substituting $P = 100$ into (1), we get

$$T = 100\sqrt{3} \approx 173 \text{ lb}$$ ■

2.8

Applications to Navigation

In some special fields, it has long been the custom to measure direction clockwise directly from a meridian; in particular, in navigation the north line is the reference.

The **course** of a ship or an aircraft is the angle measured from the north clockwise through the east to the direction in which the ship is sailing. The **bearing** of B from A is the angle measured clockwise from the north to line segment AB. In

Figure 2.28, distinguish clearly between the bearing of B from A and the bearing of A from B.

The **heading** is the direction in which the vehicle is pointed. Note that the heading and the course may not be the same, being affected by "winds aloft" in the case of aircraft and by the current in the case of ships.

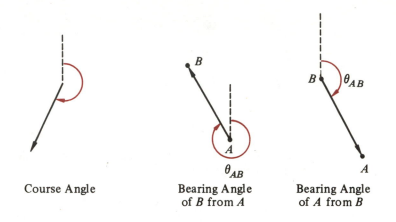

FIGURE 2.28

Course Angle   Bearing Angle   Bearing Angle
                    of B from A of A from B

The heading vector, the course vector and the wind vector comprise a right triangle as shown in Figure 2.29.

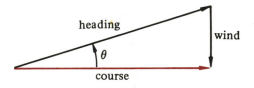

FIGURE 2.29

EXAMPLE 1 A ship sails 15.0 km on a course of 125°. It then sails 30.0 km on a course of 215°. What course must it then sail to return to its starting point? Assume that the course and heading are identical.

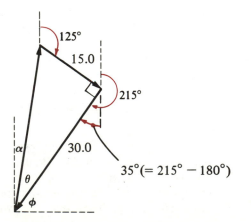

FIGURE 2.30

SOLUTION   From Figure 2.30, we conclude that

$$\tan \theta = \frac{15.0}{30.0} = \frac{1}{2} = 0.500$$

which means that

$$\theta = 26.6°$$

From Figure 2.30, $\phi = 90° - 35° = 55°$. Hence the course angle α is given by

$$\alpha = 90° - 26.6° - 55° = 8.4°$$ ∎

EXAMPLE 2   Consider a flight from Chicago to Boston to be along a west to east direction, with an airline distance of 870 statute miles (see Figure 2.31). A light plane having an airspeed of 180 mph makes the round trip. How many flying hours does it take for the round trip with a constant southerly wind of 23.0 mph? What are the headings for the two parts of the round trip?

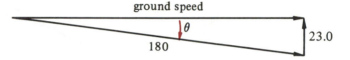

FIGURE 2.31

SOLUTION   Let θ be the angle necessary to compensate for the wind. Then

$$\sin \theta = \frac{23.0}{180} = 0.128$$

and $\theta = 7°20'$
Hence the ground speed of the plane is

$$(\cos 7°20')(180 \text{ mi/hr}) = 179 \text{ mi/hr}$$

The round trip will take

$$\frac{1,740 \text{ mi}}{179 \text{ mi/hr}} = 9.72 \text{ hr or about 9 hr 43 min}$$

The heading for the trip from west to east is $90° + 7°20'$ or $97°20'$, and the heading for the trip from east to west is $270° - 7°20'$, or $262°40'$.   ∎

$|\vec{F}|$ SIZE OR LENGTH

Exercises for Sections 2.7 and 2.8

In Exercises 1–6 find the horizontal and vertical components of the given force.

1. $|\mathbf{F}| = 25.0$ lb, $\theta = 45°10'$ 2. $|\mathbf{F}| = 100$ lb, $\theta = 65°20'$

3. $|\mathbf{F}| = 0.020$ lb, $\theta = 9°15'$ 4. $|\mathbf{F}| = 9050$ lb, $\theta = 40°17'$

5. $|\mathbf{F}| = 16.7$ lb, $\theta = 29.6°$ 6. $|\mathbf{F}| = 397.6$ lb, $\theta = 2.7°$

In Exercises 7–12 find the resultant force for the components given.

7. $F_x = 20.0$ lb, $F_y = 15.0$ lb   8. $F_x = 56.0$ lb, $F_y = 13.0$ lb

9. $F_x = 17.5$ lb, $F_y = 69.3$ lb   10. $F_x = 0.012$ lb, $F_y = 0.200$ lb

11. $F_x = 0.130$ lb, $F_y = 0.080$ lb   **12.** $F_x = 1930$ lb, $F_y = 565$ lb

13. Find the vertical and horizontal components of the force in Figure 2.32.

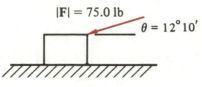

$|F| = 75.0$ lb

$\theta = 12°\,10'$

FIGURE 2.32

14. Find the vertical and horizontal components in Figure 2.32, if $|F| = 250$ lb and $\theta = 21.7°$.

15. The weight of an object is represented by a vector acting vertically downward. (See Figure 2.33.) If a 15.3-pound block rests on a plane inclined at 28.4° above the horizontal, what is the component of force acting perpendicular to the plane?

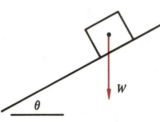

W

θ

FIGURE 2.33

16. In exercise 15, find the component of force acting parallel to the plane.

17. In Figure 2.34, find the force in the two supporting cables. *Hint*: see Example 5, Section 2.7.

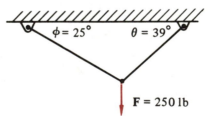

$\phi = 25°$ $\theta = 39°$

$F = 250$ lb

FIGURE 2.34

18. In Figure 2.34 find the force in each of the supporting cables if $\theta = 43.0°$ and $\phi = 32.3°$.

19. A 75.0 lb block rests on an inclined plane. If the block exerts a force of 67.0 lb perpendicular to the plane, what is the angle of inclination of the plane?

20. A horizontal force of 750 lb is applied to a block resting on a plane inclined at 17.9° above the horizontal. Find the component of the force parallel to the plane.

21. A girl weighing 100 lb sits on a swing. The swing is pulled by a horizontal rope until the swing makes an angle of 15.0° with the vertical. What is the tension in each of the supporting ropes?

22. A rocket has a velocity of 540 m/sec at 78.6° above the horizontal. How fast is the rocket rising vertically?

23. A balloon is rising at the rate of 20 ft/sec and is being carried horizontally by a wind that has a velocity of 25 mph. Find its actual velocity and the angle that its path makes with the vertical (60 mph = 88 ft/sec).

24. A boat travels at the rate of 5 kph in still water and is pointed directly across a stream having a current of 3 kph. What will be the actual speed of the boat, and in which direction will the boat go?

25. In which direction must the boat of the preceding exercise be pointed so that the boat will go straight across the stream?

26. An airplane is flying at 500 mph with a windspeed of 150 mph at right angles to it. Compute the angle between course and heading.

27. A ship is sailing due east at a constant speed. At noon, a lighthouse is observed on a bearing of 180° at a distance of 15 nautical miles. At 1 P.M. the bearing is 210°. Find the rate at which the ship is sailing and the bearing of the lighthouse at 3 P.M.

28. How far north does an airplane traveling 1000 kph fly in 1 hr if it is on a course of 300°?

29. A pilot flies a flight plan to fly from Juliette airport to Denver, which is 280 miles due north. If the airplane cruises at 160 mph and there is a constant 18.0 mph wind from the west, what heading should the pilot fly? How long will it take to make the trip?

Figure 2.35(a) shows a schematic diagram of an electric circuit containing an alternating current generator, a resistance, R, and an inductive reactance, X_L, connected in series. A quantity Z, called the impedance of the circuit and used in a–c circuit theory, is related to the resistance and the inductive reactance by the right triangle relationship shown in Figure 2.35(b). The angle θ is called the phase angle. The following exercises are illustrative of the use of the impedance triangle in a–c circuit theory.

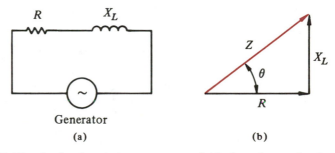

FIGURE 2.35

(a)   (b)

In Exercises 30–35, solve for the missing components in the impedance triangle.

30. $R = 30.0, \theta = 60.0°$ **31.** $R = 30.0, Z = 60.0$

32. $Z = 100, \theta = 20.0°$ **33.** $R = 200, X = 100$

34. $Z = 1000, X = 800$ **35.** $X = 25.0, \theta = 30.0°$

2.9   **Applications to Space-Related Sciences**

In space-related sciences, trigonometry has many important applications ranging from solutions of right triangles to problems of a complex analytical nature. The following material is, to a large extent, taken from *Space Mathematics*, published by the National Aeronautics and Space Administration and with permission of that organization. In Chapter 4, several similar applications are presented.

The weight of an astronaut on the moon is one-sixth his weight on Earth.

This fact has a marked effect on such simple acts as walking, running, jumping, and the like. To study these effects and to train astronauts for working under lunar gravity conditions, scientists at NASA Langley Research Center have designed an inclined plane apparatus to simulate reduced gravity.

The apparatus consists of an inclined plane and a sling that holds the astronaut in a position perpendicular to the inclined plane as shown in Figure 2.36. The sling is attached to one end of a long cable that runs parallel to the inclined plane. The other end of the cable is attached to a trolley that runs along a track high overhead. This device allows the astronaut to move freely in a plane perpendicular to the inclined plane.

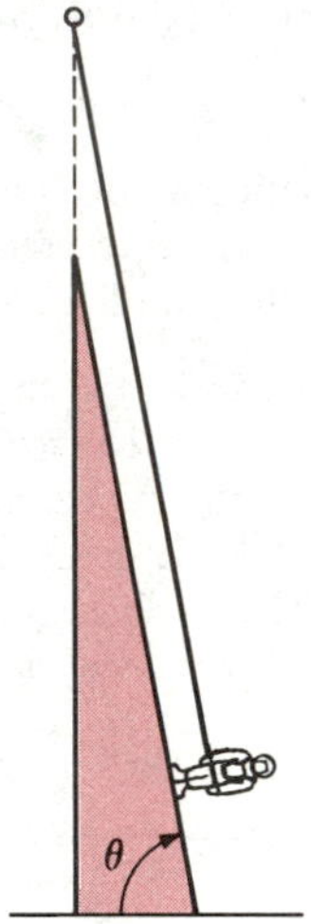

FIGURE 2.36

EXAMPLE 1   Let W be the weight of the astronaut and θ the angle between the inclined plane and the ground. Make a vector diagram to show the tension in the cable and the force exerted by the inclined plane against the feet of the astronaut.

SOLUTION   The weight of the astronaut is resolved into two components, one parallel to the inclined plane, the other perpendicular to it. These components are $W \sin \theta$ and $W \cos \theta$, respectively. To be in equilibrium, the component $W \sin \theta$ must be balanced by the tension in the cable, and the component $W \cos \theta$ must be balanced by the force exerted by the inclined plane (see Figure 2.37).

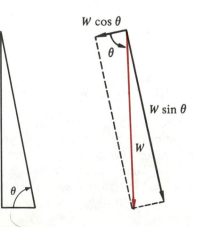

FIGURE 2.37

EXAMPLE 2   From the point of view of the astronaut in the sling, the inclined plane is the ground, the downward force against the inclined plane, is $W \cos \theta$. What is the value of θ required to simulate lunar gravity? What is the tension in the cable?

SOLUTION   To simulate lunar gravity, we must have $W \cos \theta = W/6$. Thus,

$$\cos \theta = 1/6 = 0.1667$$

$$\theta = 80°24' \text{ to the nearest minute}$$

The tension in the cable is

$$W \sin 80°24' = 0.986 \, W \qquad \blacksquare$$

A parallel of latitude on the Earth's surface is a circle in a plane parallel to the equatorial circle. The latitude angle is the angle made by the radius from the midpoint of the equator and the equator itself (see Figure 2.38).

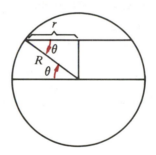

FIGURE 2.38

EXAMPLE 3   Show that the length of any parallel of latitude around Earth is equal to the equatorial distance around Earth times the cosine of the latitude angle.

SOLUTION   By the definition of the cosine function,

$$\cos \theta = r/R \text{ or } r = R \cos \theta$$

Let the length of the parallel of latitude be C_p. If C_e denotes the average circumference of Earth, then

$$C_p = 2\pi r = 2\pi R \cos \theta = C_e \cos \theta \qquad \blacksquare$$

An interesting method of measuring the height of clouds is shown in Figure 2.39. A sweeping light beam is placed at point A, and a light source detector is placed at point B. The axis of the detector is maintained vertical, and the light beam is made to sweep from the horizontal ($\alpha = 0°$) to the vertical ($\alpha = 90°$). When the beam illuminates the base of the clouds directly above the detector, as shown in the figure, the detector is activated, and the angle α is read. Since d is known, the height h can be computed.

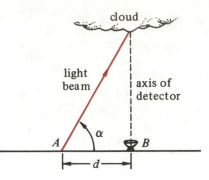

FIGURE 2.39

EXAMPLE 4   If the height of the cloud is 2050 ft and the distance d is 1000 feet, compute the angle α.

SOLUTION   Using the relation $\dfrac{h}{d} = \tan \alpha$, we have

$$2050 \text{ feet} = (1{,}000 \text{ feet})(\tan \alpha)$$

$$2.050 = \tan \alpha.$$

$$\alpha = 64°.$$

The light source of the system must be reasonably close to the detector so that the illumination of the cloud above the detector is sufficiently strong to be detected. At many U.S. National Weather Service stations two beam sources are used, one 800 ft and the other 1600 ft from the detector. To have reliable readings, α may not exceed 85°. ■

The next example deals with design of a lunar lander. Surprisingly, it is an application of trigonometry that gives the necessary design goal for construction of the legs of the lander. Since three points determine a plane, spacecraft that are designed to land on the moon are designed with three legs.

FIGURE 2.40

EXAMPLE 5   A spacecraft designed to soft land on the moon has three feet that form an equilateral triangle on level ground and each of the three legs make an angle of 37.0° with the vertical. If the impact force of 1500 lb is evenly distributed, find the force on each leg. (Figure 2.41.)

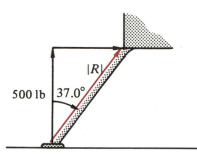

FIGURE 2.41

SOLUTION   Consider one leg. Five hundred pounds is the vertical component of force **R** in the leg acting at 37.0° from the vertical. Thus,

$$\cos 37.0° = \frac{500 \text{ lb}}{|\mathbf{R}|}$$

$$|\mathbf{R}| = \frac{500}{\cos 37.0°} = 626 \text{ lb} \qquad \blacksquare$$

The last example shows how right triangle trigonometry can be used to find vertical dimensions of objects that appear in reconnaissance and satellite photographs. All the analyst needs is the angle of elevation of the sun and the scale of the photograph.

EXAMPLE 6   A representation of an aerial photograph of a complex of buildings is shown in Figure 2.42(a). If the sun was at an angle of 26.5° when the photograph was taken, how high is the rectangular shaped building?

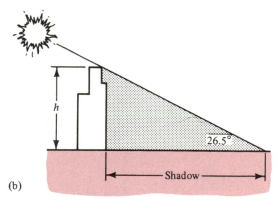

Scale: 1 cm = 250 m

(a)   (b)

FIGURE 2.42

SOLUTION   The length of the shadow measures 0.48 cm. To get the real length of the shadow multiply 0.48 cm by the scale factor given in the photograph. Thus,

$$\text{Shadow length} = 0.48(250) = 120 \text{ m}$$

Using Figure 2.42(b) as a model of the actual situation, we see that $\tan 26.5° = h/120$. Solving for h, we get

$$h = 120 \tan 26.5° = 120(0.4986) = 59.8 \text{ m}$$

as the height of the building. $\blacksquare$

Exercises for Section 2.9

1. What is the angle of the inclined plane used to simulate gravity for a 200-pound astronaut walking on an asteroid whose gravity is $\frac{1}{8}$ that of Earth?

2. If the plane used to simulate gravity were inclined at 60°, what percentage of the astronaut's weight would bear against the plane?

3. What gravity would be simulated by a plane inclined at 50°?

4. Find the length of the 30° parallel, north or south latitude. (Assume $C_e = 25,000$ mi.)

5. Determine the length of the Arctic Circle (66°33′N).

6. How far is it "around the world" along the parallel of 80°N latitude?

7. The 40° parallel of latitude passes through the United States. If a citizen of the United States were to travel due East along the 40° parallel, how far would he travel before returning to his starting point?

8. Using the cloud altitude detector, compute the height of the cloud if the light source is 100 feet from the detector and the angle is 82°.

9. Find the angle α when clouds are 1000 ft high and the light source is located 100 ft from the detector.

10. If clouds are at 1000 m and 2000 m, what angle does the light pass through in detecting the two different cloud layers? Assume $d = 800$ m.

11. Suppose that a detector which is 1 km from the light beam detects a cloud layer when $\alpha = 60°$ and another layer when $\alpha = 75°$. What is the vertical separation of the cloud layers?

12. At the U.S. National Weather Service Station, the beam, which is 1600 ft from the detector, intersects a cloud when $\alpha = 65°$. At what angle will a beam, which is 750 ft from the detector, intersect the same cloud?

13. In determining the height of a cloud, the light source 800 ft from the detector makes an angle of 85° with the horizontal. Find the corresponding angle for the light source at 1600 ft.

14. Suppose the Moonlander is to be designed for an impact force of double that mentioned in Example 5. How does this affect the force on each leg?

15. Suppose that each of the three legs of a moonlander will withstand an axial load of 1000 lb. What is the maximum angle that the legs can make with the vertical and still support a total impact force on the lander of 2500 lb?

16. Suppose a moonlander is to be designed for a total impact force of 3500 lb and that each of the three legs makes an angle of 28° with the vertical. Find the design impact force in each leg.

17. Ten holes are to be drilled in a circular cover plate of a rocket motor. The holes are equally spaced on a circle of radius 12.9 cm as shown in the figure. What is the center-to-center distance between the holes?

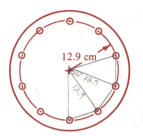

18. A 78.0-ft rocket with its base on the ground is elevated at an angle of 69°40'. What is the height of the nose of the rocket above the ground?

19. The triangular wing of a delta-wing airplane is swept back at an angle of 51.5° to the centerline of the fuselage. If the leading edge of the wing is 28.3 ft long and the fuselage is 4.20 ft wide, what is the wingspan of the airplane?

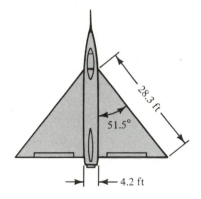

20. The shroud of a nose cone is shown in cross-section in Figure 2.43. What is the diameter of the rocket using this nose cone?

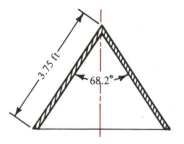

FIGURE 2.43

21. An astronomer measures the shadow of a crater on the moon and finds its length to be 0.32 cm. If the sun was at an angle of 49.1° to the horizontal when the photograph was taken, how deep is the crater? Assume the map scale is 1 cm = 2500 m.

22. Figure 2.44 represents an aerial photograph of a cliff in a remote region of Antarctica. Compute the height of the cliff if the elevation angle of the sun was 19.0° when the photograph was taken. (Figure on the next page.)

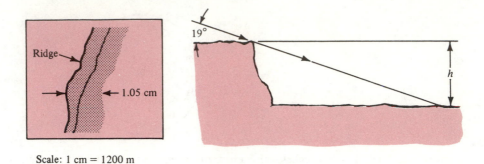

Scale: 1 cm = 1200 m

FIGURE 2.44

23. A space capsule orbits the moon at an altitude of 100 miles. A sighting from the capsule to the moon's horizon shows an angle of depression of 22.6°. Find the radius of the moon.

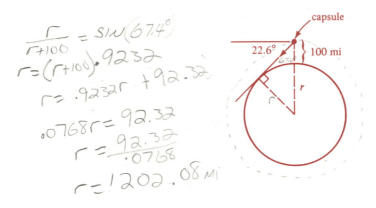

$$\frac{r}{r+100} = \sin(67.4°)$$
$$r = (r+100) \cdot .9232$$
$$r = .9232r + 92.32$$
$$.0768r = 92.32$$
$$r = \frac{92.32}{.0768}$$
$$r = 1202.08 \text{ MI}$$

Key Topics for Chapter 2

Define and/or discuss each of the following.

sin θ, cos θ, tan θ, cot θ, sec θ, csc θ

The Reciprocal Relations

The Pythagorean Relation

The Special Angles

Use of Trigonometric Tables

Interpolation

Solution of a Right Triangle

Components of a Vector

Course

Heading

Bearing

Review Exercises for Chapter 2

ODD

1. Given $\sin \theta = 5/13$, find the other five trigonometric functions.

2. Given $\tan A = 2/3$, find the other five trigonometric functions.

3. Given $\sin \phi = 1/\sqrt{3}$, find all trigonometric functions of the angle complementary to angle ϕ.

4. Use Table A to find $\sin 37°18'$.

5. Use Table A to find $\tan 56°28'$.

6. Use Table C to find $\tan 1.07$.

7. Use Table B to find $\cos 48.24°$.

8. Use Table B to find $\tan 28.32°$.

In Exercises 9–12 find θ to the nearest minute.

9. $\cos \theta = 0.6653$
10. $\tan \theta = 2.008$
11. $\sin \theta = 0.1414$
12. $\sec \theta = 1.200$

In Exercises 13–16 find θ to the nearest hundredth of a degree.

13. $\cos \theta = 0.4182$
14. $\sin \theta = 0.8214$
15. $\tan \theta = 1.2413$
16. $\cos \theta = 0.5613$

In Exercises 17–20 find ϕ to the nearest thousandth of a radian.

17. $\tan \phi = 1.5574$
18. $\sin \phi = 0.9438$
19. $\csc \phi = 2.009$
20. $\cos \phi = 0.5553$

21. Solve the right triangle ABC if $A = 32°$ and $a = 3.0$.

22. Solve the right triangle ABC if $A = 17°23'$ and $b = 5.8$.

23. Solve the right triangle ABC if $a = 29$ and $c = 41$. (c is the hypotenuse)

24. Solve the right triangle ABC if $b = 7.5$ and $c = 11.3$.

25. Find the horizontal and vertical components of F if $|\mathbf{F}| = 50$ lb and $\theta = 36.7°$.

26. Find the horizontal and vertical components of F if $|\mathbf{F}| = 8$ lb and $\theta = 17.9°$.

27. Find the resultant force for components $F_x = 43$ lb and $F_y = 72$ lb.

28. Find the resultant force for components $F_x = 9.4$ lb and $F_y = 3.7$ lb.

29. An engineers' drawing of a component of a trimming die is shown in Figure 2.45. Find the measure of angle θ.

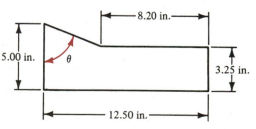

FIGURE 2.45

30. An airplane heading due east at 350 mph experiences a 40 mph crosswind out of the north. What direction is the plane moving relative to an observer on the ground?

Test 1 for Chapter 2

In Exercises 1–9 answer *true* or *false*.

1. For any acute angle θ, $\cos \theta = 1/\csc \theta$.

2. If θ is acute and $\tan \theta = \cot \theta$, then $\theta = 45°$.

3. If θ and α are complementary angles, then $\sin^2 \theta + \sin^2 \alpha = 1$.

4. If θ is acute and $\sin \theta = \frac{1}{2}$, then $\theta = 30°$.

5. For acute angles, if $\theta_1 < \theta_2$, then $\cos \theta_1 < \cos \theta_2$.

6. If $\theta = 45°$, then $\sin \theta = \cos \theta$.

7. $\sin^2 60° + \cos^2 30° = 1$.

8. $\tan^2 42° + 1 = \sec^2 42°$.

9. Vectors are quantities requiring both a magnitude and a direction for their description.

10. Use Table B to find α if $\tan \alpha = 0.7310$.

11. Find the other five trigonometric functions of θ if $\sec \theta = u/v$.

12. Solve for θ where $0 < \theta < 90°$.

    (a) $\sin \theta = 2$ \hspace{3cm} (b) $\cos \theta = \frac{1}{2}$

13. By interpolation, using Table A, approximate the value of x.

    (a) $x = \sin 33°13'$ \hspace{2.5cm} (b) $\cos x = 0.2461$

14. Solve the right triangle ABC if $A = 32°$ and $a = 3$. (The angle C is the 90° angle.)

15. The angles of elevation to a weather balloon from two tracking stations are 30° and 45°. If the balloon is at an altitude of 50,000 ft, how far apart are the two tracking stations?

Test 2 for Chapter 2

1. Given $\sin \theta = 0.2301$ and $\cos \theta = 0.8837$, find $\tan \theta$ without using a table.

2. Show that $\sin^2 30° + \cos^2 30° = 1$.

3. Find $\sin \theta$ and $\cos \theta$ if $\tan \theta = \frac{5}{12}$.

4. Find the value of   (a) $\sin 15°23'$ and   (b) $\tan 76°54'$.

5. Given $\tan \phi = 0.5679$, find the acute angle ϕ using Table A.

6. Solve the right triangle with $A = 32°20'$ and $b = 22$.

7. Solve the right triangle with $a = 5$ and $c = 8$.

8. A tunnel through a mountain ascends at an angle of 5°25′ with the horizontal. If the tunnel is 5000 ft long, what is the vertical rise of the tunnel?

9. Eight holes are equally spaced on the circumference of a circle. If the center-to-center distance between the holes is 3.8 cm, what is the radius of the circle?

10. A horizontal force of 15 lb is applied to a block resting on a plane inclined at 23° with the horizontal. Resolve the force into components — one parallel to the plane and one perpendicular to the plane.

3

Trigonometric Functions for any Angle

3.1   Extending the Basic Definitions

In the preceding chapter our use of the trigonometric functions was restricted to acute angles even though the definitions were made in general terms of the coordinates of a point on the terminal side of an angle in standard position.* Now we will consider in detail the trigonometric functions for angles of any magnitude, such as $\sin 215°$, $\tan 1042°$ and $\cos(-351°)$. First we review the basic definitions.

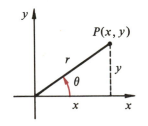

FIGURE 3.1

The generated angle θ, shown in standard position in Figure 3.1, has the point P on its terminal side. The coordinates of P are (x, y) and the distance from the origin to P is r. The trigonometric functions of the angle θ are then given by:

$$(3.1) \qquad \sin\theta = \frac{\text{ordinate of } P}{\text{radius of } P} = \frac{y}{r} \qquad\qquad \csc\theta = \frac{\text{radius of } P}{\text{ordinate of } P} = \frac{r}{y}$$

*Recall that an angle is in standard position if its initial side is along the positive x-axis.

86

$$\cos \theta = \frac{\text{abscissa of } P}{\text{radius of } P} = \frac{x}{r} \qquad \sec \theta = \frac{\text{radius of } P}{\text{abscissa of } P} = \frac{r}{x}$$

$$\tan \theta = \frac{\text{ordinate of } P}{\text{abscissa of } P} = \frac{y}{x} \qquad \cot \theta = \frac{\text{abscissa of } P}{\text{ordinate of } P} = \frac{x}{y}$$

Thus we can write the trigonometric functional values for any angle if we know the abscissa and ordinate of a point on the terminal side of the angle and its distance from the origin. Recall from Section 1.1 that r is always positive but that the abscissa and ordinate are positive or negative, depending upon the quadrant in which the terminal side lies.

EXAMPLE 1   An angle θ in standard position has the point $(-6, 3)$ on its terminal side. Find the values of the six trigonometric functions of θ. (See Figure 3.2).

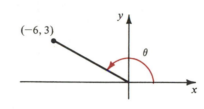

FIGURE 3.2

SOLUTION   Using (3.1) with $x = -6$, $y = 3$, and $r = \sqrt{(-6)^2 + 3^2} = \sqrt{45} = 3\sqrt{5}$, we get

$$\sin \theta = \frac{y}{r} = \frac{3}{3\sqrt{5}} = \frac{1}{\sqrt{5}} \qquad \csc \theta = \frac{r}{y} = \frac{3\sqrt{5}}{3} = \sqrt{5}$$

$$\cos \theta = \frac{x}{r} = \frac{-6}{3\sqrt{5}} = -\frac{2}{\sqrt{5}} \qquad \sec \theta = \frac{x}{r} = -\frac{3\sqrt{5}}{-6} = -\frac{\sqrt{5}}{2}$$

$$\tan \theta = \frac{y}{x} = \frac{3}{-6} = -\frac{1}{2} \qquad \cot \theta = \frac{x}{y} = \frac{-6}{3} = -2$$ ■

EXAMPLE 2   An angle θ in standard position has its terminal side passing through the point $(3, -4)$. Find the six trigonometric functions of the angle. (See Figure 3.3.)

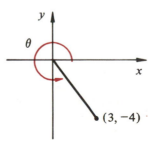

FIGURE 3.3

SOLUTION   Using Equation (3.1) with $x = 3$, $y = -4$, and $r = \sqrt{3^2 + (-4)^2} = 5$,

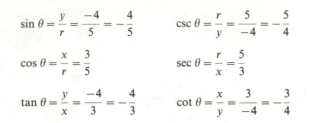

$$\sin \theta = \frac{y}{r} = \frac{-4}{5} = -\frac{4}{5} \qquad \csc \theta = \frac{r}{y} = \frac{5}{-4} = -\frac{5}{4}$$

$$\cos \theta = \frac{x}{r} = \frac{3}{5} \qquad \sec \theta = \frac{r}{x} = \frac{5}{3}$$

$$\tan \theta = \frac{y}{x} = \frac{-4}{3} = -\frac{4}{3} \qquad \cot \theta = \frac{x}{y} = \frac{3}{-4} = -\frac{3}{4}$$

In each of these examples four of the six trigonometric values are negative. This follows from the fact that the signs of the functional values depend on the signs of x and y.

The sine function is the ratio of y to r, which means that it is positive for angles in the first and second quadrant and negative for those in the third and fourth. This is because y is positive above the x-axis and negative below.

The cosine function, which is the ratio of x to r, is positive for angles in the first and fourth quadrants and negative in the second and third. This is because x is positive to the right of the y-axis and negative to the left.

The tangent function, which is the ratio of y to x, is positive in the first and third quadrants because y and x have the same signs in these quadrants. The tangent is negative in the second and fourth quadrants. The signs of the remaining three functions can be analyzed in the same way. Table 3.1 summarizes the results for all six functions.

TABLE 3.1

Quadrant	Positive Functions	Negative Functions
I	All	None
II	sine	cosine, secant
	cosecant	tangent, cotangent
III	tangent	sine, cosine
	cotangent	secant, cosecant
IV	cosine	sine, cosecant
	secant	tangent, cotangent

EXAMPLE 3

(a) $\sin \theta > 0$ in quadrants I and II.

(b) $\tan \theta < 0$ in quadrants II and IV.

(c) $\sec \theta < 0$ in quadrants II and III.

EXAMPLE 4   Show that the terminal side of θ is in quadrant II if $\sin \theta > 0$ and $\cos \theta < 0$.

SOLUTION   From Table 3.1, $\sin \theta > 0$ in quadrants I and II. Likewise $\cos \theta < 0$ in quadrants II and III. Since both of the given conditions are satisfied in quadrant II we conclude that the terminal side of θ must be in this quadrant.

EXAMPLE 5

(a) If $\sec \phi > 0$ and $\sin \phi < 0$, the terminal side of angle ϕ is in quadrant IV.

(b) If $\csc \phi > 0$ and $\cot \phi < 0$, the terminal side of angle ϕ is in quadrant II.

The definitions of the trigonometric functions of any angle show that the functional values are completely determined by the location of the terminal side when the angle is in standard position. Thus, **coterminal angles have equal functional values**. For instance, since 30° and 390° are coterminal, sin 30° = sin 390°, cos 30° = cos 390°, tan 30° = tan 390°, etc. Thus, in finding values of trigonometric functions, we need only consider angles between 0° and 360°.

An angle whose terminal side lies on a coordinate axis is called a **quadrantal angle**. Angles of 0°, ±90°, ±180° are examples. For these angles, one of the coordinates of a point on the terminal side must be zero. Since division by zero is undefinable, two of the six trigonometric functions will be undefined at each quadrantal angle.

EXAMPLE 6   Find the trigonometric functions of an angle θ whose terminal side passes through the point $(-1, 0)$ as shown in Figure 3.4.

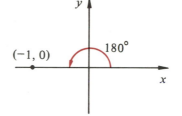

FIGURE 3.4

SOLUTION   In this case $x = -1$, $y = 0$ and $r = 1$. Since $\theta = 180°$, we write

$$\sin 180° = \frac{y}{r} = \frac{0}{1} = 0 \qquad\qquad \csc 180° = \frac{r}{y} = \frac{1}{0} \ \text{(undefined)}$$

$$\cos 180° = \frac{x}{r} = \frac{-1}{1} = -1 \qquad\qquad \sec 180° = \frac{r}{x} = \frac{1}{-1} = -1$$

$$\tan 180° = \frac{y}{x} = \frac{0}{-1} = 0 \qquad\qquad \cot 180° = \frac{x}{y} = \frac{-1}{0} \ \text{(undefined)}$$

The preceding example exhibits the values for a quadrantal angle of 180°, or one coterminal with it. The values of the other quadrantal angles can be found by a similar procedure and are listed for your reference in Table 3.2.

TABLE 3.2

Deg.	Rad.	sin θ	cos θ	tan θ	cot θ	sec θ	csc θ
0	0	0	1	0	undefined	1	undefined
90	$\pi/2$	1	0	undefined	0	undefined	1
180	π	0	-1	0	undefined	-1	undefined
270	$3\pi/2$	-1	0	undefined	0	undefined	-1

Exercises for Section 3.1

Find the values of the trigonometric functions of an angle in standard position whose terminal side passes through the points given in Exercises 1–9.

1. (2, 4)   **2.** (−1, 5)   **3.** (−9, 16)   **4.** (3, 1)

5. (2, −7)   **6.** (−1, −1)   **7.** (3, −1)   **8.** (1, −1)

9. $(-\sqrt{3}, -1)$

10. Show that the values of the trigonometric functions are independent of the choice of the point P on its terminal side.

11. In which quadrants must the terminal side of θ lie for $\sin \theta$ to be positive? $\cos \theta$? $\tan \theta$?

12. Find the six trigonometric functions of 90° by noticing that (0, 1) lies on the terminal side of the angle of 90° when it is placed in standard position.

13. Find the six trigonometric functions of 0°. Use (1, 0) as a point on the terminal side.

14. Find the six trigonometric functions of 270°. Use (0, −1) as a point on the terminal side.

15. For which values of θ is $\sin \theta = 1$? For which values of θ is $\cos \theta = 1$?

16. Show that $\sin 420° = \sin 60°$.

17. Show that $\tan 765° = \tan 45°$.

In Exercises 18–23, indicate the quadrant in which the terminal side of θ lies for the given conditions.

18. $\sin \theta > 0$ and $\tan \theta < 0$ **19.** $\sec \theta < 0$ and $\cot \theta < 0$

20. $\cos \theta > 0$ and $\sin \theta < 0$ **21.** $\tan \theta > 0$ and $\csc \theta < 0$

22. $\sin \theta < 0$ and $\cos \theta < 0$ **23.** $\sin \theta < 0$ and $\sec \theta > 0$

In Exercises 24–31 determine the numerical sign of the given functions.

24. $\cos 115°$ **25.** $\tan 300°$ **26.** $\sec(-25°)$ **27.** $\sin 170°$

28. $\sin 210°$ **29.** $\sec 124°$ **30.** $\cot 252°$ **31.** $\cos(-100°)$

3.2   Elementary Relations

From the discussion in the previous section it follows that the reciprocal relationships

$$\sin \theta = \frac{1}{\csc \theta} \qquad \cos \theta = \frac{1}{\sec \theta} \qquad \tan \theta = \frac{1}{\cot \theta}$$

are valid for angles greater than 90°. Similarly, we can extend the quotient relation

$$\tan \theta = \frac{\sin \theta}{\cos \theta}$$

Because of these relations, the values of all six trigonometric functions can be determined if you know the value of one of them and the quadrant of the terminal side of the angle. If the quadrant is not given, two sets of values are possible.

EXAMPLE 1   Given that $\tan \theta = -\frac{4}{3}$ and that θ in standard position has its terminal side in quadrant II, find the values of the other five trigonometric functions.

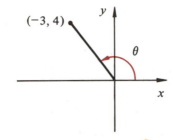

FIGURE 3.5

SOLUTION   We choose a convenient point on the terminal side, in this case $(-3, 4)$ as shown in Figure 3.5. (If we had been told to locate the point in quadrant IV, it would be the point $(3, -4)$.) The desired trigonometric functions for the given angle are

$$\sin \theta = \frac{4}{5} \qquad \cos \theta = \frac{-3}{5} \qquad \cot \theta = \frac{-3}{4} \qquad \sec \theta = \frac{5}{-3} \qquad \csc \theta = \frac{5}{4} \qquad \blacksquare$$

EXAMPLE 2   Given that $\cos \theta = -\frac{5}{13}$, find the values of the other trigonometric functions. (See Figure 3.6.)

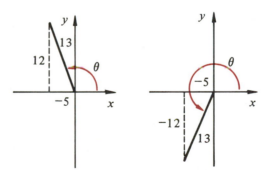

FIGURE 3.6

SOLUTION   Since the quadrant is not specified, two angles between 0 and 2π will satisfy the given condition. One is in the second quadrant and the other in the third quadrant. For the second quadrant angle,

$$\sin \theta = \frac{12}{13} \qquad \tan \theta = -\frac{12}{5} \qquad \cot \theta = -\frac{5}{12} \qquad \sec \theta = -\frac{13}{5} \qquad \csc \theta = \frac{13}{12}$$

For the third quadrant angle,

$$\sin \theta = -\frac{12}{13} \qquad \tan \theta = \frac{-12}{-5} = \frac{12}{5} \qquad \cot \theta = \frac{-5}{-12} = \frac{5}{12}$$

$$\sec \theta = -\frac{13}{5} \qquad \csc \theta = -\frac{13}{12} \qquad \blacksquare$$

3.3

Approximating the Values of the Trigonometric Functions

Suppose that r is chosen *one* unit long for the terminal side of an angle θ in standard position, then

$$\sin \theta = \frac{y}{r} = \frac{y}{1} = y$$

$$\cos \theta = \frac{x}{r} = \frac{x}{1} = x$$

That is, the values of $\sin \theta$ and $\cos \theta$ are equal to the y and x coordinates of the point $P(x, y)$. Thus, the coordinates of the points on a circle of radius 1 centered at the origin are identified with the trigonometric functions. In Figure 3.7, you see such a circle with angles from 0° to 360° marked off in 10° increments. The y coordinate of a point on the circle corresponds to $\sin \theta$ and the x coordinate corresponds to $\cos \theta$. Using this circle, you can approximate the values of the trigonometric functions for any angle θ.

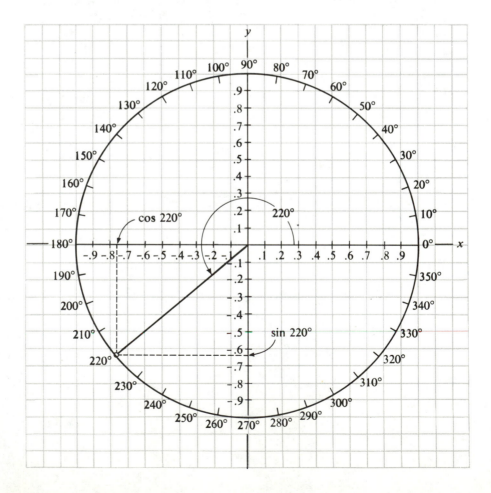

FIGURE 3.7

EXAMPLE 1   Find the values of the trigonometric functions of 220°.

SOLUTION   With reference to Figure 3.7, we estimate sin 220° to be −0.64 and cos 220° to be −0.77. Therefore,

$$\sin 220° = -0.64$$

$$\cos 220° = -0.77$$

$$\tan 220° = \frac{\sin 220°}{\cos 220°} = \frac{-0.64}{-0.77} = 0.83$$

$$\cot 220° = \frac{1}{\tan 220°} = \frac{1}{0.83} = 1.2$$

$$\sec 220° = \frac{1}{\cos 220°} = \frac{1}{-0.77} = -1.3$$

$$\csc 220° = \frac{1}{\sin 220°} = \frac{1}{-0.64} = -1.6$$

■

A circle centered at the origin of radius 1 is called a **unit circle**. What the previous discussion has shown is that the point of intersection of the terminal side of an angle with the unit circle has coordinates which are precisely the values of the sine and cosine. Figure 3.8 shows six examples. You can check these values by referring to Figure 3.7.

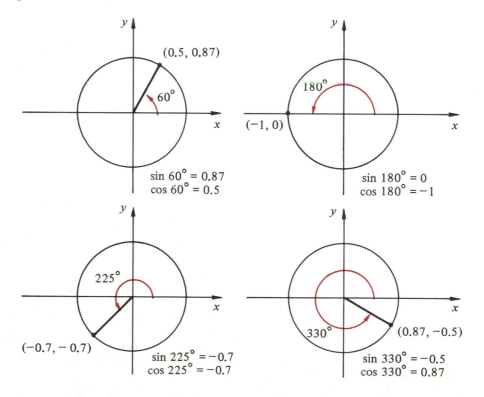

FIGURE 3.8

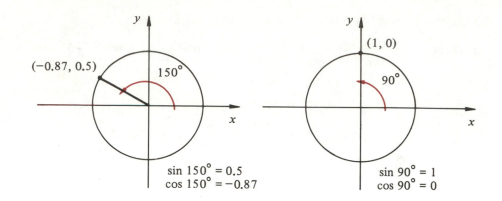

FIGURE 3.8
(Continued)

Exercises for Sections 3.2 and 3.3

In Exercises 1–16, find all of the trigonometric functions at an angle θ that satisfies the conditions.

1. $\tan \theta = \frac{3}{4}$ in Q I

2. $\sec \theta = -3$

3. $\tan \theta = \frac{3}{4}$ in Q III

4. $\tan \theta = \frac{1}{2}$ in Q III

5. $\cos \theta = -1$

6. $\cot \theta = -\pi$ in Q IV

7. $\sin \theta = 0$

8. $\sin \theta = \dfrac{\sqrt{3}}{2}$

9. $\sin \theta = -\frac{1}{2}$

10. $\sin \theta = \frac{1}{5}$

11. $\tan \theta = 10$ in Q I

12. $\csc \theta = 2$

13. $\cos \theta = \frac{12}{13}$

14. $\cos \theta = -\dfrac{\sqrt{3}}{2}$ in Q II

15. $\tan \theta = -\pi$

16. $\cot \theta = -\sqrt{2}$

Use Figure 3.7 to approximate the values of the trigonometric functions in Exercises 17–34.

17. $\sin 70°$

18. $\cos 35°$

19. $\tan 10°$

20. $\sin 125°$

21. $\sin 155°$

22. $\tan 130°$

23. $\cos 165°$

24. $\sin 110°$

25. $\sin 215°$

26. $\sin \pi/4$

27. $\tan 345°$

28. $\cos 345°$

29. $\sec 250°$

30. $\sin 300°$

31. $\cos 2\pi/3$

32. $\csc 200°$

33. $\sin 315°$

34. $\tan 5\pi/4$

35. Use the definitions of the trigonometric functions for any angle to show that $\sin^2 \theta + \cos^2 \theta = 1$.

3.4   **The Use of Tables for Angles that Are Not Acute**

As you learned in the previous section, the values of the trigonometric functions are defined, with a few exceptions, for any angle. Now you will learn how to use trigonometric tables to find values for angles greater than 90°. The process is not difficult, but it does require some explanation since trigonometric tables are traditionally tabulated from 0° to 90°.

To give you an idea of the general procedure, consider the problem of finding tan 150°. A generated angle of 150° is shown in Figure 3.9. The acute angle formed by the terminal side and the x-axis is obviously 30°. This angle is called the *reference angle of 150°*. Now, from our knowledge of the 30–60° right triangle, we know that the terminal side must pass through the point $(-\sqrt{3}, 1)$. Therefore,

$$\tan 150° = \frac{-1}{\sqrt{3}} \quad \text{and} \quad \tan 30° = \frac{1}{\sqrt{3}}$$

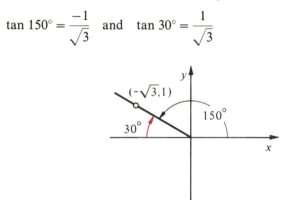

FIGURE 3.9

Of interest here is the fact that tan 150° has the same numerical value as the tangent of its reference angle. The implication of the example is that tan 150° can be found by finding tan 30° in a table and then prefixing the appropriate numerical sign. The following discussion generalizes the procedure for finding the trigonometric values for any generated angle.

Consider angles α, β, γ, and δ, in standard position, whose terminal sides each lie in a different quadrant as shown in Figure 3.10(a). To find the values of the trigonometric functions for each of these angles, we must first find the measure of the acute angle made by the terminal side of the given angle and the x-axis. This angle is called the **reference angle** for the given angle. In Figure 3.10(a), each reference angle is denoted with a prime and can be found by using the following rules.

(1)   First Quadrant Angle: $\alpha' = \alpha$

(2)   Second Quadrant Angle: $\beta' = 180° - \beta$

(3)   Third Quadrant Angle: $\gamma' = \gamma - 180°$

(4)   Fourth Quadrant Angle: $\delta' = 360° - \delta$

The reference angles found in Figure 3.10(a) are shown in standard position in Figure 3.10(b). The coordinates of the intersection of the terminal side of each angle in Figure 3.10(a) with a general circle of radius r are numerically equal (i.e., except for sign) to those obtained for the reference angles placed in standard position. It follows

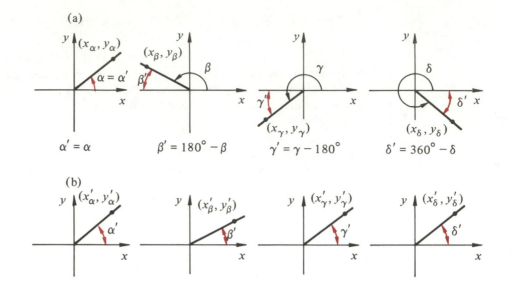

FIGURE 3.10

then that, except possibly for numerical sign, any trigonometric function of an angle has the identical value as the same function of the reference angle. For example, we find that

$$\tan \alpha = \frac{y_\alpha}{x_\alpha} = \frac{y'_\alpha}{x'_\alpha} = \tan \alpha'$$

$$\tan \beta = \frac{y_\beta}{x_\beta} = \frac{y'_\beta}{-x'_\beta} = -\tan \beta'$$

$$\tan \gamma = \frac{y_\gamma}{x_\gamma} = \frac{-y'_\gamma}{-x'_\gamma} = \tan \gamma'$$

$$\tan \delta = \frac{y_\delta}{x_\delta} = \frac{-y'_\delta}{x'_\delta} = -\tan \delta'$$

Notice that these results agree with those given in Table 3.1 with respect to the numerical sign of the tangent in each quadrant.

Thus the functional values of the reference acute angle determine completely the values of the trigonometric functions of an angle in standard position. The algebraic sign is determined by the quadrantal location of the terminal side. In summary, to determine values of the trigonometric functions for any angle, proceed as follows:

(1)   Determine the positive coterminal angle with measure between 0° and 360° corresponding to the given angle.

(2)   Sketch the angle in standard position.

(3)   Determine the reference acute angle.

(4)   Find the value of the trigonometric function of the reference acute angle, usually from the Tables in the Appendix.

(5)   The value of the trigonometric function of the reference acute angle is the same as the value of the trigonometric function of the given angle except perhaps for the sign. The sign is determined from the location of the terminal side of the given angle.

EXAMPLE 1   Find cos 145°20′. (See Figure 3.11.)

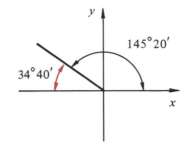

FIGURE 3.11

SOLUTION   We see that 145°20′ is a second quadrant angle, and, therefore, the reference angle is

$$\theta' = 180° - 145°20' = 34°40'$$

The reference angle is indicated in the figure. Remembering that the cosine function is negative in the second quadrant, we have

$$\cos 145°20' = -\cos 34°40' = -0.8225 \quad \text{(Table A, Appendix)}$$ ■

EXAMPLE 2   Find tan 4. (See Figure 3.12.)

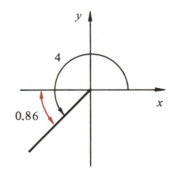

FIGURE 3.12

SOLUTION   An angle of 4 radians lies in the third quadrant, as in Figure 3.12. The reference angle is then

$$\theta' = 4 - \pi = 4 - 3.14 = 0.86$$

Since the tangent function is positive in the third quadrant,

$$\tan 4 = \tan 0.86$$

$$= 1.16 \quad \text{(Table C, Appendix)}$$ ■

EXAMPLE 3   Find sin 1000 and sin 1000°.

SOLUTION   Dividing 1000 by 6.28, we find that the angle of 1000 radians is coterminal with the angle of 1.48 radians. Since this is a first quadrant angle,

$$\sin 1000 = \sin 1.48$$

$$= 0.9959 \qquad \text{(Table C)}$$

Since $1000°/360° = 2$ with a remainder of $280°$, the angle of $1000°$ is coterminal with a fourth quadrant angle. The reference acute angle is $80°$. Hence,

$$\sin 1000° = \sin 280°$$

$$= -\sin 80°$$

$$= -0.9848 \qquad \text{(Table A)}$$ ■

While 3.14 is often used as an approximation for π large errors can occur. For instance, using 3.14 in Example 3 led to the conclusion that 1000 radians is coterminal with 1.48 radians. If the $\boxed{\pi}$ *button is used, then we find that 1000 radians is coterminal with 0.9735 radians. Quite a difference! From this you would obtain the more accurate value of* $\sin 1000 = \sin 0.9735 = 0.8269$. *Or, if your calculator accepts angles greater than π, then you can verify directly that* $\sin 1000 = 0.8269$.

EXAMPLE 4   Find $\cos(-527°)$

SOLUTION   This angle is coterminal with $-167°$, which, in turn, is coterminal with $193°$. This $193°$ angle, being in the third quadrant, has a reference acute angle of $13°$. Hence, $\cos(-527°) = \cos 193° = -\cos 13° = -0.9744$.   ■

EXAMPLE 5   Find $\tan 100\pi/3$.

SOLUTION   Dividing the given angle by 2π we obtain

$$\frac{100}{3}\pi \div 2\pi = 16 \text{ with a remainder of } \frac{4}{3}\pi$$

Since $4\pi/3$ is a third quadrant angle and since the reference acute angle is $\pi/3$.

$$\tan\frac{100}{3}\pi = \tan\frac{4}{3}\pi = \tan\frac{1}{3}\pi = \sqrt{3}$$ ■

There are always two angles on the interval $0° \leq \theta < 360°$ that have the same trigonometric ratio. Consequently, finding an angle corresponding to a given trigonometric ratio is more involved than it was for acute angle trigonometry. For example, if $\sin\theta = \frac{1}{2}$, both $\theta = 30°$ and $\theta = 150°$ are correct answers for $0° \leq \theta < 360°$. This redundancy can be avoided by stipulating certain conditions on angle θ. Thus, if we stipulate $\sin\theta = \frac{1}{2}$ and $\cos\theta < 0$, then $\theta = 150°$ is the only answer, since $\sin\theta > 0$ and $\cos\theta < 0$ implies that θ is in the second quadrant.

EXAMPLE 6   Given $\tan \beta = 0.3115$ and $\sin \beta < 0$, find β on the interval $0° \leq \beta < 360°$.

SOLUTION   Since $\tan \beta$ is positive and $\sin \beta$ is negative, β must be a third quadrant angle. From Table B, $\tan 17.3° = 0.3115$. Therefore,

$$\beta = 180° + 17.3° = 197.3°$$
■

EXAMPLE 7   Given $\sin \theta = -0.6068$ and $\cos \theta > 0$, find θ on the interval $0 \leq \theta < 2\pi$.

SOLUTION   The given conditions can be satisfied only by a fourth quadrant angle. Since $\sin 0.652 = 0.6068$ we have that

$$\theta = 2\pi - 0.652 = 5.631$$
■

By considering cases where the angle θ has, in turn, its terminal side in each of the four quadrants, the following general relations may be shown.

$$\sin \theta = -\sin(-\theta) \qquad \cos \theta = \cos(-\theta)$$
$$\sin(\pi - \theta) = \sin \theta \qquad \cos(\pi - \theta) = -\cos \theta$$
$$\sin(\pi + \theta) = -\sin \theta \qquad \cos(\pi + \theta) = -\cos \theta$$
$$\sin(2\pi - \theta) = -\sin \theta \qquad \cos(2\pi - \theta) = \cos \theta$$

You should not attempt to memorize this type of relation but rather be able to work out any of the results listed by the methods of this section. Figure 3.13 gives the idea for a second quadrant angle. From the Figure we see that $\pi - \theta$ is the reference acute angle for θ. Hence, since θ is in the second quadrant,

$$\sin \theta = \sin(\pi - \theta) \text{ and } \cos \theta = -\cos(\pi - \theta)$$

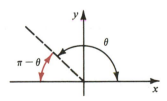

FIGURE 3.13

3.5   The Use of Calculators for Angles that Are Not Acute

Most popular calculators give the trigonometric functions of any angle by simply entering the angle and pushing the appropriate "function" button. That is, the calculator takes the drudgery out of finding trigonometric functions for angles that are not acute.

In some calculators the angle must be stated as an angle between $0°$ and $180°$ or $0°$ and $-180°$ before it can be used in the calculator. Check your particular operating manual to see if any such restrictions exist for your calculator.

EXAMPLE 1   Use a calculator to find $\sin 192°$.

SOLUTION   On many calculators sin 192° can be evaluated by simply putting the calculator in the degree mode, entering 192 and pushing the $\boxed{\text{sin}}$ button. The register will read -0.20791. Therefore,

$$\sin 192° = -0.20791$$

To solve this same problem using a calculator for which the angle must be stated between 0° and 180° or 0° and $-180°$ we proceed as follows. The angle 192° is coterminal with the angle $-168°$. Therefore, $\sin 192° = \sin(-168°)$. Enter -168 and push the $\boxed{\text{sin}}$ button. The register will read -0.20791.* ■

EXAMPLE 2   Use a calculator to find tan 53.

SOLUTION   For most calculators finding tan 53 is simply a matter of putting the calculator in radian mode, entering 53 and pushing $\boxed{\text{tan}}$ to obtain -0.43116. If your calculator requires that the angle be between $-\pi$ and π you must proceed as follows. Divide 53 by 2π to obtain 8.435212. This means that 53 radians is 8 revolutions plus 0.435212 of a revolution. Thus, $0.435212(2\pi)$ $= 2.7345175$ is coterminal with 53 radians and tan 53 = tan 2.7345175. ■

You will notice that when using a calculator to find an angle when its trigonometric function is known yields only one angle. For example, if $\sin \theta = -0.5$, your calculator will give only the one angle $\theta = -30°$. However, if $\cos \theta = -0.5$, your calculator will give the angle $\theta = 120°$. The reason we get a negative acute angle in the first case and a positive obtuse angle in the second is explained in Chapter 8. To avoid confusion at this time, we suggest that you use your calculator to find the reference angle for a given trigonometric function. The reference angle will be obtained from your calculator if you enter the absolute value of the given function. For example, if $\tan \theta = -0.3500$, the reference angle is found by entering 0.3500. That is,

$$0.3500 \boxed{\text{inv}} \boxed{\text{tan}} = 19.29° \quad \text{(Some calculators use } \boxed{\text{arc}} \text{ instead of } \boxed{\text{inv}} \text{)}$$

The desired angles are now found by the procedure described in the previous section; that is, $\theta_{\text{II}} = 180° - 19.29° = 160.71°$ and $\theta_{\text{IV}} = 360° - 19.29° = 340.71°$.

EXAMPLE 3   Find angles θ $(0 \le \theta < 360°)$ if $\sin \theta = -0.5664$.

SOLUTION   Enter the absolute value of the given function and push $\boxed{\text{inv}}$ $\boxed{\text{sin}}$. The resulting reference angle is

$$0.5664 \boxed{\text{inv}} \boxed{\text{sin}} = 34.5°$$

Now, since the sine is negative in both the third and fourth quadrants, we get

$$\theta_{\text{III}} = 180° + 34.5° = 214.5°$$

$$\theta_{\text{IV}} = 360° - 34.5° = 325.5°$$
■

EXAMPLE 4   Find angles ϕ $(0 \le \phi < 2\pi)$ if $\sec \phi = 2.992$ and $\tan \phi < 0$.

The numbers of digits displayed in the register will depend upon the brand of calculator. The functional values are rounded-off to five figures in this section.

SOLUTION

In this case the given function is positive so the reference angle is found by the following sequence.

2.992 $\boxed{1/x}$ $\boxed{\text{inv}}$ $\boxed{\cos}$ = 1.23   (Note: cos x = 1/sec x)

Since the secant is positive in both the first and fourth quadrants and tangent is negative in the second and fourth quadrants, we want the fourth quadrant angle

$$\phi_{\text{IV}} = 2\pi - 1.23 = 5.05$$

■

Exercises for Sections 3.4 and 3.5

In Exercises 1–10 express the trigonometric functions in terms of the same function of a positive acute angle.

1. cos 125°   **2.** tan 94°   **3.** sin 225°

4. csc 252°28′   **5.** tan 1243°   **6.** tan 100

7. sec 10   **8.** sin 90   **9.** cos 45

10. cos 5

Evaluate Exercises 11–20 using Table A. Check your answer with a calculator.

11. sin 154°   **12.** sin 333°   **13.** tan 96°

14. cos 205°   **15.** sec 163°20′   **16.** tan 200°35′

17. cos 285°50′   **18.** csc 261°   **19.** sin 247°36′

20. cos 108°29′

SEMMTRY

Evaluate Exercises 21–30 using Table C. Check your answer with a calculator.

21. sin 100   **22.** tan −100   **23.** sin 4.22

24. cos 5.16   **25.** csc 20   **26.** sec(−4)

27. cot 1.335   **28.** tan −2.745   **29.** cos 3.01

30. cos −7.485

Evaluate Exercises 31–40.

31. sin 13π/3   **32.** csc 41π/4   **33.** tan 69π/6

34. cos 19π/6   **35.** tan 22π/3   **36.** cot 27π/4

37. cos 61π/2   **38.** sin(−35π/6)   **39.** sec 43π/6

40. sin 1000π

In Exercises 41–45, find θ where $0 \leq \theta < 2\pi$.

41. sin θ = −0.6157,   cos θ > 0   **42.** cos θ = 0.4226,   tan θ < 0

43. cot θ = 0.9004,   sin θ < 0   **44.** cos θ = 0.8241,   sin θ > 0

45. tan θ = −1.804,   sec θ > 0

In Exercises 46–51, find θ where $0 \le \theta < 360°$.

46. $\sin \theta = 0.4331$, $\quad \cos \theta < 0$

47. $\sin \theta = -0.4253$, $\quad \tan \theta < 0$

48. $\cos \theta = -0.8635$, $\quad \cot \theta > 0$

49. $\cot \theta = 3.0326$, $\quad \csc \theta < 0$

50. $\cos \theta = 0.9012$, $\quad \sec \theta > 0$

51. $\tan \theta = -6.8269$, $\quad \csc \theta > 0$

52. Construct a table showing the exact values of the trigonometric functions for angles with radian measures of 0, $\pi/6$, $\pi/4$, $\pi/3$ and angles symmetric to those around the unit circle.

Key Topics for Chapter 3

Define and/or discuss each of the following.

Angles greater than 90°

Extended trigonometric functions

Quadrantal angles

Reciprocal relations

Quotient relation

Unit circle

Reference angle

Review Exercises for Chapter 3

1. Find the six trigonometric functions of the angle whose terminal side passes through $(-2, 5)$.

2. Determine the quadrant in which the terminal side of θ lies if $\cot \theta < 0$ and $\sec \theta < 0$.

3. Find the other five trigonometric functions of θ if $\cos \theta = -4/5$ and $\tan \theta > 0$.

4. Find the other five trigonometric functions of α if $\csc \alpha = 13/5$ and $\cos \alpha < 0$.

In Exercises 5–10 indicate the reference angle for the given angle.

5. 315°18′   **6.** 109°41′   **7.** 241.9°   **8.** 185.3°   **9.** 4.75   **10.** 2.73

In Exercises 11–18 evaluate the indicated function.

11. $\tan 203°$ **12.** $\sin 310°20′$ **13.** $\cos(-112.6°)$ **14.** $\csc(-9°15′)$

15. $\sec 98.3°$ **16.** $\cot 189.6°$ **17.** $\sin 3.07$ **18.** $\tan 5.27$

19. Given $\sin x = -0.8102$, find x $(0 \le x < 360°)$ to the nearest minute.

20. Given $\tan \theta = 1.202$, find θ $(0 \le \theta < 360°)$ to the nearest minute.

21. Given $\sec \phi = 2.603$ and $\sin \phi < 0$, find $\phi(0 \le \phi < 360°)$ to the nearest tenth degree.

22. Given $\csc \theta = -1.118$ and $\cot \theta > 0$, find θ $(0 \le \theta < 360°)$ to the nearest tenth degree.

23. Given $\tan \beta = -0.7761$ and $\cos \beta > 0$, find β $(0 \le \beta < 2\pi)$ to the nearest hundredth radian.

Test 1 for Chapter 3

In Exercises 1–10, answer *true* or *false*.

1. If $\theta_1 < \theta_2$ and both θ_1 and θ_2 are in the third quadrant, then $\cos \theta_1 < \cos \theta_2$.

2. $\sin 30° = \sin(180° - 30°)$

3. $\sin(\pi/6) = \cos(-\pi/3)$

4. $\sin(2\pi + \theta) = \sin(2\pi - \theta)$ for any θ

5. $\cos(\pi/4) = \sin(3\pi/4)$

6. $\cos(\theta - \pi) = \cos(\theta + \pi)$

7. If $45° < \theta < 90°$, then $\sin \theta > \cos \theta$.

8. If $0 < \theta < 360°$, and $\sin \theta < \cos \theta$, then $0 < \theta < 45°$.

9. If $0 < \theta < 360°$ and $\tan \theta = 1$, then $\theta = 45°$ or $225°$.

10. In the third quadrant, $\cot \theta < 0$.

11. Find the six trigonometric functions of an angle in standard position whose terminal side passes through $(2, -1)$.

12. Given that $\tan \theta = -3$ and θ is an angle in the second quadrant, find the values of the other 5 trigonometric functions.

13. Find $\tan -1030°$.   14. Find $\sin 45$.

15. Find θ if $\tan \theta = 2$ and $2\pi < \theta < 4\pi$.

16. Find θ if $\sin \theta = .6773$ and $-\pi < \theta < \pi$.

Test 2 for Chapter 3

1. Find the sine and tangent functions of the angle in standard position whose terminal side passes through $(-2, 3)$.

2. Answer true or false
   (a) $\sin \theta > 0$ in Q III   (b) $\sin \theta < 0$ in Q IV   (c) $\tan \theta < 0$ in Q III

3. Given $\tan \theta = -\sqrt{2}$ in Q II, find $\cos \theta$ and $\csc \theta$.

4. Evaluate $\sin 206°$.   5. Evaluate $\cos 495°20'$.   6. Evaluate $\tan 200$.

7. Find θ if $0 < \theta < 360°$ if $\sin \theta = -0.4564$.

8. Find θ if $-\pi < \theta < \pi$ if $\tan \theta = 1.3957$.

9. Find θ if $0 < \theta < 2\pi$ $\sin \theta = -0.5666$.

10. If you are told that θ is an angle such that $\cos \theta = -1.1231$, what do you say about θ?

4

The Solution of Oblique Triangles

4.1   Oblique Triangles

Any triangle which is not a right triangle is called **oblique**. Hence, in an oblique triangle, none of the angles are equal to 90°. In Section 2.7, you learned that a right triangle is uniquely determined by two of its five unknown parts, if at least one of the two is the length of a side. In this chapter, we study conditions under which an oblique triangle is solvable and we arrive at roughly the same conclusion: a knowledge of at least three of the six parts is necessary to solve the general triangle.

The three parts needed to solve a triangle are not completely arbitrary. For example, if the three angles of a triangle are given, no unique solution is possible. This is because many triangles can be constructed having these angles but different side lengths. All such triangles would be **similar** but not **congruent**.* As the chapter develops, you will discover other more subtle cases in which three parts assigned arbitrarily yield impossible or ambiguous situations.

From the perspective of solving oblique triangles it is convenient to group the three parts needed to describe a triangle into the categories or cases as follows:

CASE 1

Two sides and the included angle are given.

*The word **congruent** means of the same shape and size. Thus, two triangles that are congruent coincide exactly in all their parts if placed properly one upon the other.

CASE 2
> Three sides are given.

CASE 3
> Two angles and one side are given.

Of course, as was noted earlier, the given information must be sufficient to form a triangle. For instance, no triangle can be formed from a given side and two angles of 88° and 95° because of the sum of the angles of a triangle can not exceed 180°. It is important that you be aware of inconsistencies of this type when solving triangles. A fourth case that arises in solving oblique triangles is important even though the information given might give two triangles.

CASE 4
> Two sides and an angle opposite one of the sides are given.

This case is sometimes referred to as the **ambiguous** case since two triangles, one triangle, or no triangles may result from data given in this form. For instance, Figure 4.1 shows two triangles that can be obtained from the given information $a = 15$, $b = 20$, and $A = 20°$. In section 4.4, we will discuss this case in detail. For the present, you should understand that it is possible for two noncongruent triangles to have the same "Case 4" data.

FIGURE 4.1

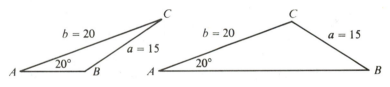

4.2   The Law of Cosines

The Law of Cosines is a formula that enables us to solve an oblique triangle when two sides and the included angle are given, as in Case I, or when three sides are given, as in Case 2. To derive the law of cosines, we subdivide a general oblique triangle into two right triangles.

Consider any oblique triangle ABC; either of the triangles shown in Figure 4.2 will do. Drop a perpendicular from the vertex B to side AC or its extension. Call the length of this perpendicular h. In either case, we obtain $h = c \sin A$, and hence,

$$h^2 = c^2 \sin^2 A$$

FIGURE 4.2
Law of Cosines

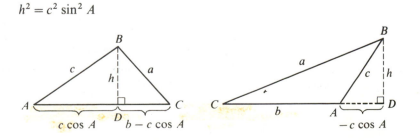

Or, using

$$\sin^2 A + \cos^2 A = 1$$

we have

$$h^2 = c^2(1 - \cos^2 A)$$

Referring to Figure 4.2, we also have, from right triangle BCD, that

$$h^2 = a^2 - (b - c \cos A)^2$$

Equating the right-hand sides of the two preceding equations, we get

$$c^2(1 - \cos^2 A) = a^2 - (b - c \cos A)^2$$

$$c^2 - c^2 \cos^2 A = a^2 - b^2 + 2bc \cos A - c^2 \cos^2 A$$

By simplifying this expression, we arrive at

(4.1) $a^2 = b^2 + c^2 - 2bc \cos A$

In a similar manner, it can be shown that

$b^2 = a^2 + c^2 - 2ac \cos B$

and

$c^2 = a^2 + b^2 - 2ab \cos C$

Each of these formulas is a statement of the Law of Cosines. In words, it says that the square of any side of a triangle is equal to the sum of the squares of the other two sides minus twice their product and the cosine of the angle between them. If the angle is 90°, the Law of Cosines reduces to the Theorem of Pythagoras so that it is quite properly considered as an extension of that famous theorem.

A calculator can be used to make the arithmetic computations that arise when the Law of Cosines is applied to a specific triangle. For example, if b, c, and A are given parts of a triangle, the following keystrokes can be used to calculate the length of side a. Referring to (4.1), we have

Some calculators require a grouping symbol before and after the expression 2bc cos A. In this case the keystroke sequence following the minus sign is

Check and see how your calculator works with (4.1).

The Law of Cosines gives the relationship between three sides and one of the angles of any triangle. Thus, if you are given any of these three parts, you can compute the remaining parts or show that such a triangle is impossible. For example, if three sides of a triangle are given, any angle may be found from Equation (4.1) by solving for $\cos A$, B, or C to obtain an alternate form of the Law of Cosines:

(4.2) $\cos A = \dfrac{b^2 + c^2 - a^2}{2bc}$, $\cos B = \dfrac{a^2 + c^2 - b^2}{2ac}$, $\cos C = \dfrac{a^2 + b^2 - c^2}{2ab}$

Note that in this form the law says that the cosine of an angle may be found by computing a fraction, whose numerator is the sum of squares of the adjacent sides minus the square of the opposite side, and whose denominator is twice the product of the adjacent sides.

To solve for angle A when sides a, b, and c are known use (4.2) *and the following keystrokes on your calculator.*

EXAMPLE 1   Solve the triangle with side $a = 5.18$, side $b = 6.00$ and angle $C = 60.0°$ as shown in Figure 4.3.

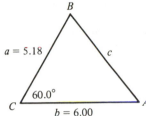

FIGURE 4.3

SOLUTION
$$c^2 = a^2 + b^2 - 2ab \cos C$$

$$= 5.18^2 + 6.00^2 - 2(5.00)(6.00) \cos 60.0°$$

$$= 31.7524$$

Therefore,

$$c = \sqrt{31.7524} = 5.63$$

To solve for angle A, we use Equation (4.2) of the Law of Cosines:

$$\cos A = \frac{b^2 + c^2 - a^2}{2bc}$$

$$= \frac{6.00^2 + 5.63^2 - 5.18^2}{2(6.00)(5.63)}$$

$$= 0.6049$$

Therefore,

$$A = 52.8°$$

The remaining angle could also be found from the Law of Cosines, but since the sum of the angles is 180°,

$$B = 180° - 60° - 52.8°$$

$$= 67.2°$$ ■

EXAMPLE 2   Two airplanes leave an airport at the same time; one going Northeast at 400 mph, and the other directly west at 300 mph. How far apart are they two hours after leaving?

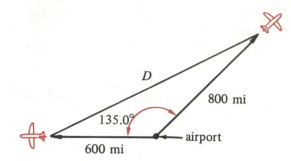

FIGURE 4.4

SOLUTION　From Figure 4.4 and from the Law of Cosines,

$$D = \sqrt{600^2 + 800^2 - 2(600)(800) \cos 135.0°}$$

Notice that $\cos 135.0° = -\cos 45°$, so

$$D = \sqrt{360,000 + 640,000 - (960,000)(-\cos 45°)}$$

$$= \sqrt{1,678,822}$$

$$= 1300 \text{ miles}$$

■

EXAMPLE 3　In a steel bridge, one part of a truss is in the form of an isosceles triangle as shown in Figure 4.5. At what angles do the sides of the truss meet?

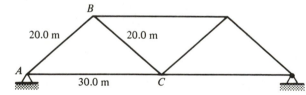

FIGURE 4.5

SOLUTION　$\cos A = \dfrac{20^2 + 30^2 - 20^2}{(2)(20)(30)} = \dfrac{900}{1200} = 0.75$

Hence,

$$A = 41.4°$$

This is also the value of angle C since the triangle is isosceles. Then,

$$B = 180° - 2(41.4°) = 97.2°$$

Suppose we had decided to use the Law of Cosines to find angle B. Then,

$$\cos B = \frac{20^2 + 20^2 - 30^2}{(2)(20)(20)} = \frac{800 - 900}{800} = -\frac{1}{8} = -0.125$$

The fact that $\cos B$ is negative tells us that angle B is greater than 90° and less than 180°. The angle whose cosine is 0.125 is 82.8°, and therefore,

$$B = 180° - 82.8° = 97.2°$$

This agrees with the previous result. Notice how the determination of angle B was affected by the fact that $\cos B$ was negative. Of course if you use a calculator to evaluate $\cos B = -0.125$, the display shows 97.2° directly. ■

EXAMPLE 4   A satellite traveling in a circular orbit 1000 mi above Earth passes directly over a tracking station at noon. Assume that the satellite takes 2.0 hr to make an orbit and that the radius of Earth is 4000 mi. Find the distance between the satellite and tracking station at 12.03 P.M.

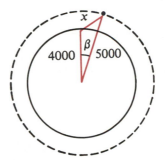

FIGURE 4.6

SOLUTION   From Figure 4.6, we see that the angle β must be computed. Since the angular velocity is $3.0°$ per minute, a total of $9.0°$ is traveled in 3 minutes. Hence, $\beta = 9.0°$. By the Law of Cosines,

$$x = \sqrt{(4000)^2 + (5000)^2 - 2(4000)(5000) \cos 9.0°}$$

$$= \sqrt{1,492,466}$$

$$= 1220$$

Thus, the distance between the satellite and the tracking station is about 1220 mi.   ■

COMMENT

> The form of the Law of Cosines as in Equation (4.2) further allows you to see that information given on the sides of triangles is not arbitrary. For example, if you are given that $a = 1$, $b = 3$, $c = 1$, then using Equation (4.2),
>
> $$\cos A = \frac{(3)^2 + (1)^2 - (1)^2}{2(3)(1)} = \frac{9}{6} = \frac{3}{2}$$
>
> However, this is impossible since $|\cos A| \leq 1$. Hence, no such triangle exists. The same conclusion is reached by noting that $a + c < b$.

Exercises for Sections 4.1 and 4.2

In each of Exercises 1–6, use the Law of Cosines to find the unknown side.

1. $a = 45.0$, $b = 67.0$, $C = 35°$
2. $a = 20.0$, $b = 40.0$, $C = 28°$
3. $a = 10.5$ $b = 40.8$, $C = 120°$
4. $b = 12.9$, $c = 15.3$, $A = 104.2°$
5. $b = 38.0$, $c = 42.0$, $A = 135.3°$
6. $a = 3.49$, $b = 3.54$, $C = 5°24'$

In each of Exercises 7–12, use the Law of Cosines to find the largest angle.

7. $a = 7.23$, $b = 6.00$, $c = 8.61$
8. $a = 16.0$, $b = 17.0$, $c = 18.0$
9. $a = 18.0$, $b = 14.0$, $c = 10.0$
10. $a = 300$, $b = 500$, $c = 600$
11. $a = 170$, $b = 250$, $c = 120$
12. $a = 56.0$, $b = 67.0$, $c = 82.0$

In each of Exercises 13–18, use the Law of Cosines to solve the triangles.

13. $a = 4.21, b = 1.84, C = 30.7°$

14. $a = 5.92, b = 7.11, C = 60.6°$

15. $a = 120, b = 145, C = 94°25'$

16. $a = 900, b = 700, c = 500$

17. $a = 2.00, b = 3.00, c = 4.00$

18. $a = 5.01, c = 5.88, B = 28°40'$

19. A boy starting from point A walks 2.50 km due west to point B. He then takes a path that is 25.4° south of west and walks 1.40 km to point C. What is the distance between A and C?

20. If the sides of the triangular sections of a geodesic dome are 1.65, 1.65 and 1.92 m, what are the interior angles of the triangular sections?

Geodesic Dome

HISTORICAL NOTE

The geodesic dome was popularized in the 1950s by architect R. Buckminster Fuller.

21. In a triangular lot ABC, the stake that marked the corner C has been lost. By consulting his deed to the property, the owner finds that $AB = 80.0$ ft, $BC = 50.0$ ft, and $CA = 40.0$ ft. At what angle with AB should she run a line so that by laying off 40.0 ft along this line she can locate corner C?

22. A solar collector is placed on a roof that makes an angle of 24.0° with the horizontal. If the upper end of the collector is supported as shown in Figure 4.7, how long is the collector?

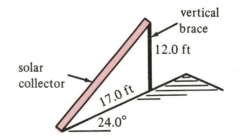

vertical brace

12.0 ft

solar collector

17.0 ft

24.0°

FIGURE 4.7

23. A reflector used in a solar furnace is composed of triangular sections, each having side lengths of 5.50, 5.50 and 1.30 ft. Find the interior angle between the sides of equal length.

24. An airplane flying directly north toward a city C alters its course toward the northeast at a point 100 km from C and heads for city B, which is 50.0 km away. If B and C are 60.0 km apart, what course should the airplane fly to get to B?

25. In planning a tunnel under a hill, an engineer lays out the triangle ABC as shown in Figure 4.8 in order to determine the course of the tunnel. If $AB = 3500$ m, $BC = 4000$ m, and angle $B = 60.0°$, what are the sizes of the angles A and C and the length of AC?

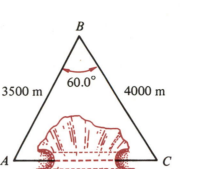

FIGURE 4.8

26. If two forces, one of 500 lb and the other of 400 lb, act from a point at an angle of 58.3° with each other, what is the size of the resultant? (See Figure 4.9.)

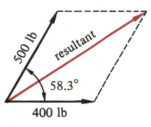

FIGURE 4.9

27. In order to measure the distance between two points, A and B, on opposite sides of a building, a third point C is chosen such that the following measurements can be made: $CA = 200$ m, $CB = 400$ m and the angle ACB measures 60.0°. What is the distance between A and B?

28. Show that for any triangle ABC,

$$\frac{a^2 + b^2 + c^2}{2abc} = \frac{\cos A}{a} + \frac{\cos B}{b} + \frac{\cos C}{c}$$

4.3   The Law of Sines

You may have noticed that the Law of Cosines cannot be used to solve triangles for which two angles and one side are given. Triangles of this type may be solved using a formula known as the **Law of Sines**. This formula in conjunction with the Law of Cosines enables us to solve any triangle for which we are given three parts, or at least to declare that no solution is possible.

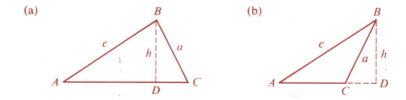

FIGURE 4.10
Law of Sines

Consider either triangle shown in Figure 4.10. By drawing a perpendicular h from the vertex B to side b or its extension, we see from Figure 4.10(a) that

$$h = c \sin A \quad \text{and} \quad h = a \sin C$$

Equating these two expressions, we get

$$c \sin A = a \sin C$$

or, rearranging terms,

$$\frac{a}{\sin A} = \frac{c}{\sin C}$$

In a similar manner, we can show that

$$\frac{a}{\sin A} = \frac{b}{\sin B}$$

Hence,

(4.3) $$\frac{a}{\sin A} = \frac{b}{\sin B} = \frac{c}{\sin C}$$

Equation (4.3) is called the **Law of Sines**. In words, it states that, in any triangle, the ratios formed by dividing the sides by the sine of the angle opposite them are equal.

Combinations of any two of the three ratios given in Equation (4.3) will yield an equation with four parts. Obviously, if we know three of these parts, we can find the fourth.

If the Law of Sines is written

$$\frac{a}{\sin A} = \frac{b}{\sin B}$$

then side a is given by

$$a = \frac{b \sin A}{\sin B}$$

The keystroke sequence for side a is then

b ⌷ x ⌷ A ⌷ sin ⌷ ÷ ⌷ B ⌷ sin ⌷ = ⌷

Similarly, if we solve the Law of Sines for sin A, *we get*

$$\sin A = \frac{a \sin B}{b}$$

Angle A can then be found by the following keystrokes

a ⌷ x ⌷ B ⌷ sin ⌷ ÷ ⌷ b ⌷ = ⌷ inv ⌷ sin ⌷

EXAMPLE 1   Given that $c = 10$, $A = 40.0°$, and $B = 60.0°$, find a, b, and C. (See Figure 4.11.)

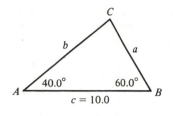

FIGURE 4.11

SOLUTION   We begin by observing that

$$C = 180° - (A + B) = 180° - (40.0° + 60.0°) = 80.0°$$

Using the Law of Sines, we have

$$\frac{a}{\sin 40°} = \frac{10.0}{\sin 80°} \quad \text{or} \quad a = \frac{10.0(0.6428)}{0.9848} = 6.53$$

and

$$\frac{b}{\sin 60°} = \frac{10.0}{\sin 80°} \quad \text{or} \quad b = \frac{10.0(0.8660)}{0.9848} = 8.79$$ ■

You should be aware of the fact that since $\sin A = \sin(180° - A)$ the formula

$$\sin A = \frac{a \sin B}{b}$$

gives two possible values for angle A. Therefore, you must be careful when using the law of sines to solve for angles. The next example illustrates the problem.

EXAMPLE 2   Given $A = 33.7°$, $b = 2.17$, and $c = 1.09$, find angle B.

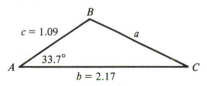

FIGURE 4.12

SOLUTION   Before we can solve for angle B we must find side a. By the law of cosines, we have

$$a = \sqrt{2.17^2 + 1.09^2 - 2(2.17)(1.09)\cos 33.7°} = 1.40$$

Now, if we use the law of sines to find angle B, we get

$$\sin B = \frac{2.17 \sin 33.7°}{1.40} = 0.860$$

$$B = 59.3°$$

However, if we use the law of cosines to find B, we get

$$\cos B = \frac{1.40^2 + 1.09^2 - 2.17^2}{2(1.40)(1.09)} = -0.5114$$

$$B = 120.7°$$

We note that the correct value 120.7° is the supplement of 59.3°. However, the law of sines alone gives us no clue that the supplement of 59.3° is the required answer. A rough drawing of the triangle will usually be sufficient to tell you which of the two angles to use. ∎

EXAMPLE 3   The unit of angular measure in the SI system is the radian. Given that $a = 17.5$ cm, $B = 0.95$ radian and $C = 1.22$ radian, find angle A and sides b and c. (See Figure 4.13.)

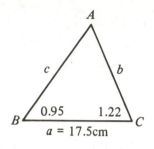

FIGURE 4.13

SOLUTION   The sum of the interior angles of a triangle is 180° or π radians. Using 3.14 to approximate π, we have

$$A = 3.14 - (0.95 + 1.22) = 0.97 \text{ radian}$$

Now using the law of sines and Table C in the appendix, we have

$$\frac{b}{\sin 0.95} = \frac{17.5}{\sin 0.97} \quad \text{or} \quad b = \frac{17.5 \sin 0.95}{\sin 0.97} = \frac{17.5(0.8134)}{0.8249} = 17.3 \text{ cm}$$

$$\frac{c}{\sin 1.22} = \frac{17.5}{\sin 0.97} \quad \text{or} \quad c = \frac{17.5 \sin 1.22}{\sin 0.97} = \frac{17.5(0.9391)}{0.8249} = 19.9 \text{ cm} \qquad ∎$$

EXAMPLE 4   A surveyor desires to run a straight line from A in the direction AB, as shown in Figure 4.14, but finds that an obstruction interferes with the line of sight. Therefore, the crew lays off the line segment $\overline{BX}$ for a distance of 150.0 m in such a way that angle $ABX = 135.0°$ and runs XY at an angle of 75.0° with $\overline{BX}$. At what distance from X on this line should a stake be placed so A, B, and C are on the same straight line?

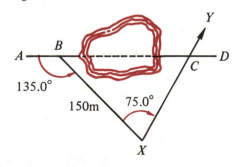

FIGURE 4.14

SOLUTION   Since angle $CBX = 45.0°$, we have, from the Law of Sines,

$$\frac{\overline{CX}}{\sin 45.0°} = \frac{150.0}{\sin 60.0°}$$

Hence,

$$\overline{CX} = \frac{\sin 45.0°}{\sin 60.0°}150 = \frac{\sqrt{2}/2}{\sqrt{3}/2}150 = \frac{\sqrt{2}}{\sqrt{3}}150 \approx \frac{1.414}{1.732}150 = 122 \text{ m}$$

The next two examples involve triangles in which two sides and an angle opposite one of them is given (Case 4). You will recall that data such as this may define one, two, or no triangles. For simplicity, we restrict the following discussion to examples in which a unique triangle is defined by the given data. The discussion of the ambiguous case is presented in the next section.

EXAMPLE 5   A satellite traveling in a circular orbit 1000 mi above Earth is due to pass directly over a tracking station at noon. Assume that the satellite takes 2 hr to make an orbit and that the radius of Earth is 4000 mi. If the tracking antenna is aimed 30.0° above the horizon, at what time will the satellite pass through the beam of the antenna? (See Figure 4.15.) Assume the beam is directed to intercept the satellite before it is overhead.

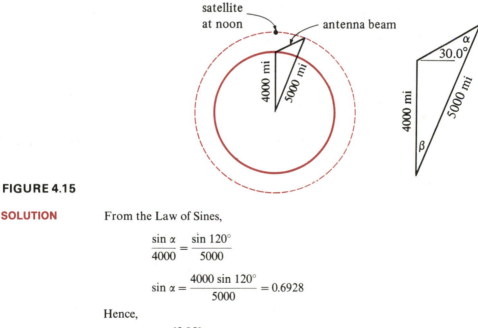

FIGURE 4.15

SOLUTION   From the Law of Sines,

$$\frac{\sin \alpha}{4000} = \frac{\sin 120°}{5000}$$

$$\sin \alpha = \frac{4000 \sin 120°}{5000} = 0.6928$$

Hence,

$$\alpha = 43.85°$$

and

$$\beta = 180° - (120° + 43.85°) = 16.15°$$

Time between $\beta = 16.5°$ and $\beta = 0.0°$ is $(16.15°/360°)(120 \text{ min}) = 5.38 \text{ min} = 5 \text{ min } 23 \text{ sec}$. Thus, the satellite passed through the beam of the antenna at $12:00 - 5.38$ min or 11:54:37 A.M.

EXAMPLE 6   Consider a flight from Chicago to Boston and a return, which is a one-way airline distance of 870 mi. A light plane having an airspeed of 180 mph makes the round trip. How many flying hours will it take for the round trip with a constant southwest wind* of 23.0 mph? What headings will the pilot use for the two parts of the trip?

SOLUTION

For the eastbound trip, shown in Figure 4.16, the Law of Sines is applied to determine θ.

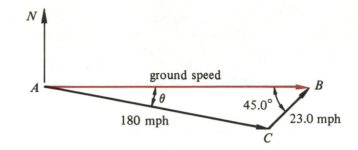

FIGURE 4.16
Eastbound trip

$$\sin \theta = \frac{23.0}{180} \sin 45.0° = 0.0904$$

$$\theta = 5.2°$$

Applying the Law of Sines again, the ground speed represented by $\overline{AB}$ is determined as follows.

$$\frac{\overline{AB}}{\sin C} = \frac{180}{\sin B} \quad \text{or} \quad \overline{AB} = \frac{\sin 129.8°}{\sin 45°} \cdot 180 = 196 \text{ mph}$$

Thus, the time required for the eastbound trip is

$$\text{Time} = \frac{\text{Distance}}{\text{Velocity}} = \frac{870 \text{ mi}}{196 \text{ mph}} = 4.44 \text{ hr}$$

For the westbound trip, shown in Figure 4.17, θ is again 5.2°, and the ground speed is found by use of the Law of Sines.

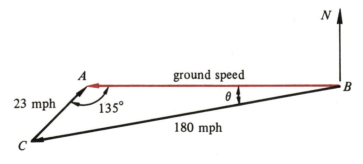

FIGURE 4.17
Westbound trip

$$\overline{AB} = \frac{\sin 39.8°}{\sin 135°} \cdot 180 = 163 \text{ mph}$$

and the time required is

$$\text{Time} = \frac{870 \text{ mi}}{163 \text{ mph}} = 5.34 \text{ hr}$$

*Wind direction is conventionally specified by the direction from which the wind is blowing. Thus, a "southwest wind" means the wind is blowing from the Southwest.

Thus, the total time for the round trip is 9.78 hr, or 9 hr 47 min. The heading for the eastbound trip is $90° + 5.2°$, or $95.2°$, and the heading for the westbound trip is $270° - 5.2°$, or $264.8°$.   ■

Exercises for Section 4.3

In Exercises 1–19 solve the given oblique triangles.

1. $A = 32.0°$, $B = 48.0°$, $a = 10.0$

2. $A = 60.0°$, $B = 45.0°$, $b = 3.00$

3. $A = 45.2°$, $a = 8.82$, $b = 5.15$

4. $A = 75.0°$, $a = 27.7$, $b = 11.8$

5. $A = 35.6°$, $b = 12.2$, $a = 17.5$

6. $A = 82.1°$, $b = 7.21$, $a = 29.0$

7. $A = 120°50'$, $a = 6.61$, $b = 5.09$

8. $A = 51°10'$, $a = 59.2$, $b = 53.5$

9. $C = 53.0°$, $b = 18.3$, $c = 30.2$

10. $C = 58.0°$, $c = 83.0$, $b = 51.0$

11. $B = 122.0°$, $b = 30.0$, $a = 25.0$

12. $B = 63.0°$, $b = 5.00$, $c = 4.00$

13. $C = 110.0°$, $B = 50.0°$, $b = 40.0$

14. $C = 73.2°$, $A = 13.7°$, $c = 20.5$

15. $B = 48.0°$, $A = 43.4°$, $c = 61.3$

16. $A = 0.88$ radian, $B = 1.29$ radians, $c = 20.7$

17. $A = 1.31$ radians, $C = 1.00$ radian, $b = 0.652$

18. $B = 0.59$ radian, $C = 1.27$ radians, $a = 274$

19. $B = 0.72$ radian, $C = 0.93$ radian, $a = 13,500$

20. The crank and connecting rod of an engine, like the one illustrated in Figure 4.18 are 30.0 cm and 100 cm long, respectively. What angle does the crank make with the horizontal when the angle made by the connecting rod is $12.0°$?

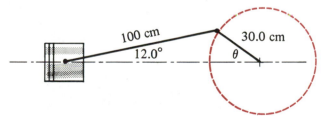

FIGURE 4.18

21. One end of a 15.5-ft plank is placed on the ground at a point 10.8 feet from the start of a $42.7°$ incline and the other end is allowed to rest on the incline. How far up the incline does the plank extend?

22. A 300-ft broadcast antenna stands at the top of a hill whose sides are inclined at $18.6°$ to the horizontal. How far down the hill will a 250-ft support cable extend if it is attached halfway up the antenna?

23. Coast guard station Bravo is located 230 mi due north of an automated search and rescue station. Station Bravo receives a distress message from an oil tanker at a bearing of $124.6°$ and the automated station receives the same message at a bearing of $52.1°$. How long will it take a helicopter from Bravo to reach the ship if the helicopter can fly at 125 mph? Bearing is defined in Section 2.8.

24. A balloon is tethered above a bridge by a cord. To find the height of the balloon above the surface of the bridge a girl measures the length of the bridge and the angles of elevation to the balloon at each end of the bridge. If she finds the length of the bridge to be 263 ft and the angles of elevation to be 64.3° and 74.1°, what is the height of the balloon?

25. A weight is attached to two vertical poles as shown in Figure 4.19. How far from the left post is the weight if $\theta = 41.0°$ and $\phi = 75.0°$.

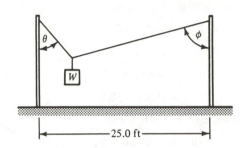

FIGURE 4.19

26. How far will the weight in Exercise 25 be from the left post if $\theta = 38.7°$ and $\phi = 69.1°$.

27. From a position at the base of a hill, an observer notes that the angle of elevation of the top of an antenna is 43.5°. See Figure 4.20. After walking 1500 ft toward the base of the antenna up a slope of 30.0° the angle of elevation is 75.4°. Find the height of the antenna and of the hill.

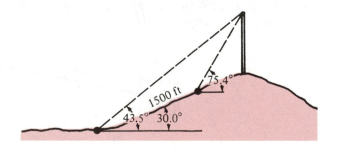

FIGURE 4.20

28. A satellite, traveling in a circular orbit 1500 mi above Earth, is due to pass directly over a tracking station at noon. Assume that the satellite takes 90 min to make an orbit and that the radius of the Earth is 4000 mi. If the tracking antenna is aimed 20.0° above the horizon, at what time will the satellite pass through the beam of the antenna?

29. In the previous exercise, determine tha angle above the horizon that the antenna should be pointed so that its beam will intercept the satellite at 12:05 P.M.

30. Consider a flight from Miami to New York to be along a north-south direction with an airline distance of approximately 1000 mi. A jetliner, having an airspeed of 500 mph, makes the round trip. If there is a constant northwest wind of 100 mph, how long will it take for the round trip? What headings will the pilot use for the two parts of the trip?

4.4   The Ambiguous Case

We now analyze how to solve those triangles for which the measure of two sides and an angle opposite one of them is given. For ease of discussion, suppose that two sides a and b and an angle A are given. As you will see, there may be one, two, or no triangles with these measurements.

$A \leq 90°$: Perhaps the best way to make the situation clear, when angle A is acute, is to draw a figure. Let us construct a line segment having a length of b units along one side of angle A to locate the vertex C. Then it is obvious from Figure 4.21 that the length of side a determines whether there are one, two, or no triangles. In Figure 4.21(a), only one triangle is possible since $a > b$. In Figure 4.21(b), two triangles are possible by swinging an arc of length a from C so that it intersects side c at two points. In Figure 4.21(c), the length of side a is such that it intersects side c at one point to form a single right triangle. Finally, in Figure 4.21(d), the length of side a is too short to intersect side c and therefore no triangle can be formed.

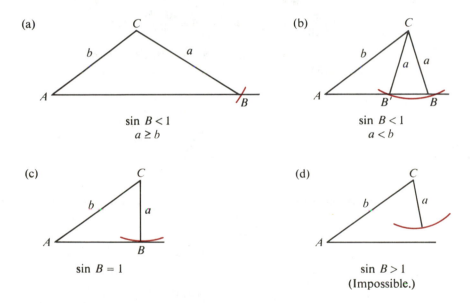

FIGURE 4.21
Ambiguous case,
$A \leq 90°$

The situations, described graphically in Figure 4.21, can be stated analytically by solving

$$\frac{a}{\sin A} = \frac{b}{\sin B}$$

for sin B, and noting that

(1)   If $a < b$ and $\sin B > 1$, then B cannot exist, so there is no triangle.

(2)   If $a < b$ and $\sin B = 1$, then $B = 90°$, so there is one right triangle.

(3)   If $a < b$ and $\sin B < 1$, then two angles at B are possible; the acute angle B from the tables and the obtuse angle $B' = 180° - B$.

(4)   If $a \geq b$, then there is only one triangle.

$A \geq 90°$: The case in which angle A is greater than or equal to 90° is much easier to analyze as shown in Figure 4.22. As you can see in Figure 4.22(a), if $a \leq b$, there is no triangle. If $a > b$, there is one triangle as shown in Figure 4.22(b).

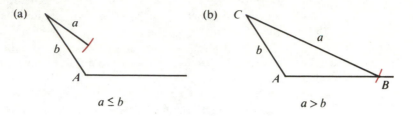

FIGURE 4.22
Ambiguous case,
$A > 90°$

EXAMPLE 1　　How many triangles can be formed if $a = 4.0$, $b = 10$, and $A = 30°$?

SOLUTION　　Using the Law of Sines, we have

$$\frac{\sin B}{10} = \frac{\sin 30°}{4.0} \quad \text{or} \quad \sin B = \frac{10(0.5)}{4.0} = 1.25$$

Since $\sin B > 1$, we conclude that no triangle corresponds to this given information. (See Figure 4.23.)

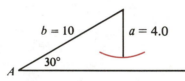

FIGURE 4.23

EXAMPLE 2　　Solve the triangle with $B = 134.7°$, $b = 526$, and $c = 481$. (See Figure 4.24.)

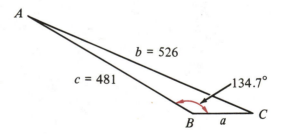

FIGURE 4.24

SOLUTION　　In this triangle the given angle is obtuse and the side opposite the given angle is greater than the side adjacent to it. This means there is only one triangle that can be formed by the given information. Using the law of sines, we have

$$\frac{\sin C}{481} = \frac{\sin 134.7°}{526} \quad \text{or} \quad \sin C = \frac{481(0.7108)}{526} = 0.6500$$

Since B is obtuse, both A and C are acute. Consequently, angle $C = 40.5°$. Angle A is then given by

$$A = 180° - (134.7° + 40.5°) = 4.8°$$

Finally, side a is given by

$$\frac{a}{\sin 4.8°} = \frac{526}{\sin 134.7°} \quad \text{or} \quad a = \frac{526(0.0837)}{0.7108} = 61.9 \quad\blacksquare$$

EXAMPLE 3   Verify that two triangles can be drawn for $a = 9.00$, $b = 10.0$, and $A = 60.0°$ and then solve each triangle.

SOLUTION   Substituting the given values in the Law of Sines, we get

$$\frac{\sin B}{10.0} = \frac{\sin 60°}{9.00} \quad \text{or} \quad \sin B = \frac{10.0(0.8660)}{9.00} = 0.9622$$

Since $\sin B < 1$ and $a < b$, we conclude that there are two possible solutions. This is shown in Figure 4.25. One triangle is ABC and the other is $AB'C'$. We find that

$$B = 74.2°$$

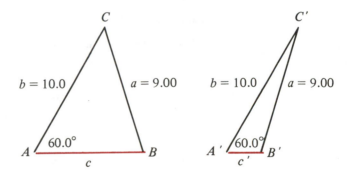

FIGURE 4.25

Therefore, angle B' is given by

$$B' = 180° - B = 180° - 74.2° = 105.8°$$

To solve triangle ABC, we note that angle C is given by

$$C = 180° - (A + B) = 180° - (60° + 74.2°) = 45.8°$$

Hence,

$$\frac{c}{\sin 45.8°} = \frac{9.00}{\sin 60°} \quad \text{or} \quad c = \frac{9.00(0.7169)}{0.8660} = 7.45$$

Similarly in triangle $AB'C'$, we have

$$C' = 180° - (A + B') = 180° - (60° + 105.8°) = 14.2°$$

So,

$$\frac{c'}{\sin 14.2°} = \frac{9.00}{\sin 60°} \quad \text{or} \quad c' = \frac{9.00(0.2453)}{0.8660} = 2.55 \quad\blacksquare$$

EXAMPLE 4   Verify that only one triangle can be drawn for the case in which $b = 10.0$, $A = 60.0°$, and $a = 11.0$.

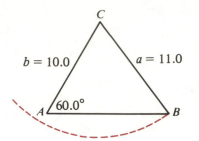

FIGURE 4.26

SOLUTION   Figure 4.26 depicts the information given. Applying the Law of Sines, we get

$$\frac{11.0}{0.866} = \frac{10.0}{\sin B}$$

from which $\sin B = 0.7873$. Since the side opposite the given angle is greater than the side adjacent, there is only one solution. Notice that had we missed the fact that $a > b$, we would have gotten

$$B = 51.9° \quad \text{and} \quad B' = 128.1°$$

However, the "solution" of $B' = 128.1°$ is impossible because a triangle with an angle of 60.0° and 128.1° cannot exist. Thus, we reach the same conclusion as before.   ■

COMMENT

> This last example shows how important it is to continually check that the information being given and being calculated is consistent and does not give impossible situations.
>
> The various subcases within the ambiguous case need not be memorized since, for any specific problem, the ambiguity or nonsolvability will become evident when the solution is attempted. For example, if in the course of solving a triangle, you find that $\sin B > 1$, this will mean that no triangle exists corresponding to the data given. If $\sin B < 1$, there are two angles which satisfy this particular inequality, but you will also have to watch for the limitation that the sum of the angles is less than 180°.

Exercises for Section 4.4

In Exercises 1–5 state whether the triangle has one solution, two solutions, or no solutions, given that $A = 30°$ and $b = 4$.

1. $a = 1$ **2.** $a = 2$ **3.** $a = 3$ **4.** $a = 4$ **5.** $a = 5$

For the triangles in Exercises 6–19, find all the unknown measurement or show that no such, triangle exists.

6. $a = 20.0, b = 10.0, A = 35°40'$ **7.** $a = 2.00, b = 6.00, A = 26°20'$

8. $a = 4.00, b = 8.00, A = 30.0°$ **9.** $a = 15.0, c = 8.00, A = 150.0°$

10. $a = 50.0, b = 19.0, B = 22°30'$ **11.** $b = 60.0, c = 74.0, B = 140.0°$

12. $a = 50.0, c = 10.0, A = 48.0°$ **13.** $C = 28.0°, a = 20.0, c = 15.0$

14. $B = 40.0°$, $a = 12.0$, $b = 10.0$ **15.** $A = 30.0°$, $b = 400$, $a = 300$

16. $B = 100.0°$, $a = 10.0$, $b = 12.0$ **17.** $C = 70.0°$, $b = 100$, $c = 100$

18. $B = 41.2°$, $a = 4.20$, $b = 3.20$ **19.** $a = 0.900$, $b = 0.700$, $A = 72°15'$

20. If $b = 12$ and $A = 30.0°$, for what values of a will two triangles result?

21. If $c = 15$ and $B = 25.0°$, for what values of b will two triangles result?

4.5   Analysis of the General Triangle

In Section 4.1, we mentioned that most of the time you could expect three parts of a triangle to be sufficient to determine it uniquely. With the aid of the two fundamental laws derived in Sections 4.2 and 4.3, we are now in a position to summarize the approach.

CASE 1

> **Two sides and an included angle given**
> Use the Law of Cosines to obtain the third side. A second angle may be obtained using either the Law of Cosines or the Law of Sines. The third angle is computed as 180° minus the sum of the other two.

CASE 2

> **Three sides given**
> Use the Law of Cosines to obtain one of the angles, preferably the largest one. A second angle may be obtained using either the Law of Cosines or Sines, and the third one may be obtained from the fact that the sum of the angles must be 180°. The sum of any two sides must exceed the length of the third side or no triangle can be formed.

CASE 3

> **Two angles and a side given**
> The two given angles must have a sum less than 180° otherwise no such triangle is possible. Use the Law of Sines to determine the two unknown sides.

CASE 4

> **Two sides and a nonincluded angle are given**
> If two sides a and b and an angle A are given, there may be two, one, or no triangles with these measurements. A carefully drawn figure will usually make the situation clear. Analytically we have,
>
> (1)   If A is acute, then there is no, one, or two solutions depending on whether $a < b \sin A$, $a = b \sin A$ or $a > b \sin A$, unless $a \geq b$ in which case there is only one solution.
>
> (2)   If A is obtuse, there is no solution, or one solution corresponding to $a \leq b$ or $a > b$.

Exercises for Section 4.5

Solve the following triangles in Exercises 1–17, or show that no such triangle exists.

1. $A = 60.0°, B = 75.0°, a = 600$
2. $A = 75.0°, a = 120, b = 75.0$
3. $B = 15.0°, C = 105.0°, a = 4.00$
4. $A = 30.0°, b = 60.0, c = 50.0$
5. $a = 8, b = 2, c = 6$
6. $C = 15.0°, b = 15.0, c = 10.0$
7. $C = 30.0°, a = 300, b = 500$
8. $a = 2000, b = 1000, c = 2500$
9. $B = 120.0°, a = 60.0, b = 25.0$
10. $B = 30.0°, a = 500, b = 400$
11. $B = 70.0°, a = 12.0, b = 6.00$
12. $A = 20.0°, a = 2.00, b = 3.00$
13. $A = 60.0°, B = 100°, b = 2.00$
14. $B = 125.2°, a = 2.20, b = 1.30$
15. $A = 100.0°, a = 2.00, b = 1.00$
16. $A = 42.3°, a = 20.0, c = 30.0$
17. $a = 2.00, b = 4.00, c = 8.00$

18. From a window 35.0 ft above the street, the angle of depression of the curb on the other side of the street is 15.0°, and, on the near side, the angle of depression is 45.0°. How wide is the street?

19. At successive milestones on a straight road leading to a mountain, readings of the angle of elevation to the top of the mountain are made of 30.0° and 45.0° respectively. What is the line of sight distance to the top of the mountain from the nearest milestone?

20. Two forces, one of 75.0 lb and the other of 100 lb act at a point. If the angle between the forces is 60.0°, find the magnitude and direction of the resultant force. Give the direction as the angle between the resultant and the 100-lb force.

21. The air speed of a plane is 400 mph, and there is a 75.0-mph wind from the northeast at a time when the heading of the plane is due east. Find the ground speed and direction of the path of the plane.

4.6   Area Formulas

As you undoubtedly know, the area of a triangle is given by one-half the product of any base with the corresponding altitude. With the use of the Law of Cosines and the Law of Sines, we can derive some equivalent expressions for area for which the height need not be specifically computed. In this section we examine three such formulas.

Consider the triangle in Figure 4.27. The altitude from the vertex B to side b is given by $h = c \sin A$. Hence, the area S is given by

(4.4) $$S = \frac{1}{2}bc \sin A$$

This procedure could be repeated for each vertex to give equivalent formulas. In general, the area of a triangle is equal to one-half the product of the lengths of any two sides and the sine of the included angle.

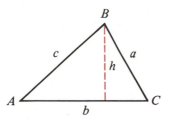

FIGURE 4.27

EXAMPLE 1   Find the area of a triangle with $a = 3.8$ in., $b = 5.1$ in., and $C = 48°$.

SOLUTION   Since C is the angle between sides a and b, the area is given by

$$S = \frac{1}{2}(3.8)(5.1) \sin 48° = 7.2 \text{ sq in.}$$ ∎

The expression for area may be altered further by using the Law of Sines. Since $c = b \sin C / \sin B$, we have from Equation (4.4),

(4.5) $$S = \frac{b^2}{2} \frac{\sin A \sin C}{\sin B}$$

Thus, the area may be calculated if one of the sides and two of the angles are given (since the third angle may then be easily found).

EXAMPLE 2   Find the area of a triangle with two angles and an included side given by $30°$, $45°$, and 2.7 cm, respectively.

SOLUTION   The remaining angle is $105°$. Hence,

$$S = \frac{(2.7)^2}{2} \frac{\sin 30° \sin 45°}{\sin 105°}$$

$$S = \frac{7.29(0.5)(0.7071)}{2(0.9659)}$$

$$= 1.3 \text{ square centimeters}$$ ∎

A formula for the area may be derived from Equation (4.4), to obtain an expression in terms of the length of the sides.

$$S^2 = \tfrac{1}{4}b^2c^2 \sin^2 A$$

$$= \tfrac{1}{4}b^2c^2(1 - \cos^2 A)$$

$$= (\tfrac{1}{2}bc)(\tfrac{1}{2}bc)(1 - \cos A)(1 + \cos A)$$

Now, from the Law of Cosines, we have that

$$\tfrac{1}{2}bc(1 + \cos A) = \tfrac{1}{2}bc\left(1 + \frac{b^2 + c^2 - a^2}{2bc}\right)$$

$$= \frac{2bc + b^2 + c^2 - a^2}{4}$$

$$= \frac{(b + c)^2 - a^2}{4}$$

$$= \frac{(b + c - a)(b + c + a)}{4}$$

Similarly, we could obtain

$$\tfrac{1}{2}bc(1 - \cos A) = \frac{(a - b + c)(a + b - c)}{4}$$

Therefore, the expression S^2 becomes

$$S^2 = \frac{(b + c - a)(a + b + c)(a - b + c)(a + b - c)}{16}$$

$$S = \tfrac{1}{4}\sqrt{(b + c - a)(a + b + c)(a - b + c)(a + b - c)}$$

Sometimes we express this formula in terms of the perimeter, $P = a + b + c$:

(4.6) $S = \tfrac{1}{4}\sqrt{P(P - 2a)(P - 2b)(P - 2c)}$

EXAMPLE 3   Find the area of the triangle whose sides are 2, 2, and 3.

SOLUTION   Since the perimeter is 7, we have

$$S = \tfrac{1}{4}\sqrt{7(7 - 4)(7 - 4)(7 - 6)} = \tfrac{1}{4}\sqrt{63} = \tfrac{3}{4}\sqrt{7} \approx 2 \text{ square units}$$ ■

Note from Equation (4.6) that the lengths of the sides of a triangle are inherently limited. For example, no triangle exists whose sides are 1, 2, and 3.

Exercises for Section 4.6

Find the area of triangles with measurements as given in Exercises 1–10.

1. $a = 15.0$, $b = 5.00$, $C = 30.0°$ 2. $a = 12.0$, $b = 10.0$, $c = 5.00$

3. $a = 10.0$, $A = 60.0°$, $B = 45.0°$ 4. $a = 10.0$, $A = 120°$, $B = 30.0°$

5. $a = 3.0$, $b = 4.0$, $c = 5.0$ 6. $a = 1.0$, $b = 4.0$, $c = 5.0$

7. $a = 1.22$, $b = 1.39$, $c = 2.51$ 8. $b = 4.30$, $B = 37.2°$, $C = 68.3°$

9. $c = 2.42$, $A = 108.3°$, $B = 31.4°$ 10. $b = 25.6$, $A = 100.3°$, $B = 30.6°$

11. Why is it impossible to express the area of a triangle only in terms of its angles?

12. How are the areas of similar triangles related?

13. Workmen need a triangular steel plate which is 12.0, 16.0, and 24.0 in., respectively, along the sides. What is the area of such a plate?

14. If the plate mentioned in Exercise 13 is made of sheet steel weighing 0.90 oz/sq in. of surface, how much does the plate weigh?

15. A farmer has a field shaped as shown in Figure 4.28. Find the area of the field.

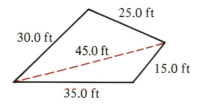

FIGURE 4.28

16. A home is built on a lot in the shape of a quadrilateral with measurements 35, 100, 28, and 83 m. If the diagonal measures 110 m, what is the area of the lot in hectares? (1 hectare = 10,000 m^2).

17. Find the perimeter of a triangle if two of the sides are 100 m and 150 m and the area is 600 m^2.

18. Find the area of a quadrilateral whose sides are successively 3, 5, 6, and 4 meters if the angle between the sides of length 3 and 5 is 100°.

19. Find all parts of a triangle whose area is 25.0 m^2 and two of whose sides are 15.3 and 11.3 m.

20. Find all possible triangles whose area is 96.5 m^2 if two of its angles are 68.2° and 58.3°.

21. A solar-reflector is composed of 36 triangular sections, each having side lengths of 6.2, 6.2 and 1.1 m. Find the total area of the reflector.

22. A triangular shaped mirror used in a solar furnace has side lengths of 25, 25 and 3.5 cm. What is the area of the mirror?

Key Topics for Chapter 4

Define and/or discuss each of the following.

Solution to the general triangle

Law of cosines

Law of sines

Ambiguous and impossible cases

Area of a general triangle

Review Exercises for Chapter 4

In Exercises 1–11 solve the triangles or show that no triangle exists.

Law of Cosines

1. $A = 29.0°, b = 17.0, c = 28.0$

2. $a = 7.80, c = 9.10, B = 38°18'$

3. $a = 11.2, b = 7.90, c = 15.4$

4. $a = 210, b = 175, c = 78.0$

5. $a = 23.0, b = 5.88, c = 17.8$

Law of Sines

6. $A = 39°12', B = 17°42', c = 20.8$

7. $C = 27.6°, A = 112.2°, a = 3120$

8. $b = 75.0, B = 0.2$ rad., $C = 0.7$ rad.

9. $a = 15.0, b = 12.0, A = 25.0°$

10. $A = 42.0°, c = 25.0, a = 17.0$

11. $B = 63°54', a = 23.0, b = 12.0$

In Exercises 12–14 find the area of the indicated triangles.

12. $a = 15.6, b = 19.2, c = 27.8$

13. $b = 7.5, c = 3.9, A = 42.5°$

14. $A = 72°, B = 37°, b = 29$

15. Solve the triangle $a = 9.06, c = 6.68, B = 138.0°$ and find its area.

16. Two light planes leave Kennedy airport at the same time, one flying a course of 265.0° at 175 mph and the other flying a course of 300.0° at 190 mph. How far apart are the two planes at the end of the two and a half hours?

17. At a certain point, the angle of elevation of the top of a tower which stands on level ground is 30.0°. At a point 100 m nearer the tower, the angle of elevation is 58.0°. How high is the tower?

18. From a helicopter the angles of depression of two successive milestones on a level road below are 15.0° and 30.0° respectively. Find the height of the helicopter.

19. In measuring the height of a bell tower with a transit set 5.00 ft above the ground, a student finds that from a point *A* the angle of elevation of the top of the tower is 45.0°. After moving the transit 50.0 ft in a straight line toward the tower, the student finds the angle to be 60.0°. Find the height of the bell tower.

20. A plane flies due East out of Atlanta at 250 mph for 1 hr and then turns and flies 35.0° North of East at 300 mph for 1.5 hr. How far was the plane from Atlanta at the end of 2.5 hr?

Test 1 for Chapter 4

In Exercises 1–5, answer *true* or *false*.

1. The law of cosines can be used to find the angles of a triangle when all three sides are given.

2. The law of sines can be used to find the angles of a triangle when all three sides are given.

3. A triangle can be found with $a = 4, b = 5$, and $c = 20$.

4. A triangle can be found with $A = 60°$, $B = 100°$ and $b = 5$.

5. Only one angle of a triangle may be obtuse.

6. Solve the triangle ABC with $b = 4.00$, $c = 5.00$ and $A = 60°$.

7. Solve the triangle ABC with $b = 5.00$, $c = 4.00$, and $B = 60°$.

8. Explain how the law of cosines may be considered as a natural generalization of the Pythagorean theorem.

9. How much would it cost to lay a 4-inch thick slab of concrete in the shape of a triangle whose sides are 15, 20 and 22 ft if the cost of the concrete is 30 dollars/cu yd.

10. A 25-ft ladder leans against a building built on a slope. If the foot of the ladder is 11 ft from the base of the building and the angle between the side of the building and the ground is 128°, how high up the side of the building does the ladder reach?

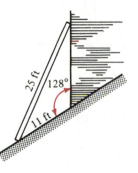

Test 2 for Chapter 4

In Exercises 1–5 solve the triangles with the given information.

1. $B = 73°20'$, $C = 15°10'$, $c = 25.0$

2. $a = 15.0$, $c = 8.00$, $B = 112.3°$

3. $a = 7.00$, $b = 3.00$, $c = 9.00$

4. $A = 46.1°$, $a = 20.0$, $b = 10.0$

5. $A = 124.0°$, $a = 5.00$, $b = 2.00$

6. An airplane is flying due East at 190 mph. Find the resultant velocity of the airplane if there is a tail wind of 50.0 mph blowing at an angle of 25.0° North of East.

7. Two forces, one of 50.0 lb and the other of 90.0 lb act from a point at an angle of 58.0° with each other. What is the size of the resultant?

5

Analytic Trigonometry

5.1   Trigonometric Functions of Real Numbers

In Chapter 3 the domain of the trigonometric functions was defined as the set of all angles, with most of the discussion being limited to the interior angles of a triangle. Because of this restriction, you may think that the trigonometric functions are important only for angles less than 180°. This is far from the case; as a matter of fact, some of the important applications of trigonometry have nothing to do with triangles.

Modern trigonometry consists of two more or less distinct branches. The study of the six ratios and their applications to problems involving triangles is one branch, called **triangle trigonometry**. The other branch is concerned with the general functional behavior of the six ratios, especially with respect to the nature of their variation and their graphs. This branch is often called **analytic trigonometry**.

In analytic trigonometry we consider the six trigonometric functions to be functions of real numbers in addition to being functions of angles. The discussion in this section shows that the extension of the domain of the trigonometric functions to include real numbers is a rather simple matter of matching real numbers with the radian measure of angles.

We begin by locating the unit circle in the rectangular plane with center at the origin and passing through (1, 0) as shown in Figure 5.1(a). The circumference of the unit circle is 2π, or approximately 6.28.

Let L denote a real number line that is parallel to the y-axis with its 0 point at (1, 0). Now if L is wrapped around the unit circle, each point on the line is mapped onto a point on the circle, as in Figure 5.1(b). The positive half line is wound in a counterclockwise direction, whereas the negative half line is wound in a clockwise direction.

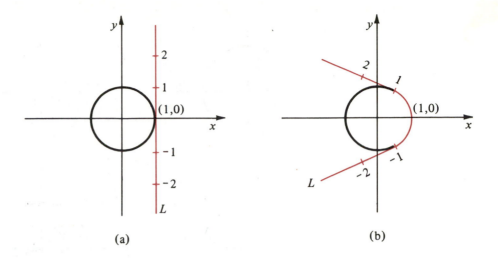

FIGURE 5.1   (a)   (b)

Each real number u mapped onto the unit circle determines both a point P in the plane and an angle α in standard position, as shown in Figure 5.2. The point P and the angle α are said to be *associated* with the real number u.

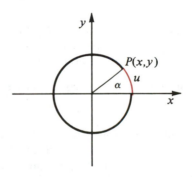

FIGURE 5.2

To establish a relationship between the number u and the angle α, remember that the radian measure of an angle is the ratio of the length of the arc of a circle subtended by the angle to the radius of the circle; that is,

$$\alpha(\text{radians}) = \frac{u \,(\text{arc length})}{r\,(\text{radius})}$$

Since $r = 1$ for the unit circle, the measure of angle α is numerically equal to the arc length u. Symbolically,

$$\boldsymbol{\alpha(\text{radians}) = u}$$

Example 1 illustrates this fact.

EXAMPLE 1   Sketch the points and angles associated with the real numbers 2, $\sqrt{13}$, -3.6, and 6.

SOLUTION   See Figure 5.3.

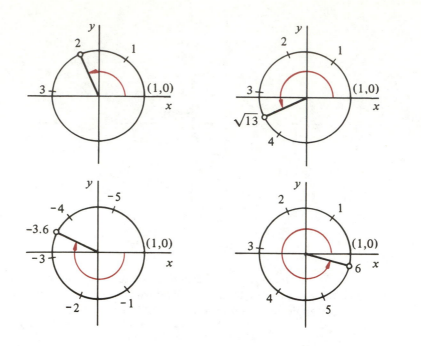

FIGURE 5.3

Using the natural association or real numbers with angles in standard position, we may define the trigonometric functions for any real number u. If α, in radians, is the angle associated with the real number u, then

$$\sin u = \sin \alpha \qquad \cos u = \cos \alpha \qquad \tan u = \tan \alpha$$

$$\sec u = \sec \alpha \qquad \csc u = \csc \alpha \qquad \cot u = \cot \alpha$$

Since it is customary not to write the dimension of radians, it will be impossible (and immaterial) to determine the distinction between an argument of a real number and an argument of an angle in radians. Take care to indicate the correct units of measurements only if the argument is an angle measured in degrees, minutes, and seconds. Otherwise, by convention, the argument is an angle measured in radians or is a real number. Thus, $\sin 30°$ means the sine of an angle of $30°$, but $\sin 30$ means either of the following two concepts, both of which give equal numerical values:

(1) the sine of an angle of 30 radians.

(2) the sine of the real number 30.

By locating the unit circle in the rectangular plane the point P associated with the number u has coordinates (x, y). This association is shown in Figure 5.2. An interesting and important relationship exists between the coordinates of the point associated with u and $\cos u$ and $\sin u$. Since we are working on the unit circle, we have

$$\cos u = \frac{x}{r} = \frac{x}{1} = x$$

$$\sin u = \frac{y}{r} = \frac{y}{1} = y$$

That is, the x- and y-coordinates of the point associated with the real number u are equal to cos u and sin u, respectively.

Figure 5.4 shows a unit circle with real numbers from 0 to 2π marked off on the circumference. Using this circle, values of cos u and sin u may be approximated for any real number u in the interval $(0, 2\pi)$. Further, as u increases the value of sin u and cos u repeat themselves every circumference or 2π units. Thus,

$$\cos u = \cos(u + 2n\pi)$$

$$\sin u = \sin(u + 2n\pi)$$

The fact that the values of the sine and cosine functions repeat every 2π units is described by saying that the sine and cosine functions are **periodic** with **period** 2π*.

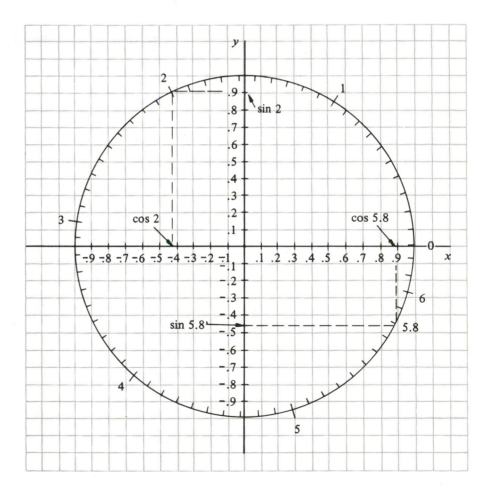

FIGURE 5.4

*More generally, a function f is periodic with period p ($p \neq 0$) if $f(x + p) = f(x)$ for all x. The smallest such p is called *the* period of the function.

EXAMPLE 2   Using Figure 5.4, find the cosine and sine of (a) 2, (b) 5.8, (c) 8.28, (d) 100, (e) −4.28.

SOLUTION   With reference to Figure 5.4 we estimate

(a) $\cos 2 = -0.42$, $\sin 2 = 0.91$.

(b) $\cos 5.8 = 0.89$, $\sin 5.8 = -0.46$.

(c)   To find $\cos 8.28$ and $\sin 8.28$, note that $8.28 = 6.28 + 2$. Thus, $\cos 8.28 = \cos 2 = -0.42$, $\sin 8.28 = \sin 2 = 0.91$.

(d)   To find $\cos 100$ and $\sin 100$, note that $100 = 15 \times 6.28 + 5.8$. Thus, $\cos 100 = \cos 5.8 = 0.89$, $\sin 100 = -0.46$.

(e)   To find $\cos -4.28$ and $\sin -4.28$, note that $-4.28 = 2 - 6.28$. Thus, $\cos -4.28 = \cos 2 = -0.42$, $\sin -4.28 = \sin 2 = 0.91$.   ■

Using multiples of π to express real numbers in connection with the argument of a trigonometric function is common practice, since it is then easy to determine the associated angle.

EXAMPLE 3   Evaluate $\sin \frac{1}{6}\pi$.

SOLUTION   Note that we need not say "$\frac{1}{6}\pi$ what?" For in this case "$\frac{1}{6}\pi$" could mean either $\frac{1}{6}\pi$ radians or merely the real number $\frac{1}{6}\pi$. Either way, the same numerical value is obtained. Hence,

$$\sin \frac{1}{6}\pi = \frac{1}{2}$$ ■

The values of the trigonometric functions for real numbers may be found from Table C in the Appendix or from a calculator by considering the real number as an angle expressed in radians. Table C may be interpreted either as a table of trigonometric functions of angles measured in radians or of trigonometric functions of real numbers. Similarly, when a calculator is being used, merely switch to "radian" mode and enter the desired real numbers.

EXAMPLE 4   Find $\cos 0.5$, $\sin 1.4$, and $\tan 0.714$ to four decimal places.

SOLUTION   From Table C, or a calculator

$$\cos 0.5 = 0.8776$$

$$\sin 1.4 = 0.9854.$$

If $\tan 0.714$ must be approximated by interpolation, we have from Table C,

$$\tan 0.71 = 0.8595$$

$$\tan 0.72 = 0.8771$$

and thus,

$$\tan 0.714 \approx (0.4)(.0176) + 0.8595 \approx 0.8665$$ ■

EXAMPLE 5   Find the values of x for which $\tan x = 1$.

SOLUTION   Some of the numbers in the solution set are $-7\pi/4$, $-3\pi/4$, $\pi/4$, and $5\pi/4$. Notice that each of these numbers is an integral multiple of $\pi/4$. Using this fact, the desired solution set can be written in the form $(4n+1)(\pi/4)$, where n is an integer.   ■

The names of the trigonometric functions are the same whether they are used in the sense of ratios or in a wider functional sense. However, by writing $y = \sin x$ or $f(x) = \sin x$, you are obviously emphasizing the functional concept of the sine function. Further, do not be misled into believing that one letter is used conventionally for the argument of the trigonometric functions. Any convenient letter or symbol suffices. Thus, $\sin x$, $\sin u$, $\sin \theta$, or $\sin y$ mean exactly the same thing. Only the application reveals if the argument is to be interpreted as an angle or as a real number.

The following example illustrates some practical applications of trigonometric functions in which the argument has nothing to do with angles.

EXAMPLE 6   (a) A weight hanging on a certain vibrating spring has a velocity described by $\sin 3t$. In this case, the argument is $3t$, where t is the time in seconds.

(b) The instantaneous voltage for certain electrical systems is given by $156 \sin 377t$, where t is the value of the time in seconds.

(c) The equation of motion of a shaft with flexible bearings is given by

$$x = x_0 \sin (\pi/2L)x$$

where x and L are given in centimeters.   ■

Exercises for Section 5.1

1. Sketch the point on the unit circle associated with each of the real numbers. Find the cosine and sine of each number using Figure 5.4 and check the answer using Table C or a calculator.

(a) 1   (b) -2 (c) 3   (d) 10

(e) 3π (f) -4 (g) -4π (h) $\frac{1}{3}\pi$

(i) $\frac{1}{3}$ (j) $\frac{1}{2}$ (k) $\sqrt{7}$ (l) 5.15

2. In calculus the ratio $\dfrac{\sin x}{x}$ is important. Using Table C or a calculator, find the values of the ratio for the following values of x.

(a) 0.3   (b) 0.2   (c) 0.1   (d) 0.05   (e) 0.01

What value do you think $\dfrac{\sin x}{x}$ approaches as x approaches 0?

What is $\dfrac{\sin x}{x}$ when $x = 0$?

3. If $f(x) = \sin x$ and $g(x) = \cos x$, show that $[f(x)]^2 + [g(x)]^2 = 1$.

If $f(x) = \sin x$, find the values in Exercises 4–10. (Use Table C or a calculator if necessary.)

4. $f(\tfrac{1}{2}\pi)$ **5.** $f(\pi)$ **6.** $f(50)$ **7.** $f(-10)$

8. $f(3\pi)$ **9.** $f(\tfrac{1}{6}\pi)$ **10.** $f(2\pi)$

If $g(x) = \cos x$, find the values in Exercises 11–17. (Use Table C or a calculator if necessary.)

11. $g(0)$ **12.** $g\left(\tfrac{1}{3}\pi\right)$ **13.** $g(\pi)$ **14.** $g(25)$

15. $g(-10)$ **16.** $g(5\pi)$ **17.** $g(5)$

18. If $f(x) = \sin x$, solve the equation $f(x) = 0$.

19. If $g(x) = \cos x$, solve the equation $g(x) = 0$.

20. For which values of x is $\sin x = 1$?

21. For which values of x is $\cos x = 1$?

22. For which values of x is $\sec x = \tfrac{1}{2}$?

23. Solve the inequality $\cos x \leq \sec x$.

24. Solve the inequality $\sin x > \csc x$.

25. Solve the inequality $\sin x \geq 1$.

A **zero** of a function $f(x)$ is a value of $x = x_0$ such that $f(x_0) = 0$.

26. What are the zeros of $\sin x$?

27. What are the zeros of $\cos x$?

5.2   Graphs of the Sine and Cosine Functions

We now examine the graphs of the sine and cosine functions considered as functions of *real numbers*. These two types of graphs have significance in many unrelated areas, and the job of graphing either the sine or the cosine is almost identical.

We first summarize some of the analytical properties.

(1)   Both $\sin x$ and $\cos x$ are **bounded**, above by 1 and below by -1. Thus, the graph of each of the functions lies between the lines $y = 1$ and $y = -1$.

(2)   Both $\sin x$ and $\cos x$ are **periodic** with period 2π, that is, the functional values repeat themselves every 2π units. Thus, only one period need be considered when graphing $\sin x$ and $\cos x$.

(3)   The sine function is **odd**; that is, $\sin x = -\sin(-x)$. Thus, the graph of $\sin x$ is symmetric about the origin.
The cosine function is **even**; that is, $\cos x = \cos(-x)$. Thus, its graph is symmetric about the y-axis.

(4) $\sin x = 0$ for $x = 0, \pm\pi, \pm 2\pi, \pm 3\pi$, and so on.
$\cos x = 0$ for $x = \pm\tfrac{1}{2}\pi, \pm\tfrac{3}{2}\pi, \pm\tfrac{5}{2}\pi$, and so on.
In each case they are the intercepts on the x-axis.

(5)   The numerical values of the sine and cosine functions for $0 \le x \le \frac{1}{2}\pi$ correspond to the values of sin x and cos x in the first quadrant. The other three quadrants yield values numerically the same with, at most, a difference in sign.

To obtain the specific graph we need a reasonable number of points for $0 < x < \frac{1}{2}\pi$. Then a smooth curve can be drawn through these points, so that the general properties may be used to obtain the remainder of the curve. Usually we emphasize certain "special" points corresponding to $x = 0, \frac{1}{2}\pi, \pi, \frac{3}{2}\pi$, and 2π.

Figure 5.5 displays the graph of one period of the sine function. To the left of the graph is a circle of radius 1. The sine function has values numerically equal to the y-coordinates of points on this circle. This figure displays the relationship between points on the unit circle and corresponding points on the graph of the function.

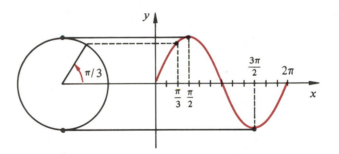

FIGURE 5.5

Figure 5.6 shows a graph of several periods of the function $y = \sin x$, and Figure 5.7 shows several periods of the cosine function. The figures show the graphs as having been terminated, but in reality they continue indefinitely. To emphasize this indefinite continuation, a statement of the period is often included with the graph.

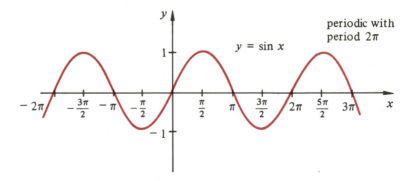

FIGURE 5.6

The graph of the sine function is sometimes called a sine wave or a **sinusoid**. In actuality, the term sinusoid is applied to any curve which has the same shape as that of the sine function. For example, the graph of the cosine function is properly called a sinusoid since (as may be shown using the techniques of the next section) the graph of the cosine function may be obtained simply by shifting the graph of the sine function $\pi/2$ units to the left.

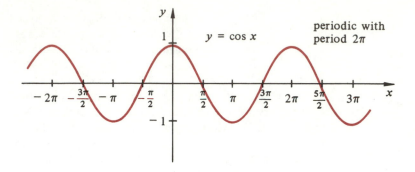

FIGURE 5.7

A **cycle** is the shortest segment of the graph that includes one period. The **frequency** of the sinusoid is defined to be the reciprocal of the period. It represents the number of cycles of the function in each unit interval. The graph of any sinusoid should clearly demonstrate the boundedness, periodicity and the intercepts.

A sketch of a sinusoid may be obtained by connecting known points on the curve with a smooth line. While there is no rule for stating how many points should be plotted for a given sinusoid, the idea is to choose just enough points to make the shape obvious. In this section we suggest you use 6 to 10 points per period of the graph. Example 1 shows the point-plotting approach to graphing $y = \cos \frac{1}{2}x$ using multiples of $\frac{1}{2}\pi$ for x.

EXAMPLE 1   Sketch the graph of $y = \cos \frac{1}{2}x$ on the interval $0 \le x \le 4\pi$ using multiples of $\pi/2$ for x.

SOLUTION   The table shows the various values for this interval. The graph (Figure 5.8) is then obtained by plotting these points. Notice that we have completed one period on the interval $[0, 4\pi]$. We conclude from this that the period of $y = \cos \frac{1}{2}x$ is 4π.

x	0	$\pi/2$	π	$3\pi/2$	2π	$5\pi/2$	3π	$7\pi/2$	4π
y	1	$\sqrt{2}/2$	0	$-\sqrt{2}/2$	-1	$-\sqrt{2}/2$	0	$\sqrt{2}/2$	1

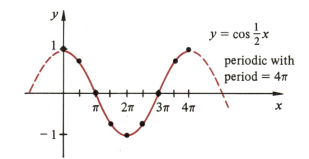

FIGURE 5.8

EXAMPLE 2   Sketch the graph of $y = 3 \sin 2x$ on the interval $0 \le x \le \pi$ using multiples of $\pi/6$ for x.

SOLUTION   To graph one period of this function we construct a table of values. The corresponding figure is shown in Figure 5.9. Notice that the plotted-points are sufficient to draw the desired curve.

x	0	$\dfrac{\pi}{6}$	$\dfrac{\pi}{3}$	$\dfrac{\pi}{2}$	$\dfrac{2\pi}{3}$	$\dfrac{5\pi}{6}$	π
y	0	$\dfrac{3\sqrt{3}}{2}$	$\dfrac{3\sqrt{3}}{2}$	0	$\dfrac{-3\sqrt{3}}{2}$	$\dfrac{-3\sqrt{3}}{2}$	0

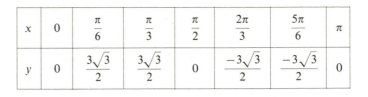

$y = 3 \sin 2x$
periodic with period $= \pi$

FIGURE 5.9

Exercises for Section 5.2

1. What are the domain and the range of $\sin x$?

2. What are the domain and the range of $\cos x$?

In Exercises 3–6, indicate the period of the sinusoids whose graphs are shown, and whether it is a sine or cosine wave.

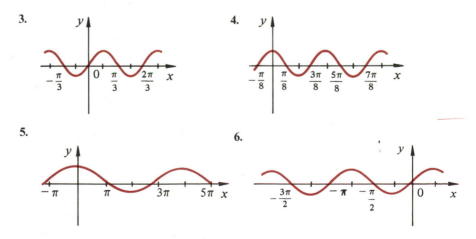

3.

4.

5.

6.

7. Sketch one period of $y = \sin x$ using multiples of $\pi/4$ for x.

8. Sketch one period of $y = \cos x$ using multiples of $\pi/4$ for x.

9. Sketch $y = 2 \sin x$ on the interval $0 \le x \le 2\pi$ using multiples of $\pi/4$ for x.

10. Sketch $y = 3 \cos x$ on the interval $0 \leq x \leq 2\pi$ using multiples of $\pi/4$ for x.

11. Sketch $y = \cos 2x$ on the interval $0 \leq x \leq \pi$ using multiples of $\pi/8$. What is its period?

12. Sketch $y = \sin 2x$ on the interval $0 \leq x \leq \pi$ using multiples of $\pi/8$. What is its period?

13. The function $f(x) = |\sin x|$ is called the **full-wave rectified sine wave**. Sketch this function on the interval $0 \leq x \leq 2\pi$ using multiples of $\pi/4$. What is its period?

14. The function defined by

$$f(x) = \begin{cases} \sin x, & 0 \leq x \leq \pi \\ 0, & \pi \leq x \leq 2\pi \end{cases}$$

and periodic with period 2π, is called the **half-wave rectified sine wave**. Sketch this function on the interval $0 \leq x \leq 4\pi$ using multiples of $\pi/4$.

15. Sketch $y = |\cos x|$ on the interval $-\pi/2 \leq x \leq 3\pi/2$ using multiples of $\pi/4$.

16. Make a sketch of $y = x$ and $y = \sin x$ on the same axes and convince yourself that $\sin x < x$ for $x > 0$.

The convention of using multiples of π to plot graphs of the trigonometric functions is convenient but not essential. We can, of course, evaluate $\sin x$ and $\cos x$ for any real number x. Sketch the functions in Exercises 17–22 on the interval $0 \leq x \leq 7$ in multiples of 1. A calculator will be helpful for these exercises.

17. $y = \cos x$ 18. $y = \sin x$ 19. $y = 1.5 \sin x$

20. $y = 3 \cos x$ 21. $y = \cos \frac{1}{2}x$ 22. $y = \sin \frac{1}{2}x$

5.3   More on the Sine and Cosine Functions

As noted at the end of the previous section you will frequently encounter variations of the functions $y = \sin x$ or $y = \cos x$, such as $\sin 2x$, $4 \cos x$, and $\cos (x + \pi)$. Modified functions of this type may be graphed by the point plotting procedure shown in Examples 1 and 2, but this is a tedious approach at best. In this section we show how, by appealing to the familiar properties of the sine or cosine function, the graphing of these modified functions may be greatly simplified.

In practice we must deal with the basic trigonometric functions altered in four ways:

    (1)   multiplication of the function by a constant.

    (2)   multiplication of the argument by a constant.

    (3)   addition of a constant to the functional value.

    (4)   addition of a constant to the argument.

We will use $y = \sin x$ to determine each of these cases with the understanding that the results apply also to $y = \cos x$. In this section we study (1) and (2) and in the next section, (3) and (4).

(1) Multiplication of the Function by a Constant

We have seen that the values of the sine function oscillate between $+1$ and -1. If we multiply $\sin x$ by a positive constant A, we then write $y = A \sin x$. Since

$$-1 \le \sin x \le 1$$

it follows that, when multiplying through by A,

$$-A \le A \sin x \le A$$

The absolute value of A is called the **amplitude** of the sine wave. If $|A|$ is greater than one, the amplitude of the basic sine wave is increased; if $|A|$ is less than one, the amplitude is decreased. Sometimes $|A|$ is called the **maximum**, or **peak** value of the function. Figure 5.10 shows the graph of $y = A \sin x$ for $A = 1, \frac{1}{2}$, and 2.

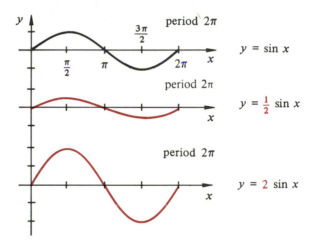

FIGURE 5.10

In Section 5.2 you learned the basic shapes of the sine and cosine graphs. Now, you can use that knowledge to help shorten the sketching process. It is easy to sketch either curve if the zeros and extreme points of the function are known. For instance, $y = \sin x$ is zero when $x = 0$, has a maximum at $x = \frac{1}{2}\pi$ ($\frac{1}{4}$ period), has a zero at $x = \pi$ ($\frac{1}{2}$ period), has a minimum at $x = \frac{3}{2}\pi$ ($\frac{3}{4}$ period), and has a zero at $x = 2\pi$ (end of one period).

EXAMPLE 1   Sketch the graph of one period of $y = \sqrt{3} \cos x$.

SOLUTION   The function $y = \sqrt{3} \cos x$ has an amplitude of $\sqrt{3}$ and a period of 2π. Also, the graph of $y = \cos x$ is at its maximum at $x = 0$, has a zero at $x = \frac{1}{2}\pi$, has a minimum at $x = \pi$, has a zero at $x = \frac{3}{2}\pi$, and is a maximum again at $x = 2\pi$. The graph of the given function is shown in Figure 5.11.

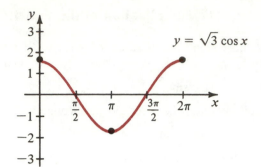

FIGURE 5.11

EXAMPLE 2   Sketch one period of $y = -3 \sin x$.

SOLUTION   The amplitude coefficient in this case is negative which means that each of the functional values will change numerical sign. Thus the graph of $y = -3 \sin x$ is the reflection in the x-axis of the graph of $y = 3 \sin x$. Both graphs are shown in Figure 5.12 to emphasize the relationship between the two curves.

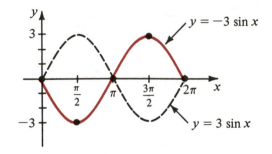

FIGURE 5.12

(2) Multiplication of the Argument by a Constant

If $\sin x$ has its argument multiplied by a constant B, the function becomes $\sin Bx$. The graph of this function remains sinusoidal in form but since the argument is Bx and since the sine function repeats itself for every increase in the argument of 2π, we can see that one period of $\sin Bx$ is contained in the interval

$$0 \le Bx \le 2\pi$$

that is, for

$$0 \le x \le \frac{2\pi}{B}$$

Therefore, multiplying the argument by a constant has the effect of altering the period to be $2\pi/B$. Thus, the period of $\sin 2x$ is π. The period of $\sin \frac{1}{2}x$ is 4π. Graphically, increasing B has the effect of squeezing the sine curve together like an accordion. Decreasing B has the effect of pulling it apart. (See Figure 5.13.) On a fundamental interval, the sine curve is 0 at $x = 0$, π, and 2π. The curve $y = \sin Bx$ is 0 at $x = 0$, π/B, and $2\pi/B$. The basic sine curve reaches a maximum at $y = \pi/2$. The curve $y = \sin Bx$ reaches a maximum at $y = \pi/(2B)$.

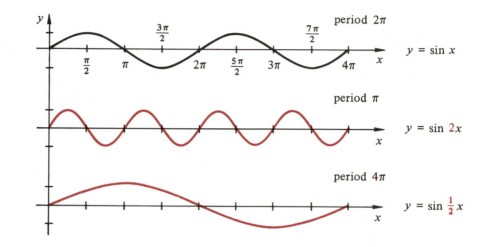

FIGURE 5.13

EXAMPLE 3   (a) $\sin 3x$ has a period of $\dfrac{2\pi}{3}$.

(b) $\cos \frac{1}{4}x$ has a period of $\dfrac{2\pi}{1/4} = 8\pi$.

(c) $\cos 10x$ has a period of $\dfrac{2\pi}{10} = \dfrac{\pi}{5}$.

(d) $\sin 4\pi x$ has a period of $\dfrac{2\pi}{4\pi} = \dfrac{1}{2}$.   ■

The graphs of the sine and cosine functions are easy to draw if you choose convenient units for the independent variable. In general, it is desirable to use units that equal one quarter of the period of the function since the zeros and the maximum and minimum values occur at multiples of $\frac{1}{4}$ period. For example, if the period is 2π, use $\frac{1}{4}(2\pi) = \frac{1}{2}\pi$ units for the graph; if the period is $\frac{1}{3}\pi$, use $\frac{1}{4}(\frac{1}{3}\pi) = \frac{1}{12}\pi$.

EXAMPLE 4   Sketch the graph of one period of $y = 2\cos 5x$.

SOLUTION   Here the amplitude is 2 and the period is $\dfrac{2\pi}{5}$. The graph is shown in Figure 5.14. Notice that the

units along the x-axis are in multiples of $\dfrac{1}{4}\left(\dfrac{2\pi}{5}\right) = \dfrac{\pi}{10}$.

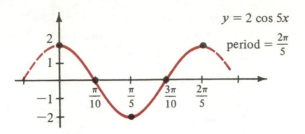

FIGURE 5.14

EXAMPLE 5 Sketch one period of the graph of $s = 2.4 \sin 3\pi t$.

SOLUTION In this case the amplitude is 2.4 and the period is $\dfrac{2\pi}{3\pi} = \dfrac{2}{3}$. The graph appears in Figure 5.15.

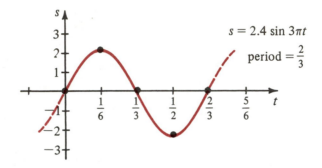

FIGURE 5.15

EXAMPLE 6 Write the equation of the sine function whose amplitude is 3 and whose period is 7.

SOLUTION The general form of the desired sine function is $y = A \sin Bx$. The amplitude 3 yields $A = 3$. Also the period is given by $\dfrac{2\pi}{B}$, so we must have $\dfrac{2\pi}{B} = 7$, from which $B = \dfrac{2\pi}{7}$. The desired function is

$$y = 3 \sin \frac{2\pi}{7} x$$

Biorhythms

An interesting application of the sine function is found in the so-called theory of biorhythms. Briefly, the theory holds that our physical, emotional and intellectual state of being, because of certain biological processes, are periodic and can be represented by sine functions. According to the theory, the cycle of each "feeling" is a constant. Specifically the periods are (1) 23 days for the physical cycle (2) 28 days for the emotional cycle and (3) 33 days for the intellectual cycle. Thus, our physical state on any day, t, since our birth is given by

$$P = \sin\left(\frac{2\pi}{23} t\right)$$

our emotional state by

$$E = \sin\left(\frac{2\pi}{28}t\right)$$

and our intellectual state by

$$I = \sin\left(\frac{2\pi}{33}t\right)$$

Using these functions, we characterize our good days as those for which P, E, and I are positive and our bad days as those for which they are negative. The closer the state to $+1$, the better the day for that particular phase of your well being. Your overall state is usually obtained by averaging the three values.

To use the formulas for the biorhythm state you must compute the number of days from the given birthdate to the present day. One convenient method is to compute the number of days from January 1, 1900 for each date and then subtract the two numbers. Recalling that leap year occurs every 4 years (but not in 1900), the number of days in the 20th century for any particular date is given by

$$T = (\text{Year of interest} - 1900)365 + (\text{Leap years since 1900})$$

$$+ (\text{Number of days from Jan. 1 in year of interest})$$

EXAMPLE 7   Find the biorhythm state of an individual on February 10, 1980 if the person was born on August 8, 1933.

SOLUTION   We first compute the number of days since 1900 for the two days in question. Notice that 1980 is a leap year but since February 2 occurs prior to the 29th we use 19 and not 20 for the number of leap years.

$$T(8/8/33) = (1933 - 1900)365 + 8 + 220 = 12273 \text{ days}$$

$$T(2/10/80) = (1980 - 1900)365 + 19 + 41 = 29260 \text{ days}$$

Thus, the number of days between the two dates is

$$t = 29260 - 12273 = 16987 \text{ days}$$

Using this value in the formulas for P, E, and I, we have

$$P = \sin\left(\frac{2\pi}{23} \cdot 16987\right) = \sin 4640.54 = -0.398$$

$$E = \sin\left(\frac{2\pi}{28} \cdot 16987\right) = \sin 3811.87 = -0.901$$

$$I = \sin\left(\frac{2\pi}{33} \cdot 16987\right) = \sin 3234.32 = -0.999$$

Check that your calculator is in the radian mode to get the indicated numbers. The average of these three numbers is -0.766. All indications are that this person should stay in bed on this date.   ■

Exercises for Section 5.3

In Exercises 1–20 sketch the graphs of the given functions. In each case give the amplitude and the period.

1. $y = 3 \sin x$

2. $y = \frac{1}{2} \sin x$

3. $y = 6 \cos x$

4. $y = \frac{1}{3} \sin x$

5. $y = \sin \frac{2}{3} x$

6. $y = 0.3 \sin 3x$

7. $y = \sin \pi x$

8. $y = \cos 0.1 x$

9. $s = \frac{1}{2} \cos 2t$

10. $y = 100 \cos 3x$

11. $y = 8.2 \sin 0.4x$

12. $v = 3 \sin \frac{7}{6} t$

13. $y = -\cos \frac{2}{5} x$

14. $y = -5 \cos 7x$

15. $p = \pi \cos 100\, t$

16. $i = -0.02 \sin \pi t$

17. $y = -12 \sin 0.2\, x$

18. $P = 10^6 \cos \frac{\pi}{1000} x$

19. $v = \dfrac{\cos 1000\pi\, x}{50}$

20. $y = \dfrac{3 \sin 25\pi t}{200}$

In Exercises 21–28 write the equation of the sine function having the indicated amplitude and period.

21. Amplitude $= \frac{1}{3}$, period $= 12$

22. Amplitude $= \frac{1}{2}$, period $= 15$

23. Amplitude $= 20$, period $= \frac{3}{8}$

24. Amplitude $= \sqrt{5}$, period $= \frac{1}{3}$

25. Amplitude $= 2.4$, period $= \frac{1}{3}\pi$

26. Amplitude $= 0.94$, period $= \frac{1}{6}\pi$

27. Amplitude $= \pi$, period $= 3\pi$

28. Amplitude $= 2/\pi$, period $= 7\pi$

29. The motion of a pendulum can be represented by the equation $x = A \sin Bt$. Write the equation of motion of a pendulum oscillating with an amplitude of 3.2 ft and a period of 2.5 sec.

30. The equation for the voltage drop across the terminals of an ordinary electric outlet is given approximately by

$$E = 156 \sin (110\, \pi t)$$

Sketch the voltage curve for several cycles.

31. If B is small the equation $y = \sin Bx$ approximates the shape of ocean waves. Sketch several cycles of an ocean wave described by

$$y = \sin \frac{1}{20}\pi x$$

Find the biorhythm state for each of the following birth dates on Jan. 25, 1983 if their birthdates are as given below. (Use $\pi = 3.1415927$.)

32. 3/11/35

33. 2/27/1963

34. 5/20/1965

35. 7/30/1966

36. 2/7/1968

37. 11/18/1969

38. Calculate your biorhythm state for todays date.

5.4   **Vertical and Horizontal Translation**

The vertical or horizontal relocation of the graph of a function without changing or distorting its shape is called a **translation**. A function is translated vertically if a constant is added to the value of the function and translated horizontally if a constant is added to the argument of the function. Both of these translations are important in applied work and are explained here in the context of sine and cosine functions.

Addition of a Constant to the Value of the Function

The graph of $y = D + \sin x$ is obtained by adding D to each value of $\sin x$. Therefore, the graph of $D + \sin x$ is simply the graph of $\sin x$ displaced D units up or down. The graph is translated up if D is positive and down if it is negative. D is called the **mean value** of the function.

EXAMPLE 1   Sketch the graph of $y = 2 + \sin x$.

SOLUTION   The graph is shown in Figure 5.16. The mean value is 2 and the period is 2π.

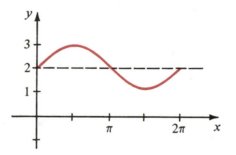

FIGURE 5.16   ■

EXAMPLE 2   Sketch the graph of $s = -1 + 3 \cos 2t$.

SOLUTION   The graph is shown in Figure 5.17. The mean value is -1 and the period is π.

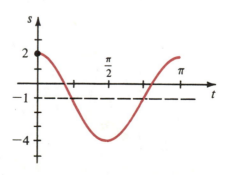

FIGURE 5.17   ■

Addition of a Constant to the Argument

The addition of a constant to the argument of sin x is written $\sin (x + C)$. The constant C has the effect of shifting the graph of the sine function to the right or to the left. Notice that $\sin (x + C)$ is zero when $x + C = 0$, that is for $x = -C$. This value of x for which the argument of the sine function is zero is called the **phase shift**. If C is positive the shift is to the left and if C is negative, the shift is to the right. Figure 5.18 shows three sine waves with phase shifts of 0, $-\frac{1}{4}\pi$, and $\frac{1}{4}\pi$.

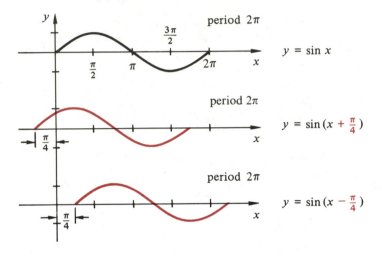

FIGURE 5.18

In the more general case, the effects of changes in amplitude, period and phase shift are all combined. The function

$$y = A \sin (Bx + C)$$

has an amplitude of A, a period of $2\pi/B$ and a phase shift corresponding to the value of x given by $Bx + C = 0$, that is $x = -C/B$. Figure 5.19 shows a graph of the basic sine curve and the graph of $y = 3 \sin (2x - \frac{1}{3}\pi)$.

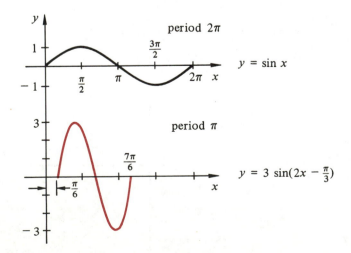

FIGURE 5.19

EXAMPLE 3   Sketch the graph of $y = 2 \sin\left(\frac{1}{3}x + \frac{1}{9}\pi\right)$.

SOLUTION   The amplitude of the graph is 2, the period is $2\pi/\left(\frac{1}{3}\right) = 6\pi$, and the phase shift is $-\frac{1}{3}\pi$. (See Figure 5.20.)

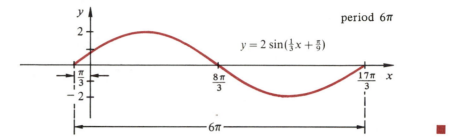

FIGURE 5.20

In all cases, the distinctive shape of the sine curve remains unaltered. This basic shape is expanded or contracted vertically by the multiplication by the amplitude constant A, expanded or contracted horizontally by the constant B, and shifted to the right or left by the constant C.

A similar analysis could be made for the cosine function. We will not discuss in detail the function $y = D + A \cos(Bx + C)$, but the constants A, B, C, and D alter the basic cosine function in a manner quite similar to that described for the sine function.

EXAMPLE 4   Sketch the graph of $y = 3 \cos\left(\frac{1}{2}x + \frac{1}{4}\pi\right)$.

SOLUTION   The amplitude is 3 since the basic cosine function is multiplied by 3. The period is $2\pi/\left(\frac{1}{2}\right) = 4\pi$. The phase shift is found from the equation $\frac{1}{2}x + \frac{1}{4}\pi = 0$, that is for $x = -\pi/2$. Hence the phase shift is $\pi/2$ units to the left. The graph of this function is shown in Figure 5.21.

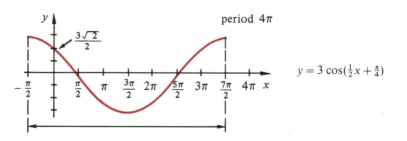

FIGURE 5.21

If the function is multiplied by a negative constant, you may use one of the relationships of Section 3.4 to put the expression into a more standard form.

EXAMPLE 5   (a) Sketch the graph of $y = -2 \sin(3x + 1)$.

(b) Express the given function in the form $y = A \sin(Bx + C)$, where A and B are positive constants.

SOLUTION   (a) The easiest way of graphing this function is to initially sketch the function $y = 2 \sin(3x + 1)$ and then reflect this graph in the x-axis. See Figure 5.22.

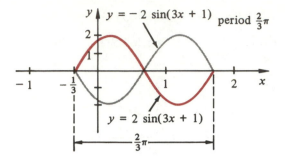

FIGURE 5.22

(b) By using the fact that $\sin x = -\sin(-x)$, we have that

$$y = -2\sin(3x+1) = 2\sin(-3x-1)$$

Then, since $\sin(x+\pi) = \sin(-x)$, we have

$$y = 2\sin[(3x+1)+\pi] = 2\sin[3x+(1+\pi)]$$

■

The Predator-Prey Problem

Certain ecological systems can be represented by periodic functions. For instance, the number of predators and the number of prey in a region vary periodically. (Coyotes are predators, rabbits are prey.) If the number of predators in a region is relatively small, the prey will increase. But, as the prey become more plentiful, the number of predators will increase because food is easy to find. As the number of predators continues to increase, the number of prey will eventually begin to decrease, which means food for the predators will become scarce, which causes the predator population to decrease, which allows more prey to survive, and so on. This cycle will repeat over and over again with the two populations oscillating about their respective mean values. The sine function is frequently used to describe the predator-prey relationship in a balanced ecological system.

EXAMPLE 6   The population of rabbits in a certain region is given by

$$N = 500 + 150\sin 2t$$

where t is time in years. Discuss the variation in the rabbit population.

SOLUTION   The constant term represents the mean population of rabbits and the coefficient of the sine function represents the variation in the population. Thus, the mean population is 500 and it varies from a high of 650 to a low of 350. The period of variation is $2\pi/2 = 3.14$ years. The population is shown graphically in Figure 5.23.

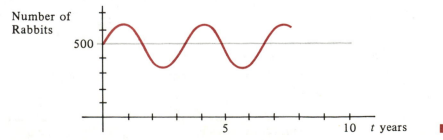

FIGURE 5.23

■

Seasonal Temperature Variation

Meteorological data provides another area where sine and cosine curves might be used to describe physical conditions. Figure 5.24 shows a representation of the daily mean temperatures for Dayton, Ohio from January, 1978 to September, 1979: the sinusoidal shape is unmistakable. The temperature variation can be approximated by the equation

$$T = T_m + A \sin\left[\frac{2\pi}{365}(t - C)\right]$$

where T_m is the mean annual temperature

A is the maximum temperature deviation from T_m

C is the phase shift found by counting the number of days from January 1 to the point at which the temperature curve crosses the mean annual temperature line.

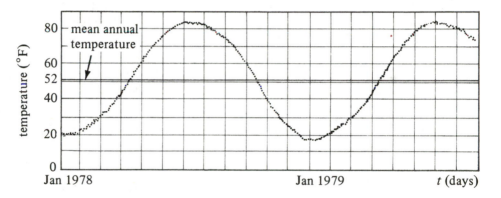

FIGURE 5.24

EXAMPLE 7   Write an equation to approximate the temperature variations shown in Figure 5.24.

SOLUTION   From Figure 5.24 we estimate that $T_m = 52°F$, $A = 32°F$, and $C = 102$ days. Figure 5.25 shows the comparison between the actual data and the graph of

$$T = 52 + 32 \sin\left[\frac{2\pi}{365}(t - 102)\right]$$

As you can see there are some specific areas of disagreement, but overall the fit appears to be good.

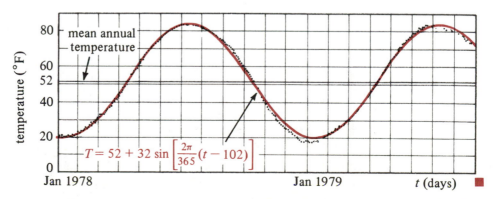

FIGURE 5.25   ■

$$y = A \sin (Bx \pm C)$$

Exercises for Section 5.4

In Exercises 1–20 indicate the mean value, amplitude, period, and phase shift. In each case sketch the graph of the function.

1. $y = 3 + \cos 2x$

2. $s = 4 + \sin 3x$

3. $v = 6 + 8 \sin t$

4. $y = -2 + 3 \cos 6x$

5. $y = -2 + \sin \frac{1}{2}x$

6. $i = 0.2 + 1.3 \cos 0.2t$

7. $M = 3 - 3 \sin 3x$

8. $y = -2 - \sin \pi x$

9. $y = \cos (x + \frac{1}{3}\pi)$

10. $y = 2 \sin \frac{1}{3}x$

11. $y = 2 \cos (\frac{1}{2}x - \frac{1}{2}\pi)$

12. $y = \sin 2(x + \frac{1}{6}\pi)$

13. $y = \cos (2x + \pi)$

14. $y = 3 \cos (3x - \pi)$

15. $y = 4 \sin \left(\frac{1}{3}x + \frac{\pi}{3} \right)$

16. $y = 0.2 \sin (0.25x - \pi)$

17. $y = \cos \left(\pi x - \frac{\pi}{4} \right)$

18. $y = \sqrt{3} \cos (\pi x + \pi)$

19. $y = 4 + \sin (4x - \pi)$

20. $y = 3 - 2 \cos \left(\frac{1}{2}x + \frac{\pi}{8} \right)$

In Exercises 21–25 write the expressions in the form $A \sin (Bx + C)$, where A and B are positive, and then sketch.

21. $-\sin (x + 1)$

22. $-\sin (-2x + 3)$

23. $-\sin (2\pi x + \frac{1}{2})$

24. $3 \cos (2\pi x + \pi)$

25. $-\cos (\pi x + 1)$

Write the equation of the sinusoids in Exercises 26–29 whose graphs are shown over one period.

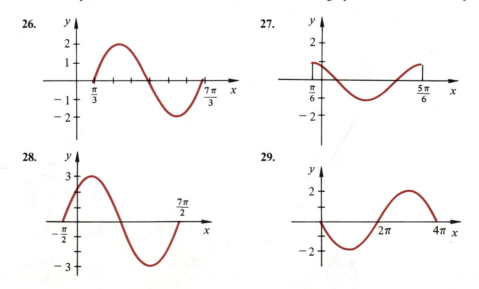

26.

27.

28.

29.

30. It is always possible to express functions of the type $A \sin(Bx + C)$ and $A \cos(Bx + C)$ in the form $A' \sin(B'x + C')$ where A', B', and C' are positive. Prove this.

31. How are the graphs of $y = \sin(-t)$ and $y = \sin(t)$ related?

32. How are the graphs of $\cos(-t)$ and $\cos(t)$ related?

33. How are the graphs of $\sin(t)$ and $\cos(\frac{1}{2}\pi - t)$ related?

34. Prove that the graph of the cosine function is the same as the sine function moved $\pi/2$ units to the left.

35. The number of deer in an ecological region is given by $D = 1500 + 400 \sin 0.4t$ and the number of pumas in the region by $P = 500 + 200 \sin(0.4t - 0.8)$, where t is time in years. Sketch the variation in these two populations on the same set of coordinates.

36. The rabbit population in an ecological region is given by $R = 1000 + 200 \sin 4t$ and the fox population by $F = 100 + 10 \sin(4t - 0.8)$, where t is the time in years. Discuss the variation and sketch the populations on the same coordinate system.

37. Draw the graph of the normal mean temperature variation for a two year period if $T = 55 + 38 \sin\left[\dfrac{2\pi}{365}(t - 100)\right]$ is the approximating equation. Assume T is in $°F$ and t is the number of days from January 1. What is the mean annual temperature of the area described by this equation?

38. Write the equation for the temperature variation shown in Figure 5.26.

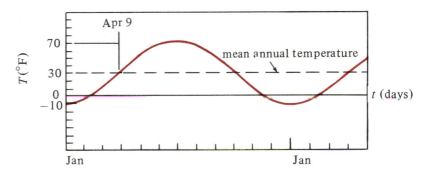

FIGURE 5.26

5.5   Harmonic Motion

Imagine an object hanging on a spring as shown in Figure 5.27. If the object is pulled down and released, it will oscillate up and down about the rest or **equilibrium** point. Assuming there is no frictional force, this oscillatory motion will continue indefinitely. Vibratory motion of this type is called **simple harmonic motion** and can be described mathematically using sine and cosine functions.

The correspondence between simple harmonic motion and the cosine function is established by referring to the circle drawn in Figure 5.27. Consider a point Q, moving at a constant angular velocity of ω rad/sec along the circumference of the circle and suppose the initial position of the point Q to be $(r, 0)$. If P is a point on the horizontal diameter of the circle, directly below Q, it is called the **projection** of Q on the diameter. As Q revolves, the point P moves back and forth along the diameter of the circle in the same way that the object attached to the spring moves, the only difference being that the movement of the object is vertical and that of P is horizontal. Thus, the point P moves with simple harmonic motion. If we can represent the motion of point P by a mathematical function, we have analytically described simple harmonic motion.

The displacement of P from the center of the circle at any time t is the distance $\overline{OP} = x$. If θ represents the angle that OQ makes with the horizontal diameter, we have

$$x = r \cos \theta$$

More commonly we write this expression as a function of time by noting that displacement = velocity·time. Thus, $\theta = \omega t$ can be used to write

$$x = r \cos \omega t$$

as the description of simple harmonic motion.

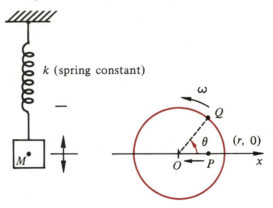

FIGURE 5.27

Simple harmonic motion

If the object is displaced through a distance y from its equilibrium position and released with an initial velocity, its motion is described by the equation

$$y = A \cos (\omega_0 t + C)$$

where ω_0 is called the **natural angular frequency** and is given by the formula $\omega_0 = \sqrt{k/M}$. In this formula, k is a constant called the **spring constant**, and M is the **mass** of the object. The period of the motion is $2\pi/\omega_0$. The amplitude and phase shift are determined from the "initial conditions," that is, the initial displacement and initial velocity. The fact that the motion is harmonic means that the mass will oscillate about its equilibrium point with bounds set by the constant A with frequency $\omega_0/2\pi$ cycles per second. In the more realistic physical case the motion would tend to decrease in amplitude due to some sort of "damping" until it eventually ceased to oscillate.

EXAMPLE 1   Find the natural angular frequency of a spring-mass system with a mass of 3 and a spring constant of 0.8.

SOLUTION   The natural frequency ω_0 is given by $\omega_0 = \sqrt{k/M} = \sqrt{0.8/3} = 0.52$.   ■

EXAMPLE 2   The motion in the spring-mass system shown in Figure 5.28 is described by the equation $y = 4 \cos 0.7t$, where y is the displacement in meters from the equilibrium point and t is the elapsed time. How long does it take for one oscillation of the mass? Where is the mass relative to the equilibrium point when $t = 0.2$ sec?

FIGURE 5.28

SOLUTION   The natural angular frequency of the system is $\omega_0 = 0.7$. Therefore, the time that it takes to complete one period is $2\pi/(0.7) = 8.98$ sec. To find the location of the mass relative to its equilibrium point when $t = 0.2$ sec, we evaluate

$$y = 4 \cos 0.7(0.2) = 4 \cos 0.14 = 3.96 \text{ m}$$ ∎

Other phenomena described by simple harmonic motion include the motion of a particle in a guitar string that has been displaced and then released, the motion of a particle of air brought about by certain sound waves, and certain radio and television devices.

Motion described by a product of a sine wave and a nonconstant factor is called **damped harmonic motion**. The nonconstant factor is called a **damping factor**, and the most relevant case is when the damping factor decreases to zero for large values of the independent variable.

EXAMPLE 3   Sketch the damped harmonic motion described by $y = 2^{-x} \sin 2x$.

SOLUTION   The graphs of 2^{-x} and -2^{-x} are shown along with the complete graph in Figure 5.29. The sine wave has varying amplitude and touches the two curves (called "envelopes") while oscillating between them.

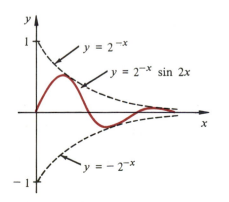

FIGURE 5.29

Exercises for Section 5.5

Find the natural angular frequency of the spring-mass system (see Figure 5.30) having the spring constant, k, and mass, M as indicated in Exercises 1–6.

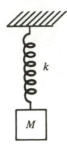

FIGURE 5.30

1. $k = 9, M = 16$ 2. $k = 1, M = 4$

3. $k = 4, M = 1$.   4. $k = 7, M = 7$

5. $k = 3.95, M = 7.61$ 6. $k = 14.5, M = 0.983$

Sketch the harmonic motion described by the functions in Exercises 7–14. Indicate the amplitude and period.

7. $x = 4 \cos 2t$ 8. $x = 3 \cos 4t$

9. $x = 0.25 \cos 5t$ 10. $x = 0.1 \cos 8t$

11. $x = -2 \cos 3t$ 12. $x = -1.5 \cos 4t$

13. $x = \cos(2t - 0.5)$ 14. $x = 0.2 \cos(4t - 0.4)$

15. How does the mass affect   (a) the amplitude of the motion   (b) the period?

16. How does the spring constant affect the period?

17. The motion in a spring-mass system is described by the equation $y = 2 \cos 3.2t$ where y is the displacement in meters from the equilibrium point and t is the elapsed time in seconds. Determine how much time it takes for one oscillation of the mass. What is its location after 5 seconds?

18. Sketch the graph of $x \sin x$ for $0 < x < 4\pi$.

19. Sketch the graph of $3^{-x} \cos x$ for $0 < x < 2\pi$.

20. Sketch the graph of $2^x \cos x$ for $0 < x < 2\pi$.

21. A piston is connected to the rim of a wheel as shown in Figure 5.31. The radius of the wheel is 2 ft and the length of the connecting rod ST is 5 ft. The wheel rotates counterclockwise at the rate of one revolution per second. Find a formula for the position of the point S, t seconds after it has coordinates $(2, 0)$. Find the position of the point S when $t = \frac{1}{2}, \frac{3}{4}$, and 2.

22. One end of a shaft is fastened to a position that moves vertically. The other end is connected to the rim of the wheel by means of prongs as shown in Figure 5.32. If the radius of the wheel is 2 ft and the shaft is 5 ft long, find a formula for the distance d ft between the bottom of the piston and the x-axis, t seconds after P is at $(2, 0)$. Assume the wheel rotates at 2 revolutions/sec.

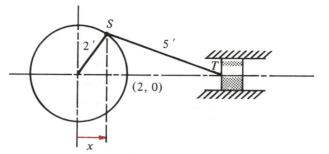

FIGURE 5.31

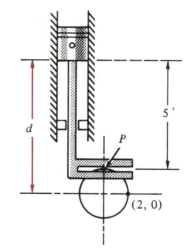

FIGURE 5.32

5.6   Addition of Ordinates

Functions such as

$$y_1 = \sin x + \cos x \quad \text{and} \quad y_2 = x + \sin x$$

which are written as a sum of more elementary functions, occur frequently. It can be a very tedious process to graph such functions if you use the method of substituting values of x and determining corresponding ordinates. Sometimes a technique called **addition of ordinates** can be useful in plotting such functions. Thus, suppose $h(x) = f(x) + g(x)$. We sketch the graphs of $f(x)$ and $g(x)$ on the same coordinate system, as in Figure 5.33. Then for particular values of x, such as x_1, we find $h(x_1)$ as the sum of $f(x_1)$ and $g(x_1)$. A vertical line is usually drawn at the point $(x_1, 0)$ and then ordinates $f(x_1)$ and $g(x_1)$ are added by using a set of dividers, or by markings on the edge of a strip of paper. This process is repeated as often as necessary to get a representation of the desired graph.

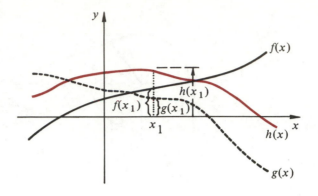

FIGURE 5.33
Addition of
ordinates

EXAMPLE 1   Use the method of addition of ordinates to sketch the function $y = \sin x + \cos 2x$.

SOLUTION   In Figure 5.34, both $\sin x$ and $\cos 2x$ are sketched along with their sum. The period of the given function is 2π even though that of $\cos 2x$ is π.

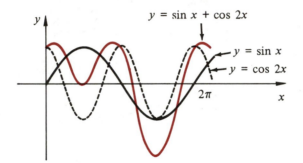

FIGURE 5.34   ■

EXAMPLE 2   Sketch the graph of $y = x + \sin x$.

SOLUTION   See Figure 5.35. In this case the basic sine curve oscillates about the curve $y = x$. Note that the given function is *not* periodic and is *not* bounded.

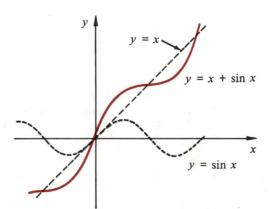

FIGURE 5.35   ■

Exercises for Section 5.6

By the method of addition of ordinates, sketch graphs of the functions in Exercises 1–20. In each case, tell if the function is periodic.

1. $y = 3 + \sin x$

2. $y = 1 + 2 \cos x$

3. $y = -1 + \cos x$

4. $y = -0.5 + \sin x$

5. $y = \sin x + 2$

6. $y = x - \sin x$

7. $y = \frac{1}{2}x + \cos x$

8. $y = \cos x - x$

9. $y = 2 - x + \sin x$

10. $y = 1 + x - \cos x$

11. $y = 0.1 x^2 + \sin 2x$

12. $y = 0.1 x^2 + \cos x$

13. $y = \sin x + \cos x$

14. $y = \sin 2x + 2 \sin x$

15. $y = \sin \frac{1}{2}x - 2 \cos x$

16. $y = \sin \frac{1}{2}x - \sin x$

17. $y = \sin x + \sin (x - \frac{1}{4}\pi)$

18. $y = \cos x + \sin (x - \frac{1}{4}\pi)$

19. $y = 2 \sin \pi x + \sin x$

20. $y = \sin \pi x - \cos 2x$

21. Compare the graph of the function $y = \sqrt{2} \sin (x + \frac{1}{4}\pi)$ with the graph of the function $y = \cos x + \sin x$.

22. Compare the graph of the function $y = \cos^2 x$ with the graph of $y = \frac{1}{2} + \frac{1}{2} \cos 2x$.

5.7   # Graphs of the Tangent and Cotangent Functions

The analytic properties of the tangent and cotangent functions discussed previously are summarized here. Each property influences the nature of the graph in a very important manner.

(1) Both $\tan x$ and $\cot x$ are periodic with period π. Thus, only an interval of length π need be analyzed for purposes of graphing the two functions, for example, $-\frac{1}{2}\pi < x < \frac{1}{2}\pi$ or $0 < x < \pi$.

(2) Both $\tan x$ and $\cot x$ are **unbounded**, which means that their values become arbitrarily large. Tan x becomes unbounded near odd multiples of $\frac{1}{2}\pi$, whereas cot x becomes unbounded near multiples of π.

(3) Both $\tan x$ and $\cot x$ are odd since they are quotients of an odd function and an even function. Thus, their graphs are symmetric with respect to the origin.

(4) Tan x is zero for $x = 0$, $\pm\pi$, $\pm 2\pi$, and so on. Cot x is zero at $x = \pm\frac{1}{2}\pi$, $\pm\frac{3}{2}\pi$, $\pm\frac{5}{2}\pi$, and so on. At these places the graph crosses the x-axis.

(5) Numerically (ignoring sign), the values of both functions are completely determined in the first quadrant, that is, for $0 < x < \frac{1}{2}\pi$.

Figure 5.36 shows a graph of several periods of the tangent function and of the cotangent function. For purposes of graphing, the places at which the graphs cross the x-axis and the places at which the functions become unbounded are emphasized. The places at which the functions become unbounded are called **vertical asymptotes**. Thus, the asymptotes for $\tan x$ are $x = \pm\frac{1}{2}\pi$, $\pm\frac{3}{2}\pi$, $\pm\frac{5}{2}\pi$, and so on. The asymptotes for $\cot x$ are $x = 0$, $\pm\pi$, $\pm2\pi$, $\pm3\pi$, and so on.

The graphs of the more general functions $y = A \tan(Bx + C)$ and $y = A \cot(Bx + C)$ are analyzed in a manner similar to that of Section 5.3.

In the case of $y = A \tan x$, we do not call A the amplitude because this would imply that the function is bounded. The constant A, then, multiplies each functional value but has no other graphical significance.

The period of $\tan Bx$ is π/B. Thus, if $B > 1$, the period is shortened; if $B < 1$, the period is larger than that of the basic tangent function.

The constant C is a phase shift constant and acts to translate the basic function to the right or left.

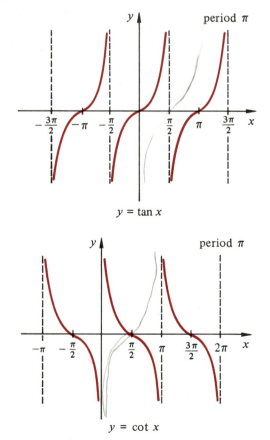

$y = \tan x$

$y = \cot x$

FIGURE 5.36

EXAMPLE 1   Sketch the function $y = \tan(4x - \frac{1}{3}\pi)$.

SOLUTION   The period of this function is $\frac{1}{4}\pi$. The phase shift is located by determining where the argument $4x - \frac{1}{3}\pi$ is equal to zero. Thus, the phase shift is $\frac{1}{12}\pi$. The graph is shown in Figure 5.37.

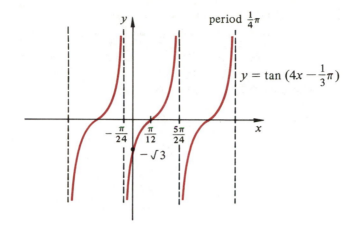

FIGURE 5.37

EXAMPLE 2   Draw the graph of $y = 3 \cot \frac{1}{2}x$.

SOLUTION   The period of this graph is $\pi/\frac{1}{2} = 2\pi$ and the phase shift is zero. The graph is shown in Figure 5.38.

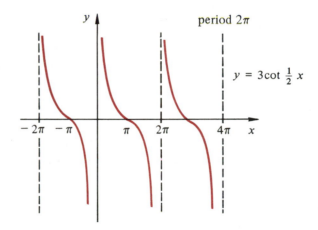

FIGURE 5.38

5.8   Graphs of the Secant and Cosecant Functions

The graphs of sec x and csc x can be sketched directly from graphs of the cos x and sin x since they are reciprocals of the respective functions. You should, of course, note their important general properties.

(1)   Both functions are unbounded. In fact, since both sin x and cos x are bounded by ± 1, the graphs of csc x and sec x lie above $y = 1$ and below $y = -1$.

(2)   Both sec x and csc x are periodic with period 2π.

(3)   The secant function is even and the cosecant function is odd.

(4)   Sec x and csc x are never 0.

(5)   Numerically, the functional values are determined for $0 < x < \pi/2$.

The graphs of both functions are sketched in Figure 5.39. In each case, their reciprocal functions are sketched lightly on the same coordinate system to show the relationship between the two.

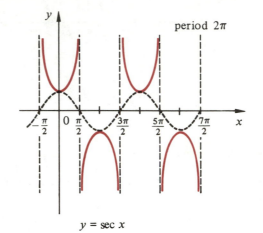

$y = \sec x$

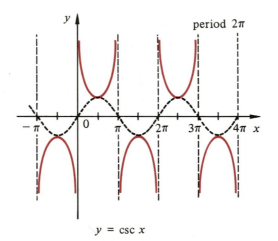

FIGURE 5.39 $y = \csc x$

The manner of sketching the more general functions

$$y = A \sec (Bx + C) \text{ and } y = A \csc (Bx + C)$$

is similar to that discussed for the other functions. Suffice it to say, that the basic waveforms remain the same, but the constants A and B exert a kind of vertical and horizontal stretching while C effects a horizontal translation.

EXAMPLE 1   Sketch the graph of $y = 0.3 \sec (x + \frac{1}{4}\pi)$.

SOLUTION   Since 0.3 is the coefficient of the secant, the graph is outside the interval $-0.3 < y < 0.3$. The period is 2π and the phase shift is $\frac{1}{4}\pi$ units to the left. The graph is shown in Figure 5.40.

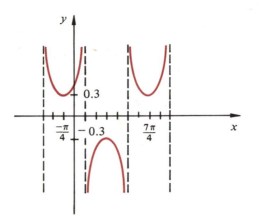

FIGURE 5.40

Exercises for Sections 5.7 and 5.8

In Exercises 1–8 find the period of each of the functions.

1. $\cot \frac{1}{2}x$ **2.** $\csc 3x$ **3.** $\sec \pi x$ **4.** $\tan 3\pi x$

5. $\tan \frac{1}{2}\pi x$ **6.** $\sec \frac{1}{3}x$ **7.** $\cot \frac{5}{6}x$ **8.** $\sec \frac{1}{\pi}x$

Sketch the graphs of the functions given in Exercises 9–17, over at least 2 periods. Give the period and the phase shift. List the asymptotes.

9. $y = \tan 2x$ **10.** $y = \tan (x + \frac{1}{2}\pi)$

11. $y = \cot (\frac{1}{4}\pi - x)$ **12.** $y = 2 \sec (x - \frac{1}{2}\pi)$

13. $y = \tan (2x + \frac{1}{3}\pi)$ **14.** $y = 2 \csc 2x$

15. $y = \csc (2x - 3\pi)$ **16.** $y = \sec (x + \frac{1}{3}\pi)$

17. $y = -\tan (x - \frac{1}{4}\pi)$

18. Does $\tan x$ exist at its asymptote?

19. How are the graphs of $\tan x$ and $\cot x$ related?

20. How are the graphs of $\sec x$ and $\csc x$ related?

21. How are the zeros of the tangent function related to the asymptotes of the cotangent function?

22. How are the zeros of the sine function and the asymptotes of the cosecant function related?

23. How are the graphs of $y = \tan x$ and $y = \tan (-x)$ related?

24. How are the graphs of $y = \tan x$ and $y = -\tan x$ related?

5.9   **A Fundamental Inequality**

The inequality

$$\sin x < x < \tan x$$

is fundamental to trigonometric analysis since it relates the arc length on the unit circle to two of the trigonometric functions. To demonstrate the validity of this inequality, construct a unit circle as shown in Figure 5.41. From the figure, it is obvious that the length of AP is less than the length of the arc OP which in turn is less than OQ.

$$\overline{AP} < \overline{OP} < \overline{OQ}$$

Since the circle has radius of 1, $\sin x = \sin \theta = \overline{AP}$, $\tan x = \tan \theta = \overline{OQ}$, and hence,

$$\sin x < x < \tan x$$

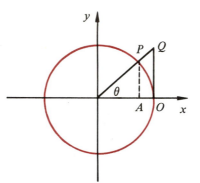

FIGURE 5.41

Figure 5.42 shows the three functions, $y = \sin x$, $y = x$, and $y = \tan x$, and, at the same time, exhibits the fact that the inequality is true only for $0 < x < \frac{1}{2}\pi$.

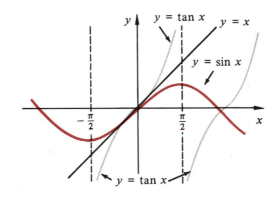

FIGURE 5.42

The ratio $\dfrac{\sin x}{x}$ is important in calculus. In the exercises for Section 5.1, you were asked to compute the values of this ratio for a few values of x near zero. You should have found that this ratio is just slightly less than 1 when x is a very small number. Using the fundamental inequality, we can show this a bit more rigorously. Divide the members of the fundamental inequality by sin x to obtain

$$1 < \frac{x}{\sin x} < \frac{1}{\cos x} \quad \text{(we are assuming } x \text{ is positive)}$$

Inverting and reversing the sense of the inequalities, we obtain

$$\cos x < \frac{\sin x}{x} < 1$$

Since $\cos x$ is near to 1 when x is small and since the ratio is trapped between two functions close to 1, it follows that it also has values close to 1. (A similar argument for negative x leads to the same conclusion.)

The function $\dfrac{\sin x}{x}$ is undefined for $x = 0$, but if we sketch its graph it will look as though its value "should be" 1. (See Figure 5.43.)

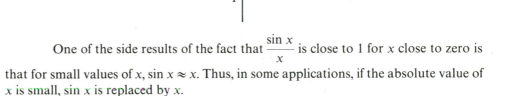

FIGURE 5.43

One of the side results of the fact that $\dfrac{\sin x}{x}$ is close to 1 for x close to zero is that for small values of x, $\sin x \approx x$. Thus, in some applications, if the absolute value of x is small, $\sin x$ is replaced by x.

Exercises for Section 5.9

1. By carefully examining the graphs of $y = 2x/\pi$ and $y = \sin x$, convince yourself that $2x/\pi \leq \sin x \leq 1$ for $0 < x < \pi/2$.

2. What are the zeros of $\dfrac{\sin x}{x}$?

3. Sketch the function $f(x) = \dfrac{\sin 2x}{x}$ for $x \neq 0$. What do you think the values of the function are approaching when x is near 0?

Key Topics for Chapter 5

Define and/or discuss each of the following.

Trigonometry of real numbers

Reference Number

Period

Amplitude

Phase Shift

Graphs of Trigonometric Functions

Bounded and unbounded trigonometric functions

Asymptotes

Odd and even functions

Cycle

Predator-Prey problem

Biorhythm

Harmonic motion

Angular frequency

Addition of ordinates

Fundamental Inequality

Review Exercises for Chapter 5

1. Let $f(x) = 2\sin(3x - 1)$. Compute

   a. $f(0)$ b. $f(\frac{1}{3})$ c. $f(\frac{2}{3}\pi)$

2. In Exercise 1 determine where the given function is equal to 0.

3. In Exercise 1 determine where the given function is equal to 1.

Sketch the graph of the following functions giving period and phase shift. Also give, where applicable, amplitude, mean value, asymptotes, and intercepts.

4. $y = -\sin(x + \frac{1}{6}\pi)$

5. $y = 2 + \sin(x - \frac{1}{3}\pi)$

6. $y = \tan(\pi x + \pi) - 1$

7. $y = 4\cot(\frac{1}{2}x + \frac{1}{8}\pi)$

8. $y = 2\cos(2 - x) + \frac{1}{2}$

9. $y = \sec\frac{1}{3}(\pi - x) + 2$

10. $y = \csc(2x + 1)$

11. Sketch the graph of the harmonic motion described by $x = 9\cos 3t$ and give the amplitude and period.

12. Sketch by using the method of addition of ordinates:

   $y = x + \cos x$

13. Locate the asymptotes for the graph of

$$y = 2 \tan (x + 3)$$

14. Completely describe the impact of the constants A, B, C, and D on the graph of $y = D + A \sin (Bx + C)$.

Test 1 for Chapter 5

Answer *true* or *false* for Exercises 1–10.

1. If x_1 and x_2 are real numbers such that $x_1 < x_2$, then $\sin x_1 < \sin x_2$.

2. The sine and cosine functions are equal to each other twice on $(0, 2\pi)$.

3. The period of $f(x) = \cos 2x$ is 4π.

4. The amplitude of $f(x) = \frac{1}{2} \sin 3x$ is $\frac{1}{2}$.

5. The graphs of $y = \sin x$ and $y = \cos (x - \pi/2)$ are identical.

6. The zeros of the tangent and cotangent function coincide.

7. $\sin x = \pi/4$ when $x = \sqrt{2}/2$.

8. If $-\pi/6 \le x \le \pi/6$, then $\cos x \ge \frac{1}{2}$.

9. If $-1 \le \tan x \le 1$, then $-\pi/4 \le x \le \pi/4$.

10. $\csc x$ is undefined for $x = 0$.

11. Sketch the graph of $y = \sqrt{13} \sin (x - \pi/4)$. Determine the amplitude, phase shift, intercepts, and period.

12. Sketch the graph of $y = -\tan 2x$. What is its period?

13. Sketch the graph of $y = -3 \csc \frac{1}{2}x$.

14. Sketch the graph of $y = 2 \cos 3x$.

15. Solve for x: $\sin x = \tan 10$.

16. Determine the zeros for $0 \le x \le 2\pi$: $f(x) = \sin (2x + 5)$.

Test 2 for Chapter 5

1. Evaluate   (a) $\cos 2$ (b) $\tan (-1.1)$ (c) $\sec 3$.

2. What are the zeros of $\tan x$? $\sec x$?

Discuss and sketch Exercises 3–8 (indicate period, amplitude and phase shift).

3. $y = 5 \cos \frac{1}{2}x$

4. $y = \sin \left(2x - \dfrac{\pi}{3} \right)$

5. $y = 3 \sec\left(x - \dfrac{\pi}{2}\right)$ **6.** $y = 2.5 \cos \pi t$

7. $y = 2 \tan 0.4x$ **8.** $y = -\sin(-x + 2)$

9. Evaluate $\tan(1 + \cos 1)$

10. Graphically solve the following system of equations

$$y = \cos \tfrac{1}{3}x$$

$$3x - 4y = 6$$

6
Identities, Equations and Inequalities

Fundamental Trigonometric Relations

Any combination of trigonometric functions such as $3 \sin x + \cos x$ or $\sec^2 x + \tan^2 x + 2 \sin x$ is called a **trigonometric expression**. One of the important things you will learn in this chapter is how to simplify or alter the form of trigonometric expressions using certain fundamental trigonometric relations.

There are eight *fundamental* relations or identities that you should know if you are to work effectively in the remainder of this book. You are already familiar with most of these relations but they are listed here for completeness. The fundamental relations fall into three groups; The Reciprocal Relations, The Quotient Relations, and The Pythagorean Relations.

The Reciprocal Relations

(6.1) $\sin \theta = \dfrac{1}{\csc \theta}$

(6.2) $\cos \theta = \dfrac{1}{\sec \theta}$

(6.3) $\tan \theta = \dfrac{1}{\cot \theta}$

169

We establish Equation (6.1) by observing for any angle θ in standard position and (x, y) on the terminal side with length of length r:

$$\sin \theta = \frac{y}{r} = \frac{1}{r/y} = \frac{1}{\csc \theta}$$

The other two relations are established in a similar manner.

The Quotient Relations

(6.4)　　$\tan \theta = \dfrac{\sin \theta}{\cos \theta}$

(6.5)　　$\cot \theta = \dfrac{\cos \theta}{\sin \theta}$

To establish Equation (6.4) we note that

$$\cos \theta \tan \theta = \left(\frac{x}{r}\right)\left(\frac{y}{x}\right) = \frac{y}{r} = \sin \theta$$

Therefore,

$$\tan \theta = \frac{\sin \theta}{\cos \theta}$$

The Pythagorean Relations

(6.6)　　$\sin^2 \theta + \cos^2 \theta = 1$

(6.7)　　$\tan^2 \theta + 1 = \sec^2 \theta$

(6.8)　　$\cot^2 \theta + 1 = \csc^2 \theta$

We prove Equation (6.6) by dividing $x^2 + y^2 = r^2$ by r^2 to get

$$\frac{x^2}{r^2} + \frac{y^2}{r^2} = 1$$

Then, since

$$\sin \theta = \frac{y}{r} \text{ and } \cos \theta = \frac{x}{r}$$

we have

$$\cos^2 \theta + \sin^2 \theta = 1$$

Note that Equation (6.7) is derived from Equation (6.6). By dividing both sides of (6.6) by $\cos^2 \theta$, we get

$$\frac{\sin^2 \theta}{\cos^2 \theta} + 1 = \frac{1}{\cos^2 \theta}$$

from which, after using the fact that $\dfrac{\sin \theta}{\cos \theta} = \tan \theta$ and $\dfrac{1}{\cos \theta} = \sec \theta$, we have

$$\tan^2 \theta + 1 = \sec^2 \theta$$

Similarly Equation (6.8) is derived from Equation (6.6) by first dividing both sides by $\sin^2 \theta$ and then applying Equations (6.1) and (6.5).

These eight relations are called the *fundamental identities* of trigonometry; they are valid for all values of the argument for which the functions in the expression have meaning. Also, as before, the variable (often the letter x is chosen instead of θ) may be regarded as either a real number or an angle, the interpretation being chosen from the context.

Using the fundamental identities you can (sometimes ingeniously) manipulate trigonometric expressions into alternative forms.

EXAMPLE 1   Write the following expression as a single trigonometric term.

$$\frac{\tan x \, \csc^2 x}{1 + \tan^2 x}$$

SOLUTION   From Equation (6.7) the denominator may be written as $\sec^2 x$. Thus,

$$\frac{\tan x \, \csc^2 x}{1 + \tan^2 x} = \frac{\tan x \, \csc^2 x}{\sec^2 x}$$

We now express $\tan x$, $\csc x$, and $\sec x$ in terms of the sine and cosine functions

$$\frac{\tan x \, \csc^2 x}{1 + \tan^2 x} = \frac{\dfrac{\sin x}{\cos x} \cdot \dfrac{1}{\sin^2 x}}{\dfrac{1}{\cos^2 x}}$$

$$= \frac{\cos^2 x \, \sin x}{\sin^2 x \, \cos x}$$

$$= \frac{\cos x}{\sin x}$$

$$= \cot x \qquad\qquad \blacksquare$$

As you can see by the preceding example, a large part of the process is algebraic. The series of steps used in the simplification procedure is not unique. For example, we could have initially expressed the complete expression in terms of the sine and cosine functions. Experience with the use of the eight fundamental relations in simplifying trigonometric expressions will give you some facility to choose a reasonable approach. Writing the entire expression in terms of the sine and cosine function is often appropriate but not necessarily the most "economical."

EXAMPLE 2   Simplify the expression

$$(\sec x + \tan x)(1 - \sin x)$$

SOLUTION   We write each of the functions in terms of the sine and cosine functions

$$(\sec x + \tan x)(1 - \sin x) = \left(\frac{1}{\cos x} + \frac{\sin x}{\cos x} \right)(1 - \sin x)$$

$$= \frac{(1 + \sin x)(1 - \sin x)}{\cos x}$$

$$= \frac{(1 - \sin^2 x)}{\cos x}$$

$$= \frac{\cos^2 x}{\cos x}$$

$$= \cos x$$

(What is the domain of this expression?)   ■

EXAMPLE 3   Expand and simplify the expression

$$(\sin x + \cos x)^2$$

SOLUTION   Note that this is *not* the same expression as $\sin^2 x + \cos^2 x$. By squaring the expression, we obtain:

$$(\sin x + \cos x)^2 = \sin^2 x + 2 \sin x \cos x + \cos^2 x$$

$$= 1 + 2 \sin x \cos x$$ ■

EXAMPLE 4   Simplify the expression

$$\sin^4 x - \cos^4 x + \cos^2 x$$

SOLUTION   We write the expression in a form involving only the cosine function. To make this simplification we note that $\sin^4 x = (\sin^2 x)^2 = (1 - \cos^2 x)^2$. Thus,

$$\sin^4 x - \cos^4 x + \cos^2 x = (1 - \cos^2 x)^2 - \cos^4 x + \cos^2 x$$

$$= 1 - 2 \cos^2 x + \cos^4 x - \cos^4 x + \cos^2 x$$

$$= 1 - \cos^2 x$$

$$= \sin^2 x$$ ■

Certain algebraic expressions encountered in calculus are often transformed into trigonometric expressions in which, after simplification, "hard to handle" terms such as radicals disappear.

EXAMPLE 5   By using the substitution $x = 2 \sin \theta$, simplify the expression $\sqrt{4 - x^2}$ and determine an interval for the variable θ which corresponds to $0 \leq x \leq 2$ in a one-to-one manner. What is $\tan \theta$?

SOLUTION   Substituting $x = 2 \sin \theta$ into the radical, we have

$$\sqrt{4 - x^2} = \sqrt{4 - 4 \sin^2 \theta}$$

$$= \sqrt{4(1 - \sin^2 \theta)}$$

$$= |2 \cos \theta| = 2|\cos \theta|$$

When $x = 0$, $\theta = 0$ and when $x = 2$, $\theta = \frac{1}{2}\pi$ so that the interval $0 \le x \le 2$ corresponds to $0 \le \theta \le \frac{1}{2}\pi$. On this interval $\cos \theta \ge 0$ so that $|\cos \theta| = \cos \theta$. Hence,

$$\sqrt{4 - x^2} = 2 \cos \theta \text{ for } 0 \le x \le 2 \text{ and } 0 \le \theta \le \frac{1}{2}\pi$$

Since $\sin \theta = x/2$, the right triangle in Figure 6.1 shows the relations necessary to establish that

$$\tan \theta = \frac{x}{\sqrt{4 - x^2}}$$

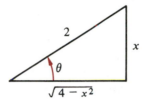

FIGURE 6.1   ■

Exercises for Section 6.1

Reduce the expressions in Exercises 1–20 to a single trigonometric function.

1. $\cos \theta + \tan \theta \sin \theta$

2. $\csc \theta - \cot \theta \cos \theta$

3. $(\tan x + \cot x)\sin x$

4. $\dfrac{1 + \cos x}{1 + \sec x}$

5. $\dfrac{(\tan x)(1 + \cot^2 x)}{(1 + \tan^2 x)}$

6. $\sec x - \sin x \tan x$

7. $\cos x \csc x$

8. $\cos x(\tan x + \cot x)$

9. $(\cos^2 x - 1)(\tan^2 x + 1)$

10. $\dfrac{(\sec^2 x - 1)}{\sec^2 x}$

11. $\dfrac{\sec x - \cos x}{\tan x}$

12. $\dfrac{1 + \tan^2 x}{\tan^2 x}$

13. $(\sin^2 x + \cos^2 x)^3$

14. $\dfrac{1 + \sec x}{\tan x + \sin x}$

15. $(\csc x - \cot x)^4(\csc x + \cot x)^4$

16. $\dfrac{\sec x}{(\tan x + \cot x)}$

17. $(\tan x)(\sin x + \cot x \cos x)$

18. $1 + \dfrac{\tan^2 x}{1 + \sec x}$

19. $\dfrac{\tan x \sin x}{(\sec^2 x - 1)}$

20. $(\cos x)(1 + \tan^2 x)$

By using the substitutions in Exercises 21–26, reduce the given expression to one involving only trigonometric functions. $(0 \leqslant \theta \leqslant \frac{1}{2}\pi)$

21. $\sqrt{a^2 + x^2}$, let $x = a \tan \theta$. What is $\sin \theta$?

22. $\sqrt{36 + 16x^2}$, let $x = \frac{3}{2} \tan \theta$. What is $\sin \theta$?

23. $\dfrac{\sqrt{x^2 - 4}}{x}$, let $x = 2 \sec \theta$.

24. $x^2\sqrt{4 + 9x^2}$, let $x = \frac{2}{3} \tan \theta$.

25. $\sqrt{3 - 5x^2}$, let $x = \sqrt{3/5} \sin \theta$.

26. $\sqrt{2 + x^2}$, let $x = \sqrt{2} \tan \theta$.

6.2 Trigonometric Identities

We noted earlier that when the two sides of an equation are equal for all values of the variable for which the equation is defined, the equation is called an **identity**. The eight fundamental relations (6.1)–(6.8) are trigonometric identities. These fundamental identities are of immediate importance, but you should be aware of the fact that there are many more trigonometric identities that arise in applications of mathematics. In most cases you are required to verify or **prove** that the given relation is an identity. There are several techniques that can be used to prove an identity, but the most common one involves using other known identities to transform one side of the equation into precisely the same form as the other. In this section you will learn to use the fundamental identities (6.1)–(6.8) to prove other identities.

   We note that it is incorrect to verify an identity by beginning with the assumption that it *is* an identity. Thus the expressions on either side of the equality are not equated initially. Operations valid for conditional equations (such as transposing) are not valid. While there is no general approach to proving identities you may find it desirable to write the given expressions in terms of sines and cosines only. Then you can often see the necessary manipulations to complete the verification.

EXAMPLE 1   Verify the identity $\cot x + \tan x = \csc x \sec x$.

SOLUTION   Here, we express the left-hand side in terms of sines and cosines. Thus,

$$\cot x + \tan x = \frac{\cos x}{\sin x} + \frac{\sin x}{\cos x} \qquad \textit{Change to sine and cosine}$$

$$= \frac{\cos^2 x + \sin^2 x}{\sin x \cos x} \qquad \textit{Adding fractions}$$

$$= \frac{1}{\sin x \cos x} \qquad \textit{cos}^2\, x + \textit{sin}^2\, x = 1$$

$$= \csc x \sec x \qquad \frac{1}{\textit{sin }x} = \textit{csc x; } \frac{1}{\textit{cos x}} = \textit{sec x}$$

Therefore, we have shown that $\cot x + \tan x = \csc x \sec x$. ■

EXAMPLE 2   Show that $\sin \theta(\csc \theta - \sin \theta) = \cos^2 \theta$ is an identity.

SOLUTION   Here the most expedient approach is to expand the left-hand side. Thus,

$$\sin \theta(\csc \theta - \sin \theta) = \sin \theta \csc \theta - \sin^2 \theta \quad \bigg| \quad \textit{multiplication}$$

$$= \sin \theta \frac{1}{\sin \theta} - \sin^2 \theta \quad \bigg| \quad \csc \theta = \frac{1}{\sin \theta}$$

$$= 1 - \sin^2 \theta \quad \bigg| \quad \textit{cancellation}$$

$$= \cos^2 \theta \quad \bigg| \quad 1 - \sin^2 \theta = \cos^2 \theta$$

Therefore, we have shown that $\sin \theta(\csc \theta - \sin \theta) = \cos^2 \theta$.   ■

EXAMPLE 3   Verify the identity

$$\frac{\cos x}{1 - \sin x} = \frac{1 + \sin x}{\cos x}$$

SOLUTION   Here, we start on the left side. One way to get $1 + \sin x$ into the numerator of the left side is to multiply the left side by $\dfrac{1 + \sin x}{1 + \sin x}$ as shown below.

$$\frac{\cos x}{1 - \sin x} = \frac{\cos x}{1 - \sin x} \cdot \frac{1 + \sin x}{1 + \sin x} \quad \bigg| \quad \frac{1 + \sin x}{1 + \sin x}$$

$$= \frac{\cos x(1 + \sin x)}{1 - \sin^2 x} \quad \bigg| \quad \textit{multiplication}$$

$$= \frac{\cos x(1 + \sin x)}{\cos^2 x} \quad \bigg| \quad 1 - \sin^2 x = \cos^2 x$$

$$= \frac{1 + \sin x}{\cos x} \quad \bigg| \quad \textit{cancellation}$$

This completes the proof.   ■

EXAMPLE 4   Verify the identity $(\csc x + \cot x)^2 = \dfrac{1 + \cos x}{1 - \cos x}$.

SOLUTION   We start on the left side by squaring the binomial.

$$(\csc x + \cot x)^2 = \csc^2 x + 2 \csc x \cot x + \cot^2 x \quad \bigg| \quad \textit{expand}$$

$$= \frac{1}{\sin^2 x} + \frac{2 \cos x}{\sin^2 x} + \frac{\cos^2 x}{\sin^2 x} \quad \bigg| \quad \textit{change to } \sin x \textit{ and } \cos x$$

$$= \frac{1 + 2 \cos x + \cos^2 x}{\sin^2 x} \quad \bigg| \quad \textit{add fractions}$$

$$= \frac{(1 + \cos x)^2}{\sin^2 x} \quad \bigg| \quad \textit{factor numerator}$$

$$= \frac{(1 + \cos x)^2}{1 - \cos^2 x} \quad \bigg| \quad \sin^2 x = 1 - \cos^2 x$$

$$= \frac{(1 + \cos x)^2}{(1 + \cos x)(1 - \cos x)} \qquad \text{\textit{factor denominator}}$$

$$= \frac{1 + \cos x}{1 - \cos x} \qquad \text{\textit{cancellation}}$$

This completes the proof. ■

Sometimes an identity is verified by manipulating the left-hand side and the right-hand side into forms that are precisely the same.

EXAMPLE 5 Verify the identity

$$\cos^2 x \tan^2 x + 1 = \sec^2 x + \sin^2 x - \sin^2 x \sec^2 x$$

SOLUTION We transform each side of the given expression into precisely the same form. First, the left-hand side becomes

$$\cos^2 x \tan^2 x + 1 = \cos^2 x \left(\frac{\sin^2 x}{\cos^2 x} \right) + 1$$

$$= \sin^2 x + 1$$

The right-hand side may be transformed as follows

$$\sec^2 x + \sin^2 x - \sin^2 x \sec^2 x = \frac{1}{\cos^2 x} + \sin^2 x - \frac{\sin^2 x}{\cos^2 x}$$

$$= \frac{1 - \sin^2 x}{\cos^2 x} + \sin^2 x$$

$$= \frac{\cos^2 x}{\cos^2 x} + \sin^2 x$$

$$= 1 + \sin^2 x$$

Since the right- and left-hand sides have been transformed into the same expression, the identity is verified. ■

EXAMPLE 6 Verify the identity

$$\frac{\cos x \cot x}{\cot x - \cos x} = \frac{\cot x + \cos x}{\cos x \cot x}$$

SOLUTION Expressing the left-hand side in terms of sines and cosines we have

$$\frac{\cos x \cot x}{\cot x - \cos x} = \frac{\cos x \dfrac{\cos x}{\sin x}}{\dfrac{\cos x}{\sin x} - \cos x}$$

$$= \frac{\dfrac{\cos^2 x}{\sin x}}{\dfrac{\cos x - \cos x \sin x}{\sin x}}$$

$$= \frac{\cos^2 x}{\sin x} \cdot \frac{\sin x}{\cos x (1 - \sin x)}$$

$$= \frac{\cos x}{1 - \sin x}$$

We now manipulate the right-hand side to agree with this expression, thus,

$$\frac{\cot x + \cos x}{\cos x \cot x} = \frac{\dfrac{\cos x}{\sin x} + \cos x}{\cos x \dfrac{\cos x}{\sin x}}$$

$$= \frac{\dfrac{\cos x + \cos x \sin x}{\sin x}}{\dfrac{\cos^2 x}{\sin x}}$$

$$= \frac{1 + \sin x}{\cos x}$$

Recognizing that

$$(1 + \sin x)(1 - \sin x) = 1 - \sin^2 x = \cos^2 x$$

we multiply both numerator and denominator by $1 - \sin x$, to get

$$\frac{1 + \sin x}{\cos x} \cdot \frac{1 - \sin x}{1 - \sin x} = \frac{1 - \sin^2 x}{\cos x (1 - \sin x)}$$

$$= \frac{\cos^2 x}{\cos x (1 - \sin x)} = \frac{\cos x}{1 - \sin x}$$

Since we have transformed both sides into the same expression, the identity is proved. ■

To show that a relation is not an identity, we only have to show that the two expressions are unequal for a particular value of x.

EXAMPLE 7   Show that the expression $\sin x = \sqrt{\sin^2 x}$ is not an identity.

SOLUTION   Note that there are many values of x for which the two expressions $\sin x$ and $\sqrt{\sin^2 x}$ *are* equal. But consider $x = -\pi/4$. The value of $\sin(-\pi/4)$ is $-\sqrt{2}/2$ while $\sqrt{\sin^2(-\pi/4)} = \sqrt{(-\sqrt{2}/2)^2}$ $= \sqrt{1/2} = \sqrt{2}/2$. Hence, the two expressions are not equal for this value of x, (other values of x could have been used) and the given expression is not an identity. ■

Exercises for Section 6.2

In Exercises 1–82 verify the identities.

1. $\sin x \cot x = \cos x$

2. $\cos x \tan x = \sin x$

3. $\sec x \cot x = \csc x$

4. $(1 + \tan^2 x) \sin^2 x = \tan^2 x$

5. $\sin^2 x(1 + \cot^2 x) = 1$

6. $\cot^2 x - \cos^2 x = \cot^2 x \cos^2 x$

7. $\csc x - \sin x = \cot x \cos x$

8. $\sec^2 x \csc^2 x = \sec^2 x + \csc^2 x$

9. $(\sin^2 x - 1)(\cot^2 x + 1) = 1 - \csc^2 x$

10. $\dfrac{\sin^2 x + \cos^2 x}{\cos^2 x} = \sec^2 x$

11. $\dfrac{2 + \sec x}{\csc x} - 2\sin x = \tan x$

12. $\dfrac{\sin^4 x - \cos^4 x}{\sin x - \cos x} = \sin x + \cos x$

13. $\dfrac{\sin x}{1 - \cos x} = \csc x + \cot x$

14. $\dfrac{\tan x - 1}{\tan x + 1} = \dfrac{1 - \cot x}{1 + \cot x}$

15. $\dfrac{\cot x + 1}{\cot x - 1} = -\dfrac{\tan x + 1}{\tan x - 1}$

16. $\dfrac{\cos x}{\sec x} + \dfrac{\sin x}{\csc x} = \sec^2 x - \tan^2 x$

17. $\dfrac{1 + \sec x}{\sin x + \tan x} = \csc x$

18. $\sec^2 x - \csc^2 x = \tan^2 x - \cot^2 x$

19. $\dfrac{1 - \sin x}{1 + \sin x} = (\sec x - \tan x)^2$

20. $\cos^2 x - \sin^2 x = 2\cos^2 x - 1$

21. $(\sin^2 x + \cos^2 x)^4 = 1$

22. $\dfrac{\csc^2 x - \cot^2 x}{\sec^2 x} = \cos^2 x$

23. $\dfrac{\tan x + \cot x}{\tan x - \cot x} = \dfrac{\sec^2 x}{\tan^2 x - 1}$

24. $\sin^2 x \sec^2 x + 1 = \sec^2 x$

25. $\dfrac{\sin x}{\csc x(1 + \cot^2 x)} = \sin^4 x$

26. $\sec^2 x - (\cos^2 x + \tan^2 x) = \sin^2 x$

27. $1 - \tan^4 x = 2\sec^2 x - \sec^4 x$

28. $\sec x + \cos x = \sin x \tan x$

29. $(\cot x + \csc x)^2 = \dfrac{1 + \cos x}{1 - \cos x}$

30. $\sin^2 x(\csc^2 x - 1) = \cos^2 x$

31. $\sec x \csc x - 2\cos x \csc x = \tan x - \cot x$

32. $\sec^4 x + \tan^4 x = 1 + 2\sec^2 x \tan^2 x$

33. $\tan^2 x - \cot^2 x = \sec^2 x - \csc^2 x$

34. $\dfrac{1 - \tan^2 x}{1 - \cot^2 x} = 1 - \sec^2 x$

35. $(1 - \sin^2 x)(1 + \tan^2 x) = 1$

36. $\dfrac{\tan x + \sin x}{\tan x - \sin x} = \dfrac{\sec x + 1}{\sec x - 1}$

37. $\dfrac{\cos x + \tan x}{\sin x \cos x} = \csc x + \sec^2 x$

38. $\cos^2 x \tan x = \dfrac{2\sin x}{\cos x + \sec x + \sin^2 x \sec x}$

39. $(\sin x - \cos x)^2 = 1 - 2\cot x \sin^2 x$

40. $\dfrac{\cos x}{\cos x - \sin x} = \dfrac{1}{1 - \tan x}$

41. $\cos^2 x - \sin x \tan x = \cos x \cot x \sin x$
$- \tan x \sin x$

42. $\dfrac{\tan^2 x}{\sin^4 x} = \dfrac{1 + \tan^2 x}{1 - \cos^2 x}$

43. $\tan x - \cot x = -\dfrac{\cos x - \sin^2 x \sec x}{\sin x}$

44. $1 - \sin x = \dfrac{\cot x - \cos x}{\cot x}$

45. $\csc^4 x + \cot^4 x = 1 + 2 \csc^2 x \cot^2 x$

46. $\dfrac{\sec^2 x + 2 \tan x}{1 + \tan x} = 1 + \tan x$

47. $(1 + \cos x)^2 = \sin^2 x \dfrac{\sec x + 1}{\sec x - 1}$

48. $(\sec x - \tan x)^2 = \dfrac{1 - \sin x}{1 + \sin x}$

49. $(\sec x + \tan x)^2 = \dfrac{\sec x + \tan x}{\sec x - \tan x}$

50. $(\csc x - \cot x)^2 = \dfrac{\csc x - \cot x}{\csc x + \cot x}$

51. $\dfrac{\csc x}{\csc x - \tan x} = \dfrac{\cos x}{\cos x - \sin^2 x}$

52. $(\tan x - 1)\cos x = \sin x - \cos x$

53. $\dfrac{1}{1 - \sin x} - \dfrac{1}{1 + \sin x} = 2 \tan x \sec x$

54. $\dfrac{\tan x - \csc x}{\tan x + \csc x} = \dfrac{\sin^2 x - \cos x}{\sin^2 x + \cos x}$

55. $\sec^4 x - \tan^4 x = \dfrac{1 + \sin^2 x}{\cos^2 x}$

56. $\dfrac{\tan x}{\sec x - \cos x} = \csc x$

57. $(\csc x - \cot x)(\sec x + 1) = \tan x$

58. $\dfrac{\cos^2 x}{1 + \sin x} = 1 - \sin x$

59. $(1 + \sin x)(\sec x - \tan x) = \cos x$

60. $\cos^4 x - \sin^4 x = 1 - 2 \sin^2 x$

61. $\sec x \csc x - 2 \cos x \csc x = \tan x - \cot x$

62. $\dfrac{\sin x}{\sin x + \cos x} = \dfrac{\tan x}{1 + \tan x}$

63. $\dfrac{\sin x}{1 + \cos x} + \dfrac{1 + \cos x}{\sin x} = 2 \csc x$

64. $2 \sin^4 x - 3 \sin^2 x + 1 = \cos^2 x(1 - 2 \sin^2 x)$

65. $\dfrac{\csc x}{\tan x + \cot x} = \cos x$

66. $\dfrac{1 - \sin x}{1 + \sin x} = \left(\dfrac{\cos x}{1 + \sin x}\right)^2$

67. $\dfrac{1 + \cos x}{1 - \cos x} = (\csc x + \cot x)^2$

68. $\dfrac{\sec^3 x - \cos^3 x}{\sec x - \cos x} = 1 + \cos^2 x + \sec^2 x$

69. $\dfrac{\cos^2 x}{1 - \sin x + \cos^2 x} = \dfrac{1 + \sin x}{2 + \sin x}$

70. $(1 + \tan x)^2 = \sec^2 x(1 + 2 \cos x \sin x)$

71. $(1 + \cot x)^2 = \csc^2 x(1 + 2 \cos x \sin x)$

72. $(\sin x + \cos x)^2 = (\sec x \csc x + 2)/\sec x \csc x$

73. $(\cos x + \sin x + \tan x)^2 = \sec^2 x(1 + 2 \sin^2 x \cos x) + 2 \sin x(1 + \cos x)$

74. $\dfrac{1 + \cos x}{2 - \cos x} = \dfrac{\sin^2 x}{2 + 3 \cos x^2 x}$

75. $(1 + \sin^2 x)^2 = 16 - 32 \cos^2 x + 24 \cos^4 x - 8 \cos^6 x + \cos^8 x$

76. $\dfrac{\tan x - \cot x}{\sec^2 x - \csc^2 x} = \sin x \cos x$

77. $\dfrac{\tan x - \tan y}{\cot x - \cot y} = \dfrac{1 + \tan x \tan y}{1 - \cot x \cot y}$

78. $\dfrac{\cos x \cos y - \sin x \sin y}{\cos x \sin y + \cos y \sin x} = \dfrac{1 - \tan x \tan y}{\tan x + \tan y}$

79. $\tan x + \cot x = \sec x \csc x$

80. $\dfrac{\tan x}{1 - \cot x} + \dfrac{\cot x}{1 - \tan x} - 1 = \sec x \csc x$

81. $(2 \cos x - \sin x)^2 + (2 \sin x + \cos x)^2 = 5$

82. $(a \cos x - b \sin x)^2 + (a \sin x + b \cos x)^2 = a^2 + b^2$

In Exercises 83–94 show that the expressions given are not identities.

83. $\cos t = \sqrt{\cos^2 t}$

84. $1 = \tan(\cot x)$

85. $1 = \sec (\cos x)$

86. $\sin x = (1 - \cos x)^2$

87. $\cos^2 x = \dfrac{1 - \sin x}{\cdot\; 2}$

88. $\sin x + \cos x = \sqrt{\sin^2 x + \cos^2 x}$

89. $\sin \frac{1}{2}x = \frac{1}{2}\sin x$

90. $\tan 2x = 2 \tan x$

91. $\sin (x + \pi) = \sin x$

92. $\cos (x + \pi) = \cos x - 1$

93. $\cos x^2 = \cos^2 x$

94. $\sin x^2 = \cos (1 - x^2)$

Determine which of the following expressions are identities in Exercises 95–104.

95. $(\cos x - \sin x)(\cos x + \sin x) = 2 \cos^2 x - 1$

96. $\sin x \sec x = \tan x$

97. $\cos x = \cot x$

98. $1 - \cot x = \cot x \tan x - \cot x$

99. $1 - \dfrac{2}{\sec^2 x} = \sin^2 x - \cos^2 x$

100. $\cos x + 1 = \sin x$

101. $\dfrac{\cos x}{1 - \sin x} = \dfrac{1 + \sin x}{\cos x}$

102. $\sin x \cot x \tan^2 x = \sec x - \sin x \cot x$

103. $\sin x \tan x + \cos x = \sec x$

104. $\sin x \tan^2 x = \sin x$

6.3   Trigonometric Equations

A **trigonometric equation** is any statement involving a conditional equality of two trigonometric expressions. A **solution** to the trigonometric equation is a value of the variable (within the domain of the function) that makes the statement true. The **solution set** is the set of all values of the variable that are solutions. To solve a trigonometric equation means to find the solution set for some indicated domain. If no domain is specifically mentioned, the domain is assumed to be all values of the independent variable for which the terms of the equation have meaning.

To solve trigonometric equations, we proceed in a series of steps until a point is reached that allows an explicit determination of the solution set. Usually, some specific knowledge about certain values of the trigonometric functions is necessary to make this determination.

We say that two trigonometric equations are **equivalent** if they have the same solution sets. Any operation on a given equation is **allowable** if the consequence of the operation is an equivalent equation. It can be shown that the permissible operations are: (1) adding or subtracting the same expression to both sides of an equality, and (2) multiplying or dividing both sides by the same nonzero expression.

The most basic type of trigonometric equation is one which is linear in a *single* trigonometric function of θ. Generally such trigonometric equations have infinitely many solutions. However, it is traditional to list only those roots within some fundamental interval. Unless otherwise stated the fundamental interval is chosen to be $[\phi, \phi + p)$, where ϕ is the phase shift and p is the period of trigonometric function. The roots on the fundamental interval are sufficient since the other roots can be obtained from these by simply adding multiples of the period.

EXAMPLE 1   Solve the equation $\cos x = \frac{1}{2}$.

SOLUTION   The period of $\cos x$ is 2π. Recalling that $\cos \frac{1}{3}\pi = \frac{1}{2}$ we have that the only solutions to this equation on the interval $0 \leq x < 2\pi$ are $x = \frac{1}{3}\pi$ and $\frac{5}{3}\pi$. (We note that the complete solution set is comprised of those values which can be written in the form $\frac{1}{3}\pi + 2n\pi$ and $\frac{5}{3}\pi + 2n\pi$ where n is an integer.) Figure 6.2 illustrates the nature of the solution set as the points of intersection of the curve $y = \cos x$ with the line $y = \frac{1}{2}$.

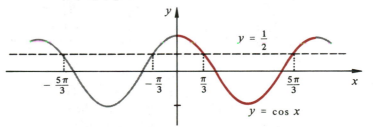

FIGURE 6.2

EXAMPLE 2   Solve the trigonometric equation $\tan 2x = 1$.

SOLUTION   Since the period of this function is $\frac{1}{2}\pi$, it is sufficient to examine for roots on the interval $0 \leq x < \frac{1}{2}\pi$. The value of θ at which $\tan \theta = 1$ is $\theta = \frac{1}{4}\pi$. Hence $\tan 2x = 1$ has the solution $x = \frac{1}{8}\pi$. Figure 6.3 shows a graphical interpretation of the solution set as the intersection of the curve $y = \tan 2x$ with the line $y = 1$. (Note that $\frac{1}{8}\pi + \frac{1}{2}\pi$ represents all solutions.)

EXAMPLE 3   Determine the fundamental interval and write the solution for $\sin (2x - \frac{1}{4}\pi) = -1$.

SOLUTION   The period of $\sin (2x - \frac{1}{4}\pi)$ is π and the phase angle is $\phi = \frac{1}{8}\pi$. Therefore the fundamental interval is $[\frac{1}{8}\pi, \frac{9}{8}\pi)$. Since $\sin \theta = -1$ implies $\theta = \frac{3}{2}\pi$ for θ in the fundamental interval, we have

$$2x - \frac{1}{4}\pi = \frac{3}{2}\pi, \quad x = \frac{7}{8}\pi$$

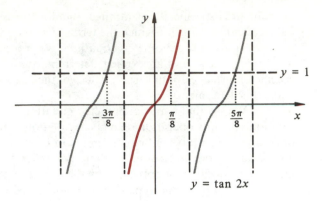

FIGURE 6.3

Note that all solutions are given by $x = \frac{7}{8}\pi + n\pi$. ■

EXAMPLE 4   Find the solution to the equation $|\sin x| = \frac{1}{2}$.

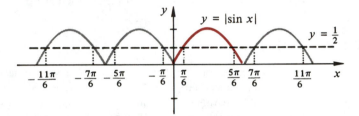

FIGURE 6.4

SOLUTION   Figure 6.4 shows the intersection points of the curve $y = |\sin x|$ with $y = \frac{1}{2}$. Since $|\sin x|$ is periodic with period π, we need only find those values of x for which the equation is true on the interval $0 \le x < \pi$. From your knowledge of the sine function, these values are $x = \frac{1}{6}\pi$ and $x = \frac{5}{6}\pi$. Thus, all solutions are given by $x = \frac{1}{6}\pi + n\pi$ and $\frac{5}{6}\pi + n\pi$. ■

Trigonometric equations that are quadratic in one of the functions can be factored into a product of linear factors. The total solution set is found by finding the solution to each of the resulting linear equations.

EXAMPLE 5   Solve the equation $2\cos^2\theta + 3\cos\theta + 1 = 0$ on $0 \le \theta < 2\pi$.

SOLUTION   This is a quadratic equation in $\cos\theta$ and may be factored as

$$(2\cos\theta + 1)(\cos\theta + 1) = 0$$

Equating each factor to zero and solving for θ, we get

$$
\begin{array}{c|c}
2\cos\theta + 1 = 0 & \cos\theta + 1 = 0 \\
\cos\theta = -\frac{1}{2} & \cos\theta = -1 \\
\theta = \frac{2}{3}\pi, \frac{4}{3}\pi & \theta = \pi
\end{array}
$$

Hence the solution is $\theta = \frac{2}{3}\pi, \pi, \frac{4}{3}\pi$. ■

If more than one function occurs in the equation, you may find it helpful to use trigonometric identities to yield an equivalent equation involving only one function.

EXAMPLE 6   Solve the equation $2 \cos^2 x - \sin x - 1 = 0$ on $0 \le x < 2\pi$.

SOLUTION   Since $\cos^2 x = 1 - \sin^2 x$, we have

$$2(1 - \sin^2 x) - \sin x - 1 = 0$$

$$2 - 2 \sin^2 x - \sin x - 1 = 0$$

$$2 \sin^2 x + \sin x - 1 = 0$$

Factoring,

$$(2 \sin x - 1)(\sin x + 1) = 0$$

The solution set of the original equation is found by separately solving the two equations

$$2 \sin x - 1 = 0 \qquad \text{and} \qquad \sin x + 1 = 0$$

The solution to $2 \sin x - 1 = 0$ is $x = \frac{1}{6}\pi$ and $x = \frac{5}{6}\pi$, and to $\sin x + 1 = 0$ is $\frac{3}{2}\pi$. Hence, the solution set is

$$x = \frac{1}{6}\pi, \; \frac{5}{6}\pi, \; \frac{3}{2}\pi \qquad \blacksquare$$

In solving trigonometric equations, the following is normally a good procedure.

(1)   Gather the entire expression to one side of the equality.

(2)   Use the fundamental identities to express this conditional equality in terms of one function, or, failing this, as a product of two expressions each involving one function.

(3)   Use some knowledge from algebra, such as the quadratic formula or techniques of factoring, to write as a product of linear factors.

(4)   The zeros (if there are any) of each of the linear factors should be recognizable by inspection. The total solution set consists of all values which are zeros of one of these linear factors.

EXAMPLE 7   Solve the equation $2 \tan \theta \sec \theta - \tan \theta = 0$ on $0 < \theta < 360°$.

SOLUTION   The given equation can be factored into

$$\tan \theta \, (2 \sec \theta - 1) = 0$$

Equating each factor to zero, we get

$\tan \theta = 0$	$2 \sec \theta - 1 = 0$
$\theta = 0°$ and $180°$	$\sec \theta = \frac{1}{2}$
	No solution possible since $\sec \theta \ge 1$.

Thus, the solutions are $\theta = 0°$ and $\theta = 180°$. $\blacksquare$

Note that squaring both sides of an equality is not an allowable operation since it does not necessarily yield an equivalent equation. In practice you need not restrict yourself to allowable operations, but when other operations, such as squaring both sides, are used, you should be aware that nonequivalent equations may result. Consequently, when nonallowable operations are used to solve an equation, each apparent solution must be checked for validity. Of course, it is *always* good practice to check your work.

EXAMPLE 8   Solve the equation $\sin x + \cos x = 1$ on $[0, 2\pi)$.

SOLUTION   If we square both sides of this equation we get

$$\sin^2 x + 2 \sin x \cos x + \cos^2 x = 1$$

from which, after using the fact that $\sin^2 x + \cos^2 x = 1$, we have

$$\sin x \cos x = 0$$

The solution to this equation consists of the values of x for which $\sin x = 0$, (that is, $x = 0$ and π), and the values of x for which $\cos x = 0$ (that is, $x = \frac{1}{2}\pi$, and $\frac{3}{2}\pi$). Hence, the possible solutions are

$$x = 0, \tfrac{1}{2}\pi, \pi, \tfrac{3}{2}\pi$$

Since the squaring operation does not yield an equivalent equation, you must check the values in this set to determine if they are truly solutions to the original equation. (You can be sure that the solution set is a *subset* of this set). It is easy to show that only $x = 0$ and $x = \frac{1}{2}\pi$ are valid solutions.   ■

EXAMPLE 9   Solve the equation $\sin^2 x + 3 \sin x - 2 = 0$ on $0 \le x \le 2\pi$.

SOLUTION   Since the given quadratic does not factor we use the quadratic formula. Thus,

$$\sin x = \frac{-3 \pm \sqrt{3^2 - 4(1)(-2)}}{2(1)} = \frac{-3 \pm \sqrt{17}}{2}$$

From this we can write

$$\sin x = \frac{-3 + \sqrt{17}}{2} = 0.5616 \quad \text{and} \quad \sin x = \frac{-3 - \sqrt{17}}{2} = -3.5616$$

The second expression has no solution since the $\sin x$ cannot be less than -1. Using a calculator in the radian mode to solve $\sin x = 0.5616$, we find that $x = 0.596$. Since the $\sin x$ is also positive in the second quadrant, another solution is $\pi - 0.596 = 2.546$. Therefore, the desired solutions are $x = 0.596$ and 2.546.   ■

Exercises for Section 6.3

In Exercises 1–10 solve the equations over the fundamental interval of the function. Make a sketch showing the solution set as the intersection of a line with the graph of some trigonometric function.

1. $\sin x = \frac{1}{2}$

2. $\cos 2x = \frac{1}{2}\sqrt{2}$

3. $\tan x = \sqrt{3}$

4. $\cos x = 1$

5. $\sin x = \frac{1}{2}\sqrt{3}$

6. $\cos (3x + \frac{1}{6}\pi) = \frac{1}{2}$

7. $\sin (2x - \frac{1}{4}\pi) = \frac{1}{2}\sqrt{3}$

8. $\tan (3x - \pi) = 1$

9. $\sin (\frac{1}{3}x - \frac{1}{12}\pi) = -\frac{1}{2}$

10. $\sin (\frac{1}{2}x + \frac{1}{8}\pi) = -1$

In Exercises 11–37 solve the trigonometric equations over the interval $[0, 2\pi)$ unless indicated otherwise.

11. $2 \sin x + 1 = 0$

12. $\sin 2x + 1 = 0$, $[0, \pi)$

13. $\cos 3x = 1$, $[0, 2\pi/3)$

14. $\tan 2x + 1 = 0$, $[0, \pi/2)$

15. $\cos^2 x + 2 \cos x + 1 = 0$

16. $\tan^2 x - 1 = 0$, $[0, \pi)$

17. $2 \sin^2 x = \sin x$

18. $\sec^2 2x = 1$, $[0, \pi)$

19. $\sec^2 x + 1 = 0$

20. $\cos^2 x = 2$

21. $\cos x = \sin x$

22. $2 \sec x \tan x + \sec^2 x = 0$

23. $\sec^2 x - 2 = \tan^2 x$

24. $4 \sin^2 x - 1 = 0$

25. $2 \cos^2 x - \sin x = 1$

26. $2 \sec x + 4 = 0$

27. $\sin^2 x - 2 \sin x + 1 = 0$

28. $\cot^2 x - 5 \cot x + 4 = 0$, $[0, \pi)$

29. $\tan^2 x - \tan x = 0$

30. $\cos 2x + \sin 2x = 0$, $[0, \pi)$

31. $\sin x \tan^2 x = \sin x$

32. $\cos x + 2 \sin^2 x = 1$

33. $\tan x + \sec x = 1$

34. $\tan x + \cot x = \sec x \csc x$

35. $\cos x + 1 = \sin x$

36. $2 \tan x - \sec^2 x = 0$, $[0, \frac{1}{2}\pi)$

37. $\csc^5 x - 4 \csc x = 0$

In Exercises 38–51 solve the equations for all values of θ over the interval $[0, 360°)$.

38. $4 \sin^2 \theta = 3$

39. $2 \cos^2 \theta = 1$

40. $\tan^2 \theta - 3 = 0$

41. $\csc^2 \theta = 1$

42. $(2 \cos \theta - \sqrt{3})(\sqrt{2} \sin \theta + 1) = 0$

43. $(\sin \theta - 1)(2 \cos \theta + \sqrt{3}) = 0$

44. $\csc \theta = 1 + \cot \theta$

45. $2 \sin \theta \tan \theta + \sqrt{3} \tan \theta = 0$

46. $2 \cos^3 \theta = \cos \theta$

47. $\tan^2 \theta - \tan \theta = 0.$

48. $2 \sin^2 \theta = 21 \sin \theta + 11$

49. $2 \cos^2 \theta + 7 \sin \theta = 5$

50. $5 \tan^2 \theta - 11 \sec \theta + 7 = 0$

51. $\sin \theta + \sqrt{3} \cos \theta = 0$

In Exercises 52–60 solve the equations for x, $0 \le x < 2\pi$. Note: For these problems, the solutions are not necessarily multiples of π.

52. $\tan^2 x = 3.2$

53. $\sin^2 x + 2 \cos^2 x = 1.7$

54. $\cos x + 2 \sin x \tan x = 1$

55. $3 \sin^2 x + \sin x = 0$

56. $\tan^2 x + 2 \sec^2 x = 1$

57. $2 \sin x - \cot x \cos x = 1$

58. $\csc^2 x + 2 \cot^2 x - 1 = 0$

59. $\sin x - \csc x + 1 = 0$

60. $2 \tan^2 x + 3 \sec x - 5 = 0$

6.4   Parametric Equations

A convenient way to define the locus* of a point in the plane is by using two equations, one for x and one for y in terms of some variable, say t. Thus, two functions of t determine the location of the set of points,

$$x = f(t) \qquad y = g(t)$$

As t varies, the point describes a curve in the plane. The equations are called **parametric equations** of the curve; the variable t is called the **parameter**.

The determination of the locus of the point described by parametric equations is usually rather awkward, and we may find it useful to *eliminate the parameter* to obtain an equation in x and y. If the parametric functions f and g contain trigonometric terms, as they do in some very typical cases, then the use of trigonometric identities is frequently helpful in this procedure. There is no general technique to describe how to eliminate the parameter, so we will be satisfied with a few examples.

EXAMPLE 1   Sketch the curve in the x-y plane described parametrically by $x = 2 \sin t$, $y = 3 \cos t$.

SOLUTION   Since $\frac{1}{2}x = \sin t$ and $\frac{1}{3}y = \cos t$, we may square both of these equations and add the corresponding sides to obtain

$$(\tfrac{1}{2}x)^2 + (\tfrac{1}{3}y)^2 = \sin^2 t + \cos^2 t = 1.$$

This is an ellipse and is sketched in Figure 6.5.

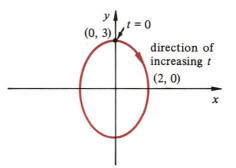

FIGURE 6.5

Note that the parametric representation gives a sense of direction to the curve since you can think of the point beginning at the point (0, 3) when $t = 0$ and going around the curve in a

---

*The **locus** of a point in the plane is the path taken by the point in satisfying a given condition, such as a mathematical formula.

clockwise sense as t increases. The point completes one revolution of the ellipse every 2π units. A curve with "direction," such as the one in this example, is said to be **oriented**. ■

EXAMPLE 2   Sketch the graph of the curve described by $x = \sin t$, $y = \sin t$.

SOLUTION   The elimination of the parameter is a simple matter of noticing that $y = x$. The graph of $y = x$ is shown in Figure 6.6 as the line that splits the first and third quadrants.

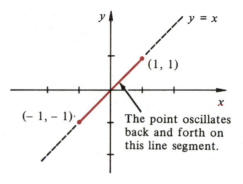

FIGURE 6.6

However, by the nature of the parametric equations, the values of x and y are limited between 1 and -1. Thus the point moves on the line $y = x$, but it oscillates between the points $(1, 1)$ and $(-1, -1)$. ■

The previous example should carry with it a general word of caution; when eliminating the parameter be careful that the process of elimination does not tend to include more points (or less) than the parametric equations themselves give.

EXAMPLE 3   Eliminate t from the parametric equations

$$x = a + b \sec t \qquad y = c + d \tan t$$

where a, b, c, and d are positive constants.

SOLUTION   We rewrite the two equations as

$$\frac{x - a}{b} = \sec t \qquad \frac{y - c}{d} = \tan t$$

Squaring and subtracting, and noting that $\sec^2 t - \tan^2 t = 1$, we have

$$\left(\frac{x - a}{b}\right)^2 - \left(\frac{y - c}{d}\right)^2 = 1$$

This is the equation of a hyperbola. It is studied in more detail in analytic geometry. ■

6.5   Graphical Solutions of Trigonometric Equations

Equations containing a mixture of trigonometric and other functions may be quite difficult to solve by analytic methods. A graphical analysis will usually yield at least an approximation to the roots and will often give helpful information even when a problem can be solved analytically.

EXAMPLE 1   Graphically solve the equation $x = \sin \frac{1}{2}\pi x$.

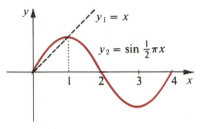

FIGURE 6.7

SOLUTION   The functions $y_1 = x$ and $y_2 = \sin \frac{1}{2}\pi x$ are sketched in Figure 6.7. Since both functions are odd, the graph is drawn for positive values of x only. The solution set to the equation is the set of x coordinates of the points of intersection of the two curves. The figure shows that the values are $x = 0$ and $x = 1$; hence by the symmetry of the graph, the solution set is $\{-1, 0, 1\}$. ∎

EXAMPLE 2   Graphically solve the equation $x \tan x = 1$.

SOLUTION   We write the equation in the form $\tan x = 1/x$ and then graph the two functions $y_1 = \tan x$ and $y_2 = 1/x$, as is shown in Figure 6.8. As in the previous example, we may, without loss of generality, sketch only the part of the graphs for $x \geq 0$. Figure 6.8 shows that there are infinitely many solutions of which the first positive one is approximately 0.86.

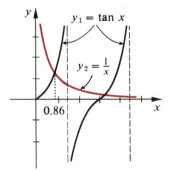

FIGURE 6.8   ∎

6.6   Trigonometric Inequalities

A **trigonometric inequality** is any conditional statement of inequality involving trigonometric expressions. The solution set is defined, analogous to those defined for trigonometric equations, as the set of values for which the conditional statement is true. Other related terminology is defined to be consistent with that which is used for trigonometric equations. Note that the term "allowable operation" is slightly more restrictive when used with inequalities in that multiplication or division of both sides is permitted only by positive expressions. Multiplication of both sides by a negative quantity results in a reversal of the inequality.

Aside from the very basic kinds of inequalities such as $\sin x < 1$ and $\cos x > 0$, most trigonometric inequalities are best solved by some combination of graphical and analytical methods.

EXAMPLE 1   Solve the inequality, $\sin x > \cos x$.

SOLUTION   In Figure 6.9, both the sine and cosine functions are sketched. The points of intersection occur at the real numbers that can be written in the form $\frac{1}{4}\pi + n\pi$. The graph shows that the sine function is greater than the cosine function over those intervals for which the left-hand end point is given for even values of n. In Figure 6.9, the solution set for one period is shown as the interval $\frac{1}{4}\pi < x < \frac{5}{4}\pi$.

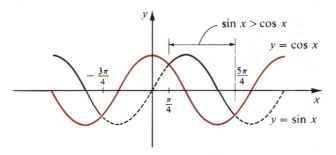

FIGURE 6.9

Inequalities that are not strictly trigonometric but which include other functions are also best analyzed graphically.

EXAMPLE 2   Solve the inequality $x > \cos x$.

SOLUTION   From Figure 6.10, we see that the point where $x = \cos x$ is approximately $x = 0.74$. Hence, from the graph, the solution is $x > 0.74$.

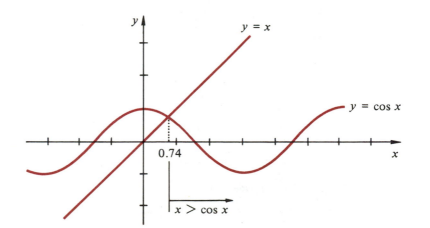

FIGURE 6.10

Exercises for Sections 6.4 through 6.6

In Exercises 1–10 eliminate t from the parametric equations to obtain one equation in x and y.

1. $x = \sin t, y = \cos t$

2. $x = 1 + \sin t, y = 1 + \cos t$

3. $x = -\cos t, y = \sin t$

4. $x = \sec t + 1, y = \tan t - 1$

5. $x = a \cos t, y = a \sin^2 t$

6. $x = \cos t, y = \sin t \cos t$

7. $x = \cos t, y = \cos t$

8. $x = \sin^2 t, y = 1$

9. $x = \cos^2 t, y = \cos^2 t$

10. $x = 2 \sin t + \cos t, y = 2 \cos t - \sin t$

Use graphical methods to approximate at least some of the solutions to the equations on $[0, 2\pi)$ in Exercises 11–20.

11. $x = \sin x$

12. $x = \sin 2x$

13. $x \sin x = 1$

14. $x = \tan x$

15. $x = \cos x$

16. $x^2 = \cos 2\pi x$

17. $x^2 - \tan x = 0$

18. $\sin x = \cos x$

19. $\tan x = \cos x$

20. $\tan x = \sin x$

In Exercises 21–26 solve the trigonometric inequalities.

21. $\sec x > \csc x$

22. $x > \sin x$

23. $x > \sin 2\pi x$

24. $x \sin x > 1$

25. $|\sin x| < |\cos x|$

26. $\sec^2 x < 1$

Key Topics for Chapter 6

Define and/or discuss each of the following.

Fundamental Identities

Verifying and proving Identities

Trigonometric equations

Parametric equations

Graphical solutions to trigonometric equations

Trigonometric Inequalities

Review Exercises for Chapter 6

Prove the following identities.

1. $1 + \sec x = \csc x(\tan x + \sin x)$

2. $\cos^2 x = \dfrac{\csc^2 x - 1}{\csc^2 x}$

3. $\sin x = \dfrac{1 - \cos^2 x}{\sin x}$

4. $\cos x = \sec x - \sin x \tan x$

5. $\cos x + \sin x = \dfrac{2\cos^2 x - 1}{\cos x - \sin x}$

6. $\sec x + \tan x = \dfrac{1}{\sec x - \tan x}$

7. $1 - \sin x = \dfrac{\cos x}{\sec x + \tan x}$

8. $\dfrac{\csc x + 1}{\cot x} = \dfrac{\cot x}{\csc x - 1}$

9. $\sec x - \cos x = \dfrac{\tan^2 x}{\sec x}$

10. $2\sin^2 x = \sin^4 x - \cos^4 x + 1$

In Exercises 11–15 solve the following trigonometric equations over one fundamental interval.

11. $\sin x = -\sqrt{3}/2$

12. $\cos 2x = 0.6789$

13. $\sin x + 2\cos^2 x = 1$

14. $\sin^2 x - \sin x = 0$

15. $\sec^5 x - 4\sec x = 0$

In Exercises 16–20 eliminate the parameter and sketch.

16. $x = 3\sin t$

$\quad y = \cos t$

17. $x = \tan t$

$\quad y = 2\sec t$

18. $x = 1 + \sin t$

$\quad y = 2\cos t$

19. $\left. \begin{array}{l} x = \tan t \\ y = \tan t \end{array} \right\} 0 \le t \le \pi$

20. $x = 1$

$\quad y = \sin t$

21. Using graphical methods find approximate solutions to $x^2 = \sin x$ on $(-\pi, \pi)$.

22. Approximate solutions to $x \cos x = 1$ on $[0, 2\pi]$.

23. Solve the inequality $\cos x > \sin x$ on $[0, 2\pi]$.

24. Solve the inequality $\sec x > 2$ on $[-\pi, \pi]$.

Test 1 for Chapter 6

In Exercises 1–10 answer *true* or *false*.

1. $\sin^4 x + \cos^4 x = 1$

2. $|\tan^2 420° - \sec^2 420°| = 1$

3. $\tan\dfrac{x}{y} = \dfrac{\tan x}{\tan y}$

4. $(\sin x + \cos x)^2 = 1$

5. $\tan^2 x = 1$, where $\cos x = \pm \sin x$

6. $\sin x = 1/(\sec x)$

7. On $0 < x < \pi/2$, $\cos x > \sin x$

8. The curve defined parametrically by $x = \cos t$, $y = \sin t$ is the same set of points as the curve $x^2 + y^2 = 1$.

9. The curve defined parametrically by $x = \sin t$, $y = \cos t$ is the same set of points as the curve $x = \cos t$, $y = \sin t$.

10. The reciprocal of $\tan x$ is $\cot x$.

11. Solve the following equation on $0 \le x < 2\pi$.

$2 \sin^2 x + 5 \sin x - 3 = 0$.

12. Solve the following equation on $0 \le x < 360°$.

$5 \sin^2 x + \cos^2 x + 4 \sin x = 0$.

13. Graph the function $f(x) = \tan^2 x - \sec^2 x$.

14. Verify the identity: $\cos x \tan x \cot^2 x = \csc x - \cos x \tan x$.

15. Eliminate the parameter and sketch: $x = \sin t$, $y = 2 \cos t$.

16. By using the substitution $x = 3 \sin \theta$, reduce the expression

$(9 - x^2)^{1/2}$

to one with one trigonometric function. What is $\tan \theta$?

Test 2 for Chapter 6

In Exercises 1–3 solve the equations on $0 \le x < 2\pi$.

1. $\tan x + 1 = 0$

2. $\cos^2 x + 5 \sin^2 x = 0$

3. $2 \sin^2 x + 3 \sin x + 1 = 0$

Verify the identities in Exercises 4–7.

4. $\dfrac{\cot^2 x - 1}{1 + \cot^2 x} = 2 \cos^2 x - 1$

5. $\dfrac{\sin x \tan x}{\tan x - \sin x} = \csc x + \cot x$

6. $\cos^2 x (1 + \tan^2 x) = 1$.

7. $\dfrac{\cos x}{\sec x (1 + \tan^2 x)} = \cos^4 x$.

8. Eliminate the parameter and sketch $x = 1 - \cos t$, $y = 2 + \sin t$.

9. Graphically solve the equation $\sin x = x^3$, on $0 \le x < \frac{1}{2}\pi$.

10. Graphically solve the inequality $x \cos x > 1$, on $0 \le x < 2\pi$.

Composite Angle Identities

7.1   The Linearity Property

In mathematics, functions that obey the laws

$$f(x + y) = f(x) + f(y) \quad \text{and} \quad f(ax) = af(x)$$

are said to have the **linearity property**. You may fail to see the importance of this property until you have to work with functions that do not have it. Beginning students tend to think that all functions have the linearity property, but this is far from the case, as the next two examples illustrate.

EXAMPLE 1   Show that the square root function does not have the linearity property.

SOLUTION   Here we show that $\sqrt{x + y} \neq \sqrt{x} + \sqrt{y}$. Letting $x = 9$ and $y = 16$, we have

$$\sqrt{x + y} = \sqrt{9 + 16} = \sqrt{25} = 5$$

But,

$$\sqrt{x} + \sqrt{y} = \sqrt{9} + \sqrt{16} = 3 + 4 = 7$$

Since $5 \neq 7$, we conclude that $\sqrt{x + y} \neq \sqrt{x} + \sqrt{y}$, and, therefore, $\sqrt{x + y}$ does not have the linearity property. ■

EXAMPLE 2   Show that the cosine function does not have the linearity property.

193

SOLUTION

We will show that $\cos(x + y) \neq \cos x + \cos y$. To this end, we let $x = \frac{1}{3}\pi$ and $y = \frac{1}{6}\pi$. Then,

$$\cos(x + y) = \cos\left(\tfrac{1}{3}\pi + \tfrac{1}{6}\pi\right) = \cos\tfrac{1}{2}\pi = 0$$

But,

$$\cos x + \cos y = \cos\tfrac{1}{3}\pi + \cos\tfrac{1}{6}\pi = \tfrac{1}{2} + \tfrac{1}{2}\sqrt{3} \neq 0$$

Therefore, we have shown that the cosine function does not have the linearity property.   ■

The implication of Example 2 is that none of the trigonometric functions have the linearity property, which is indeed the case. The purpose of this chapter is to show precisely which formulas the trigonometric functions of sums and differences of angles *do* obey. The formulas will be derived from the viewpoint of trigonometric functions of angles, with the understanding that they are also valid when the domain is the set of real numbers. The most basic of these so-called addition formulas is the formula for $\cos(A - B)$. This formula is derived in the next section.

7.2   The Cosine of the Difference of Two Angles

The formula for $\cos(A - B)$ is so basic that it is the only one of the sum and difference formulas which must be derived directly from the definitions of the trigonometric functions.

Let A and B represent angles in standard position superimposed on a circle of radius 1. Figure 7.1(a) is a picture of the general situation. The terminal side of A intersects the unit circle in the point $(x_A, y_A) = (\cos A, \sin A)$.

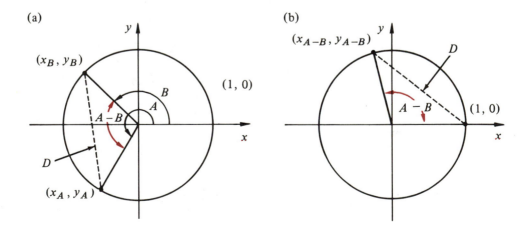

(a)   (b)

FIGURE 7.1

Similarly, the terminal side of B intersects the circle at $(x_B, y_B) = (\cos B, \sin B)$. The distance D between these two points is given by

$$D^2 = (x_A - x_B)^2 + (y_A - y_B)^2$$

$$= (\cos A - \cos B)^2 + (\sin A - \sin B)^2$$

$$= (\cos^2 A - 2 \cos A \cos B + \cos^2 B)$$

$$+ (\sin^2 A - 2 \sin A \sin B + \sin^2 B)$$

Using the fact that $\cos^2 A + \sin^2 A = 1$ and $\cos^2 B + \sin^2 B = 1$, we have

$$D^2 = 2(1 - \cos A \cos B - \sin A \sin B)$$

Now rotate the angle $A - B$ until it is in standard position as shown in Figure 7.1(b). The coordinates of the point of intersection of the terminal side of the angle $A - B$ and the unit circle are

$$(x_{A-B}, y_{A-B}) = [\cos (A - B), \sin (A - B)]$$

D is now the distance connecting $(1, 0)$ to this point. Using the distance formula,

$$D^2 = (x_{A-B} - 1)^2 + (y_{A-B})^2$$

$$= (\cos (A - B) - 1)^2 + \sin^2 (A - B)$$

$$= \cos^2 (A - B) - 2 \cos (A - B) + 1 + \sin^2 (A - B)$$

Since $\cos^2 (A - B) + \sin^2 (A - B) = 1$, we can write D^2 as

$$D^2 = 2(1 - \cos (A - B))$$

Equating this result to the first expression we derived for D^2, we have

$$1 - \cos A \cos B - \sin A \sin B = 1 - \cos(A - B)$$

Simplifying this expression, we have the fundamental formula

(7.1) $\cos (A - B) = \cos A \cos B + \sin A \sin B$

Formula (7.1) was derived under the conditions that $A > B$ and that A and B are between 0 and 2π. However, since

$$\cos (A - B) = \cos (B - A) = \cos (A - B + 2n\pi)$$

it follows that the formula is perfectly general. Since it is true for all angles A and B and, consequently for all real numbers, it is an **identity**.

The principal use of this formula is to derive other important relations. However, it can also be used to obtain the value of the cosine function at a particular angle (or real number) if that angle can be expressed as the difference of two angles for which the exact value of the cosine is known.

EXAMPLE 1   Without the use of tables, find the exact value of $\cos \frac{1}{12}\pi$.

SOLUTION   First notice that $\frac{1}{12}\pi = \frac{1}{3}\pi - \frac{1}{4}\pi$. Hence,

$$\cos \tfrac{1}{12}\pi = \cos \left(\tfrac{1}{3}\pi - \tfrac{1}{4}\pi\right)$$

$$= \cos \tfrac{1}{3}\pi \cos \tfrac{1}{4}\pi + \sin \tfrac{1}{3}\pi \sin \tfrac{1}{4}\pi$$

$$= \frac{1}{2}\left(\frac{\sqrt{2}}{2}\right) + \left(\frac{\sqrt{3}}{2}\right)\left(\frac{\sqrt{2}}{2}\right)$$

$$= \frac{\sqrt{2} + \sqrt{6}}{4}$$

■

If, in Formula (7.1), we replace B by $-B$, we obtain

$$\cos (A - (-B)) = \cos A \cos (-B) + \sin A \sin (-B)$$

Since the cosine function is even, $\cos (-B) = \cos B$; since the sine function is odd, $\sin (-B) = -\sin B$. Hence,

(7.2) $\cos (A + B) = \cos A \cos B - \sin A \sin B$

EXAMPLE 2

(a) $-\sin 2\alpha \sin 3\alpha + \cos 2\alpha \cos 3\alpha = \cos (2\alpha + 3\alpha) = \cos 5\alpha$

(b) $7 \cos (2x + 4y) = 7 \cos 2x \cos 4y - 7 \sin 2x \sin 4y$

■

EXAMPLE 3   Find the exact value of $\cos 75°$.

SOLUTION   Since $75° = 30° + 45°$,

$$\cos 75° = \cos (30° + 45°)$$

$$= \cos 30° \cos 45° - \sin 30° \sin 45°$$

$$= \frac{\sqrt{3}}{2} \cdot \frac{\sqrt{2}}{2} - \frac{1}{2} \cdot \frac{\sqrt{2}}{2}$$

$$= \frac{\sqrt{6} - \sqrt{2}}{4}$$

■

EXAMPLE 4   Find the value of $\cos (A - B)$ given that $\sin A = \tfrac{3}{5}$ in quadrant II and $\tan B = \tfrac{1}{2}$ in quadrant I.

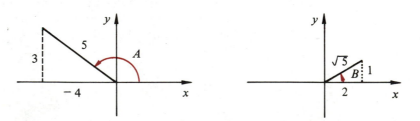

FIGURE 7.2

SOLUTION   From Figure 7.2, we see that $\cos A = -4/5$, $\sin B = 1/\sqrt{5}$, and $\cos B = 2/\sqrt{5}$. Hence,

$$\cos(A - B) = \cos A \cos B + \sin A \sin B$$

$$= \left(-\frac{4}{5}\right)\left(\frac{2}{\sqrt{5}}\right) + \left(\frac{3}{5}\right)\left(\frac{1}{\sqrt{5}}\right)$$

$$= -\frac{5}{5\sqrt{5}} = -\frac{\sqrt{5}}{5}$$ ∎

In certain applied problems, we encounter functions of the form $c_1 \cos Bx + c_2 \sin Bx$. In analyzing the properties of this function, it is easy to see that it is periodic with a period of $2\pi/B$; however, in its present form, it is not easy to recognize the amplitude and phase shift of the oscillation. To obtain these properties, as well as the period, we make use of the identity

$$c_1 \cos Bx + c_2 \sin Bx = A \cos(Bx - C)$$

where $A = \sqrt{c_1^2 + c_2^2}$ and $\tan C = c_2/c_1$.

To verify this identity we note that by Equation (7.2), we have

$$A \cos(Bx - C) = A \cos C \cos Bx + A \sin C \sin Bx$$

Therefore,

$$c_1 \cos Bx + c_2 \sin Bx = A \cos C \cos Bx + A \sin C \sin Bx$$

if and only if, $c_1 = A \cos C$ and $c_2 = A \sin C$. Squaring c_1 and c_2 and adding, we get

$$c_1^2 + c_2^2 = A^2 \cos^2 C + A^2 \sin^2 C = A^2 (\cos^2 C + \sin^2 C)$$

or

$$A = \sqrt{c_1^2 + c_2^2}$$

Also, the ratio of c_2 to c_1 yields

$$\frac{c_2}{c_1} = \frac{A \sin C}{A \cos C} = \tan C$$

It should be noted that C is not just *any* angle for which $\tan C = c_2/c_1$: The angle must be chosen so that its terminal side passes through the point (c_1, c_2).

EXAMPLE 5   Express $f(x) = \sin x + \cos x$ as a cosine function and sketch.

SOLUTION   Using the above formulas, we see that

$$f(x) = A \cos(x - C) \quad \text{where} \quad A = \sqrt{1^2 + 1^2} = \sqrt{2} \quad \text{and} \quad \tan C = 1$$

Since $c_1 = c_2 = 1$ and $(1, 1)$ is in the first quadrant, we let $C = \frac{1}{4}\pi$. Therefore,

$$f(x) = \sqrt{2} \cos(x - \tfrac{1}{4}\pi)$$

This is a function with amplitude $\sqrt{2}$, period 2π and phase shift $\frac{1}{4}\pi$. The graph is shown in Figure 7.3 along with the graphs of the sine and cosine functions.

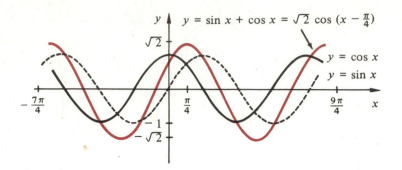

FIGURE 7.3

Exercises for Sections 7.1 and 7.2

1. Show that $\sin(x + y) \neq \sin x + \sin y$. (Let $x = \frac{1}{3}\pi$ and $y = \frac{1}{6}\pi$.)

2. Show that $\tan(x + y) \neq \tan x + \tan y$. (Let $x = \frac{1}{3}\pi$ and $y = \frac{1}{6}\pi$.)

3. Show that $\sin 2x \neq 2 \sin x$. (Let $x = \frac{1}{4}\pi$.)

4. Show that $\cos 2x \neq 2 \cos x$. (Let $x = \frac{1}{4}\pi$.)

5. Show that $\tan 2x \neq 2 \tan x$. (Let $x = \frac{1}{4}\pi$.)

In Exercises 6–10 find the *exact* values of the trigonometric functions.

6. $\cos 105°$ 7. $\cos \frac{1}{12}\pi$ 8. $\cos \frac{11}{12}\pi$ 9. $\cos 195°$ 10. $\cos 345°$

Use Equation (7.1) to show that Exercises 11–13 are true.

11. $\cos(\pi - \theta) = -\cos \theta$ 12. $\cos(\frac{1}{2}\pi - \theta) = \sin \theta$ 13. $\cos(\frac{1}{2}\pi + \theta) = -\sin \theta$

14. Using Equation (7.1), give a proof that the cosine is an even function.

Verify the identities in Exercises 15–19.

15. $\cos(\frac{1}{3}\pi - x) = \dfrac{\cos x + \sqrt{3}\sin x}{2}$ 16. $\cos(\frac{1}{4}\pi + \theta) = \dfrac{\cos \theta - \sin \theta}{\sqrt{2}}$

17. $\cos(\frac{3}{2}\pi + x) = \sin x$

18. $\cos(x + y)\cos(x - y) = \cos^2 x - \sin^2 y$

19. $\cos(x + y) + \cos(x - y) = 2 \cos x \cos y$

Reduce Exercises 20–22 to a single term.

20. $\cos 2x \cos 3x + \sin 2x \sin 3x$

21. $\cos 7x \cos x - \sin 7x \sin x$

22. $\cos \frac{1}{6}x \cos \frac{5}{6}x - \sin \frac{1}{6}x \sin \frac{5}{6}x$

Find the value of $\cos(A + B)$ for the conditions in Exercises 23–26.

23. $\cos A = \frac{1}{3}$, $\sin B = -\frac{1}{2}$, A in quadrant I, B in quadrant IV

24. $\cos A = \frac{3}{5}$, $\tan B = \frac{12}{5}$, both A and B acute.

25. $\tan A = \frac{24}{7}$, $\sec B = \frac{5}{3}$, A in quadrant III, B in quadrant I.

26. $\sin A = \frac{1}{4}$, $\cos B = \frac{1}{2}$, both A and B in quadrant I.

Let A and B both be positive acute angles.

27. Find $\cos A$ if $\cos(A+B) = \frac{5}{6}$ and $\sin B = \frac{1}{3}$.

28. Find $\cos A$ if $\cos(A-B) = \frac{3}{4}$ and $\cos B = \frac{2}{3}$.

Graph the functions in Exercises 29–32. What is the amplitude and phase shift of each?

29. $f(x) = \cos x - \sin x$

30. $f(x) = 2\cos x + 2\sin x$

31. $f(x) = \cos 2x + \sqrt{3}\sin 2x$

32. $f(x) = -\cos 2x + \sqrt{3}\sin 2x$

In Exercises 33–38 show that $\cos(A+B) \neq \cos A + \cos B$. Then show that $\cos(A+B) = \cos A \cos B - \sin A \sin B$. Use a calculator when necessary.

33. $A = 0$, $B = 0$

34. $A = 30°$, $B = 30°$

35. $A = 0.987$, $B = 0.111$

36. $A = -0.912$, $B = 0.912$

37. $A = 23.1$, $B = 14.14$

38. $A = 1.57$, $B = -1.57$

7.3   Other Addition Formulas

From the formula for the cosine of the difference, it is easy to establish that

(7.3) $\cos(\tfrac{1}{2}\pi - \theta) = \sin\theta$ and $\sin(\tfrac{1}{2}\pi - \theta) = \cos\theta$

Note that this is a general statement of the identity relating cofunctions of complementary angles. In Chapter 2 this formula was derived with the condition that the two angles were acute. We now see that this is a general formula, true for all values of the argument.

EXAMPLE 1   (a) $\cos(-10°) = \cos(90° - 100°) = \sin 100°$,

(b) $\sin(\tfrac{5}{6}\pi) = \sin(\tfrac{1}{2}\pi - (-\tfrac{1}{3}\pi)) = \cos(-\tfrac{1}{3}\pi) = \cos\tfrac{1}{3}\pi$.   ■

You know the fundamental Pythagorean relationship $\sin^2 x + \cos^2 x = 1$. A formula involving *two* numbers x and y that looks quite similar is also true if x and y have a sum of $\tfrac{1}{2}\pi$.

EXAMPLE 2   If x and y are complementary numbers (that is their sum is $\tfrac{1}{2}\pi$), show that $\sin^2 x + \sin^2 y = 1$.

SOLUTION   Since $x + y = \tfrac{1}{2}\pi$, $y = \tfrac{1}{2}\pi - x$. Therefore, $\sin y = \sin(\tfrac{1}{2}\pi - x) = \cos x$. Hence, from the Pythagorean relation,

$$\sin^2 x + \cos^2 x = 1$$

we have, using $\sin y = \cos x$,

$$\sin^2 x + \sin^2 y = 1 \text{ (Assuming } x + y = \tfrac{1}{2}\pi)$$ ■

By letting $\theta = A + B$ in Equation (7.3), we can write,

$$\sin(A + B) = \cos(\tfrac{1}{2}\pi - (A + B))$$

$$= \cos((\tfrac{1}{2}\pi - A) - B)$$

$$= \cos(\tfrac{1}{2}\pi - A)\cos B + \sin(\tfrac{1}{2}\pi - A)\sin B$$

(7.4) $\sin(A + B) = \sin A \cos B + \cos A \sin B$

Similarly, if $\theta = A - B$, we have

$$\sin(A - B) = \sin A \cos(-B) + \cos A \sin(-B)$$

and, since the cosine is an even function and the sine function is odd, this becomes

(7.5) $\sin(A - B) = \sin A \cos B - \cos A \sin B$

EXAMPLE 3   Find the exact value of $\sin(\tfrac{1}{12}\pi)$.

SOLUTION   Since $\tfrac{1}{12}\pi = \tfrac{1}{3}\pi - \tfrac{1}{4}\pi$, we have that

$$\sin(\tfrac{1}{12}\pi) = \sin(\tfrac{1}{3}\pi)\cos(\tfrac{1}{4}\pi) - \cos(\tfrac{1}{3}\pi)\sin(\tfrac{1}{4}\pi)$$

$$= \frac{\sqrt{3}}{2}\frac{\sqrt{2}}{2} - \frac{1}{2}\frac{\sqrt{2}}{2}$$

$$= \frac{1}{4}(\sqrt{6} - \sqrt{2})$$ ■

EXAMPLE 4   Show that $\sin x + \cos x = \sqrt{2}\sin(x + \tfrac{1}{4}\pi)$.

SOLUTION   If $\sin x + \cos x = A\sin(x + C) = A\sin x \cos C + A\cos x \sin C$, then $A\cos C = 1$ and $A\sin C = 1$. Squaring these two equations and adding the results, we get

$$A^2\cos^2 C + A^2\sin^2 C = 1^2 + 1^2 = 2$$

or

$$A = \sqrt{2}$$

Also,

$$\frac{A\sin C}{A\cos C} = \frac{1}{1}$$

so,

$$\tan C = 1$$

We choose C to be a first quadrant angle since its terminal side must pass through (1, 1). Thus, $C = \frac{1}{4}\pi$ and hence,

$$\sin x + \cos x = \sqrt{2} \sin\left(x + \tfrac{1}{4}\pi\right)$$

The sketch of this function is precisely that given in Figure 7.3. Can you verify this?   ■

Sum and difference formulas for the tangent follow directly from those for the sine and cosine.

$$\tan(A + B) = \frac{\sin(A + B)}{\cos(A + B)}$$

$$= \frac{\sin A \cos B + \cos A \sin B}{\cos A \cos B - \sin A \sin B}$$

Now divide both the numerator and denominator by $\cos A \cos B$:

$$\tan(A + B) = \frac{\dfrac{\sin A \cos B}{\cos A \cos B} + \dfrac{\cos A \sin B}{\cos A \cos B}}{\dfrac{\cos A \cos B}{\cos A \cos B} - \dfrac{\sin A \sin B}{\cos A \cos B}}$$

Simplifying, we get

(7.6) $$\tan(A + B) = \frac{\tan A + \tan B}{1 - \tan A \tan B}$$

Similarly,

(7.7) $$\tan(A - B) = \frac{\tan A - \tan B}{1 + \tan A \tan B}$$

EXAMPLE 5   Verify that $\tan(\theta + \pi) = \tan\theta$.

SOLUTION   Using (7.6) and the fact that $\tan\pi = 0$, we have

$$\tan(\theta + \pi) = \frac{\tan\theta + \tan\pi}{1 - \tan\theta\,\tan\pi} = \tan\theta$$ ■

EXAMPLE 6   Reduce $\dfrac{\tan(x + y) - \tan(x - y)}{1 + \tan(x + y)\tan(x - y)}$ to a single term.

SOLUTION   We recognize this expression as the righthand side of (7.7) with $A = x + y$ and $B = x - y$. Thus,

$$\frac{\tan(x + y) - \tan(x - y)}{1 + \tan(x + y)\tan(x - y)} = \tan[(x + y) - (x - y)] = \tan 2y$$ ■

In summary, we have the following sum and difference formulas which have been derived:

$$\sin(A \pm B) = \sin A \cos B \pm \cos A \sin B$$

$$\cos(A \pm B) = \cos A \cos B \mp \sin A \sin B$$

$$\tan(A \pm B) = \frac{\tan A \pm \tan B}{1 \mp \tan A \tan B}$$ $\tan(A-B)$

By convention, the symbols $\pm$ and $\mp$ used in the same formula mean to use the topmost signs together and the bottommost signs together.

Exercises for Section 7.3

Find the exact value of Exercises 1–5.

1. $\sin\left(\frac{5}{12}\pi\right)$ **2.** $\tan 15°$ **3.** $\sin\left(\frac{7}{12}\pi\right)$

4. $\sin(345°)$ **5.** $\cot\left(\frac{5}{12}\pi\right)$

In Exercises 6–13 verify the identities.

6. $\sin\left(A + \frac{1}{4}\pi\right) = \dfrac{\sqrt{2}}{2}(\sin A + \cos A)$ **7.** $\tan\left(A + \frac{1}{2}\pi\right) = -\cot A$

8. $\tan\left(A + \frac{1}{4}\pi\right) = \dfrac{1 + \tan A}{1 - \tan A}$ **9.** $\cot(A + B) = \dfrac{\cot A \cot B - 1}{\cot A + \cot B}$

10. $\dfrac{\sin(A + B)}{\sin(A - B)} = \dfrac{\tan A + \tan B}{\tan A - \tan B}$

11. $\sin(A + B)\sin(A - B) = \sin^2 A - \sin^2 B$

12. $\sin(A + B) + \sin(A - B) = 2\sin A \cos B$

13. $\tan A + \tan B = \dfrac{\sin(A + B)}{\cos A \cos B}$

In Exercises 14–17 reduce the expressions to a single term.

14. $\sin 2x \cos 3x + \sin 3x \cos 2x$ **15.** $\dfrac{\tan 3x - \tan 2x}{1 + \tan 3x \tan 2x}$

16. $\sin \frac{1}{3}x \cos \frac{2}{3}x + \sin \frac{2}{3}x \cos \frac{1}{3}x$ **17.** $\dfrac{\tan(x + y) + \tan z}{1 - \tan(x + y)\tan z}$

Find the values of $\sin(A + B)$ and $\tan(A + B)$ if:

18. $\sin A = \frac{3}{5}$, $\cos B = \frac{4}{5}$, both A and B in quadrant I.

19. $\tan A = -\frac{7}{24}$, $\tan B = \frac{5}{12}$, A in quadrant II, B in quadrant III.

20. $\cos A = \frac{1}{3}$, $\cos B = -\frac{1}{3}$, A in quadrant IV, B in quadrant III.

Express Exercises 21–24 as a sine function with a phase shift and sketch.

21. $\sin 2x + \cos 2x$ **22.** $\cos x$

23. $\sqrt{3}\sin \pi x + \cos \pi x$ **24.** $7\sin 2x - 24\cos 2x$

In Exercises 25–30, use a calculator to show that the left member is equal to the right member.

25. $\sin(27° + 96°) = \sin 27° \cos 96° + \cos 27° \sin 96°$

26. $\cos(12.6° + 8.7°) = \cos 12.6° \cos 8.7° - \sin 12.6° \sin 8.7°$

27. $\cos(1.1 - 0.3) = \cos 1.1 \cos 0.3 + \sin 1.1 \sin 0.3$

28. $\sin(2.8 - 1.6) = \sin 2.8 \cos 1.6 - \cos 2.8 \sin 1.6$

29. $\tan(0.4 + 0.3) = \dfrac{\tan 0.4 + \tan 0.3}{1 - \tan 0.4 \tan 0.3}$

30. $\tan(2.9 - 1.2) = \dfrac{\tan 2.9 - \tan 1.2}{1 + \tan 2.9 \tan 1.2}$

7.4   Multiple and Half Angle Formulas

In the two previous sections, we have been primarily interested in expanding trigonometric functions whose arguments were $A \pm B$. Now we will derive formulas for functions of $2A$ and $\frac{1}{2}A$. If A represents an angle, the formulas are called the double and half angle formulas, respectively.

The double angle formulas are easily proved by choosing $B = A$ in the formulas for the sum of two angles. Thus,

$$\sin 2A = \sin(A + A)$$

$$= \sin A \cos A + \sin A \cos A$$

(7.8) $\sin 2A = 2 \sin A \cos A$

and

$$\cos 2A = \cos(A + A)$$

$$= \cos A \cos A - \sin A \sin A$$

(7.9) $\cos 2A = \cos^2 A - \sin^2 A$

By use of the Pythagorean relationship, this last formula may also be expressed in the equivalent forms

(7.9a) $\cos 2A = 2\cos^2 A - 1$

and

(7.9b) $\cos 2A = 1 - 2\sin^2 A$

Similarly,

$$\tan 2A = \frac{\tan A + \tan A}{1 - \tan A \tan A}$$

(7.10) $\tan 2A = \dfrac{2 \tan A}{1 - \tan^2 A}$

EXAMPLE 1   Find $\sin 2A$ if $\sin A = \frac{1}{3}$ and A is in quadrant II.

SOLUTION   Since $\sin A = \frac{1}{3}$, we have from Figure 7.4 that $\cos A = -\frac{1}{3}\sqrt{8}$ and hence,

$$\sin 2A = 2 \sin A \cos A$$

$$= 2\left(\frac{1}{3}\right)\left(\frac{-\sqrt{8}}{3}\right) = \frac{-2\sqrt{8}}{9}$$

Note that $\sin 2A \neq \frac{2}{3}$.

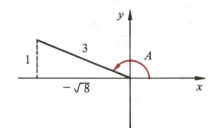

FIGURE 7.4

EXAMPLE 2   Sketch the graph of $y = \sin x \cos x$. Where does the maximum value of this function occur?

SOLUTION   Initially you may think that you must graph this function by "point plotting," but, by multiplying and dividing the right-hand side by 2, you obtain,

$$y = \frac{2 \sin x \cos x}{2} = \frac{1}{2} \sin 2x$$

Thus, the graph of this function is a sine wave with amplitude $\frac{1}{2}$ and period π.

Its maximum value of $\frac{1}{2}$ occurs at $x = \frac{1}{4}\pi + n\pi$. (See Figure 7.5.)

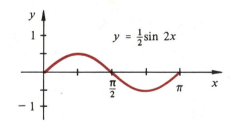

FIGURE 7.5

The half-angle formulas are direct consequences of the formulas for the cosine of a double angle. Since

$$\cos 2A = 2 \cos^2 A - 1$$

we have, upon solving for $\cos A$,

$$\cos A = \pm \sqrt{\frac{1 + \cos 2A}{2}}$$

Using the formula $\cos 2A = 1 - 2 \sin^2 A$ and solving for $\sin A$,

$$\sin A = \pm \sqrt{\frac{1 - \cos 2A}{2}}$$

By letting $A = \frac{1}{2}x$ in both of these formulas, we have

(7.11) $\cos \frac{1}{2}x = \pm \sqrt{(1 + \cos x)/2}$

and

(7.12) $\sin \frac{1}{2}x = \pm \sqrt{(1 - \cos x)/2}$

To get the formula for $\tan \frac{1}{2}x$, we write

$$\tan \frac{1}{2}x = \frac{\sin \frac{1}{2}x}{\cos \frac{1}{2}x}$$

$$= \frac{\pm \sqrt{(1 - \cos x)/2}}{\pm \sqrt{(1 + \cos x)/2}}$$

$$= \pm \sqrt{(1 - \cos x)/(1 + \cos x)}$$

Multiplying numerator and denominator by this expression by $(1 + \cos x)$, we get

$$\tan \frac{1}{2}x = \pm \sqrt{(1 - \cos^2 x)/(1 + \cos x)^2}$$

$$= \pm \sqrt{\sin^2 x/(1 + \cos x)^2}$$

To simplify this expression we note that $1 + \cos x$ is never negative and that $\tan \frac{1}{2}x$ and $\sin x$ can be shown to have the same sign for all values of x. Therefore, $\tan \frac{1}{2}x$ always has the same sign as $\sin x/(1 + \cos x)$ and we write

(7.13) $\tan \frac{1}{2}x = \dfrac{\sin x}{1 + \cos x}$

EXAMPLE 3   If $\tan \theta = -\frac{4}{3}$ and $-\frac{1}{2}\pi < \theta < 0$, find $\sin \frac{1}{2}\theta$ and $\cos \frac{1}{2}\theta$.

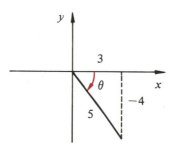

FIGURE 7.6

SOLUTION   Figure 7.6 shows the angle θ in standard position. From this, $\cos \theta = \frac{3}{5}$, and hence,

$$\sin \frac{1}{2}\theta = -\sqrt{\frac{1 - (3/5)}{2}} = -\frac{\sqrt{5}}{5}$$

(The minus sign is chosen because $\sin \frac{1}{2}\theta$ is negative on the indicated range). Similarly,

$$\cos \tfrac{1}{2}\theta = \sqrt{\frac{1+(3/5)}{2}} = \sqrt{4/5} = \frac{2\sqrt{5}}{5}$$

∎

EXAMPLE 4   Find an expression for $\sin 3\theta$ in terms of $\sin \theta$.

SOLUTION
$$\sin 3\theta = \sin (2\theta + \theta)$$
$$= \sin 2\theta \cos \theta + \sin \theta \cos 2\theta$$
$$= 2 \sin \theta \cos^2 \theta + \sin \theta (1 - 2 \sin^2 \theta)$$
$$= 2 \sin \theta (1 - \sin^2 \theta) + \sin \theta (1 - 2 \sin^2 \theta)$$
$$= 3 \sin \theta - 4 \sin^3 \theta$$

∎

EXAMPLE 5   Solve the equation $\cos 2x = \cos x$ on the interval $0 \le x \le 2\pi$.

SOLUTION   We first use the identity for $\cos 2x$ to transform the equation into one involving $\cos x$.

$$2 \cos^2 x - 1 = \cos x$$

Transferring $\cos x$ to the left-hand side,

$$2 \cos^2 x - \cos x - 1 = 0$$

Factoring,

$$(2 \cos x + 1)(\cos x - 1) = 0$$

From this we have the two separate equations

$$2 \cos x + 1 = 0 \quad \text{and} \quad \cos x - 1 = 0$$

The solution set to the first of these on $0 \le x \le 2\pi$ is $x = \tfrac{2}{3}\pi$ and $\tfrac{4}{3}\pi$ and, to the second one, is $x = 0$ and $x = 2\pi$. Hence, the complete solution is

$$x = 0, \tfrac{2}{3}\pi, \tfrac{4}{3}\pi, 2\pi$$

∎

EXAMPLE 6   If $\cot 2\theta = -\tfrac{7}{24}$ and θ is in quadrant I find $\sin \theta$ and $\cos \theta$.

SOLUTION   A sketch such as the one shown in Figure 7.7 is often helpful for this kind of problem. From this figure we can see immediately that $\cos 2\theta = -\tfrac{7}{25}$. Hence,

$$\sin \theta = \left(\frac{1 - \cos 2\theta}{2} \right)^{1/2}$$
$$= \left(\frac{1 + \tfrac{7}{25}}{2} \right)^{1/2}$$
$$= \frac{4}{5}$$

Similarly, $\cos \theta = \dfrac{3}{5}$.

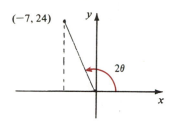

FIGURE 7.7

Exercises for Section 7.4

Express each of the functions in Exercises 1–6 as a single trigonometric term.

1. $2 \sin 3x \cos 3x$

2. $6 \sin \frac{1}{2}x \cos \frac{1}{2}x$

3. $\sin^2 4x - \cos^2 4x$

4. $4 \sin^2 x \cos^2 x$

5. $\dfrac{2 \tan \frac{1}{6}x}{1 - \tan^2 \frac{1}{6}x}$

6. $\dfrac{\sin 6x}{1 + \cos 6x}$

7. Sketch the graph of the function $f(x) = \sin 2x \cos 2x$. What is the maximum value of the function? Where does it occur?

8. Sketch the graph of the two functions $f(x) = \cos^2 x - \sin^2 x$ and $g(x) = \cos^2 x + \sin^2 x$. In each case tell what the maximum values are.

9. Sketch the graph of the function $f(x) = \sec x \csc x$. Does this function have a maximum? Where is this function undefined?

10. Sketch the graph of the function $f(x) = (2 \tan x)/(1 + \tan^2 x)$. What is the maximum value of this function? What is the period?

11. Sketch the graph of the function $f(x) = \tan x + \cot x$. Where is this function undefined? What is the period?

12. Sketch the graph of the function $f(x) = \cot x - \tan x$. What are the zeros of this function? Is it bounded or unbounded? What is the period?

In Exercises 13–16 find $\sin 2A$, $\cos 2A$, and $\tan 2A$ given that

13. $\sin A = \frac{3}{5}$, A in quadrant I

14. $\cos A = -\frac{12}{13}$, A in quadrant III

15. $\tan A = \frac{7}{24}$, A in quadrant III

16. $\sec A = -\frac{13}{5}$, A in quadrant II

Determine the exact values in Exercises 17–21.

17. $\sin \frac{1}{8}\pi$

18. $\cos \frac{5}{8}\pi$

19. $\tan 157.5°$

20. $\sin 67.5°$

21. $\cos \frac{1}{12}\pi$

Verify the identities in Exercises 22–42.

22. $(\sin x + \cos x)^2 = 1 + \sin 2x$

23. $\cos 3x = 4 \cos^3 x - 3 \cos x$

24. $\sin 4x = 4 \cos x \sin x(1 - 2 \sin^2 x)$

25. $\tan x + \cot x = 2 \csc 2x$

26. $\cos 4x = 8 \cos^4 x - 8 \cos^2 x + 1$

27. $\cot^2 \frac{1}{2}x = \dfrac{\sec x + 1}{\sec x - 1}$

28. $\cos^4 x = \frac{3}{8} + \frac{1}{2} \cos 2x + \frac{1}{8} \cos 4x$

29. $\sec^2 x = \dfrac{4 \sin^2 x}{\sin^2 2x}$

30. $\tan 2x = \dfrac{2}{\cot x - \tan x}$

31. $2 \cot 2x \cos x = \csc x - 2 \sin x$

32. $\dfrac{1 - \tan x}{1 + \tan x} = \dfrac{1 - \sin 2x}{\cos 2x}$

33. $\tan^2 x + \cos 2x = 1 - \cos 2x \tan^2 x$

34. $\dfrac{2 \tan x}{\tan 2x} = 1 - \tan^2 x$

35. $\dfrac{\cos 3x}{\sec x} - \dfrac{\sin x}{\csc 3x} = \cos^2 2x - \sin^2 2x$

36. $\dfrac{\csc x - \sec x}{\csc x + \sec x} = \dfrac{\cos 2x}{1 + \sin 2x}$

37. $1 + \cot x \cot 3x = \dfrac{2 \cos 2x}{\cos 2x - \cos 4x}$

38. $\tan 3x = \dfrac{\tan^3 x - 3 \tan x}{3 \tan^2 x - 1}$

39. $\sin 6x \tan 3x = 2 \sin^2 3x$

40. $\dfrac{\cos^3 x + \sin^3 x}{2 - \sin 2x} = \frac{1}{2}(\sin x + \cos x)$

41. $\cos 4x \sec^2 2x = 1 - \tan^2 2x$

42. $16 \cos^5 x = 5 \cos 3x + 10 \cos x + \cos 5x$

In Exercises 43–45 find the indicated functional value. (Assume 2θ is in quadrant I.)

43. Find $\tan \theta$ if $\sin 2\theta = \frac{5}{13}$

44. Find $\sin \theta$ if $\sin 2\theta = \frac{3}{5}$

45. Find $\cos \theta$ if $\cos 2\theta = \frac{24}{25}$

In Exercises 46–54 solve the equations for those numbers which belong to the interval $0 \le x \le 2\pi$.

46. $\sin 2x = \sin x$

47. $\sin x = \cos x$

48. $\sin 2x \sin x + \cos x = 0$

49. $\tan 2x = \tan x$

50. $\cos x - \sin 2x = 0$

51. $\sin 2x + \cos 2x = 0$

52. $\sin 2x - 2 \cos x + \sin x - 1 = 0$

53. $2(\sin^2 2x - \cos^2 2x) = 1$

54. $\sin 2x \cos x - \frac{1}{2} \sin 3x = \frac{1}{2} \sin x$

Show that each of the expressions in Exercises 55–59 reduces to 1.

55. $(\sin x + \cos x)^2 - \sin 2x$

56. $\dfrac{\sin 2x \sin x}{2 \cos x} + \cos^2 x$

57. $\sec^4 x - \tan^4 x - 2 \tan^2 x$

58. $\left[\dfrac{\sin 2x}{\sin x} - \dfrac{\cos 2x}{\cos x} \right] \sec x - \tan^2 x$

59. $\cos 2x + \sin 2x \tan x$

In Exercises 60–64 use a calculator to show that the left member is equal to the right member.

60. $\sin 2(1.05) = 2 \sin 1.05 \cos 1.05$

61. $\cos^2 0.47 = \frac{1}{2}(1 + \cos 2(0.47))$

62. $\sin^2 1.3 = \frac{1}{2}(1 - \cos 2(1.3))$

63. $\cos 2(0.22) = \cos^2 0.22 - \sin^2 0.22$

64. $\tan \frac{1}{2}(3.0) = \dfrac{\sin 3.0}{1 + \cos 3.0}$

7.5   Sums and Differences of Sines and Cosines

Sometimes you will want to write a sum of sines and cosines as a product. A scheme, based on the addition formulas, follows.

EXAMPLE 1   Factor $\sin 7x + \sin 3x$.

SOLUTION   We write

$$\sin 7x + \sin 3x = \sin (5x + 2x) + \sin (5x - 2x)$$

$$= \sin 5x \cos 2x + \sin 2x \cos 5x + \sin 5x \cos 2x - \sin 2x \cos 5x$$

$$= 2 \sin 5x \cos 2x$$

■

The method, illustrated in Example 1, is called the **average angle method**. In the general case, we can proceed:

$$\sin A + \sin B = \sin \left(\frac{A+B}{2} + \frac{A-B}{2} \right) + \sin \left(\frac{A+B}{2} - \frac{A-B}{2} \right)$$

$$= \sin \frac{A+B}{2} \cos \frac{A-B}{2} + \cos \frac{A+B}{2} \sin \frac{A-B}{2}$$

$$+ \sin \frac{A+B}{2} \cos \frac{A-B}{2} - \cos \frac{A+B}{2} \sin \frac{A-B}{2}$$

(7.14) $$\sin A + \sin B = 2 \sin \frac{A+B}{2} \cos \frac{A-B}{2}$$

The following formulas are derived analogously. If you followed the technique, above, you will not have to memorize them.

(7.15) $$\sin A - \sin B = 2 \cos \frac{A+B}{2} \sin \frac{A-B}{2}$$

(7.16) $$\cos A + \cos B = 2 \cos \frac{A+B}{2} \cos \frac{A-B}{2}$$

(7.17) $$\cos A - \cos B = -2 \sin \frac{A+B}{2} \sin \frac{A-B}{2}$$

EXAMPLE 2   The difference quotient of a function is defined to be

$$\Delta f = \frac{f(x + \Delta x) - f(x)}{\Delta x}$$

Find the difference quotient for the sine function.

SOLUTION   We first compute $\sin(x + \Delta x) - \sin x$.

$$\sin(x + \Delta x) - \sin x = 2\cos\frac{(x + \Delta x) + x}{2}\sin\frac{(x + \Delta x) - x}{2}$$

$$= 2\cos\frac{2x + \Delta x}{2}\sin\frac{\Delta x}{2}$$

Thus, the difference quotient for the sine function is

$$\frac{\sin(\Delta x/2)\cos(x + (\Delta x/2))}{\Delta x/2}$$

∎

EXAMPLE 3   Prove the identity $\dfrac{\sin 6x - \sin 4x}{\cos 6x + \cos 4x} = \tan x$.

SOLUTION

$$\frac{\sin 6x - \sin 4x}{\cos 6x + \cos 4x} = \frac{2\cos\frac{1}{2}(6x + 4x)\sin\frac{1}{2}(6x - 4x)}{2\cos\frac{1}{2}(6x + 4x)\cos\frac{1}{2}(6x - 4x)}$$

$$= \frac{\cos 5x \sin x}{\cos 5x \cos x}$$

$$= \tan x$$

∎

EXAMPLE 4   Solve the equation $\sin 3x + \sin x = 0$ on the interval $0 \le x \le \pi$.

SOLUTION   To put the given equation in factored form we use (7.14) with

$$A = 3x \text{ and } B = x.$$

Therefore,

$$\sin 3x + \sin x = 2\sin 2x \cos x$$

The given equation is then

$$\sin 3x + \sin x = 2\sin 2x \cos x = 0$$

Now, since $\sin 2A = 2\sin A \cos A$, we can write

$$2(2\sin x \cos x)\cos x = 0$$

$$4\sin x \cos^2 x = 0$$

Equating each factor to zero, we have

$$\sin x = 0 \qquad \bigg| \qquad \cos^2 x = 0$$

$$x = 0, \pi \qquad \bigg| \qquad x = \tfrac{1}{2}\pi$$

The desired solution set is $x = 0, \tfrac{1}{2}\pi, \pi$.

∎

7.6   Product Formulas

Our objective in this section is to find formulas to express products of trigonometric functions as sums of trigonometric functions. The derivations are made from the formulas for the sine and cosine of the sum and difference. If we add the formulas for sin $(A + B)$ and sin $(A - B)$ we obtain

(7.18) $\sin A \cos B = \frac{1}{2}\{\sin (A + B) + \sin (A - B)\}$

Upon subtracting sin $(A - B)$ from sin $(A + B)$ and simplifying:

(7.19) $\cos A \sin B = \frac{1}{2}\{\sin (A + B) - \sin (A - B)\}$

In like manner, by first adding and then subtracting the formulas for cos $(A + B)$ and cos $(A - B)$, we obtain

(7.20) $\cos A \cos B = \frac{1}{2}\{\cos (A + B) + \cos (A - B)\}$

(7.21) $\sin A \sin B = \frac{1}{2}\{\cos (A - B) - \cos (A + B)\}$

EXAMPLE 1   Express sin mx cos nx as a sum of functions.

SOLUTION   Using Equation (7.18) with $A = mx$ and $B = nx$, we have

$$\sin mx \cos nx = \frac{1}{2}\{\sin (mx + nx) + \sin (mx - nx)\}$$
$$= \frac{1}{2}\{\sin (m + n)x + \sin (m - n)x\}$$ ■

EXAMPLE 2   In the analysis of some types of harmonic motion the governing equation is $y(t) = A(\cos \omega t - \cos \omega_0 t)$ where the difference of ω and ω_0 is considered to be very small. Make a sketch of the graph of this function.

SOLUTION   By using Equation (7.17), we write

$$y(t) = 2A \sin \frac{\omega_0 - \omega}{2} t \sin \frac{\omega_0 + \omega}{2} t$$

If ω is close to ω_0, the resultant oscillation can be interpreted to have a frequency close to $\omega_0/2\pi$ (and of course close to $\omega/2\pi$) with variable amplitude given by

$$2A \sin \frac{\omega_0 - \omega}{2} t$$

which fluctuates with frequency $(\omega - \omega_0)/4\pi$. Oscillations of this type are called **beats**. (See Figure 7.8.)

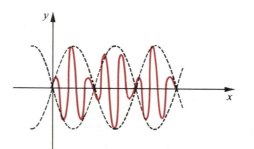

FIGURE 7.8   ■

Exercises for Sections 7.5 and 7.6

Express each of Exercises 1–7 as a product.

1. $\sin 3\theta + \sin \theta$

2. $\cos 3\alpha - \cos 8\alpha$

3. $\sin 8x + \sin 2x$

4. $\sin \frac{1}{2}x - \sin \frac{1}{4}x$

5. $\cos 50° - \cos 30°$

6. $\sin \frac{3}{4}\pi - \sin \frac{1}{4}\pi$

7. $\sin \frac{3}{4} - \sin \frac{1}{4}$

Express each of Exercises 8–12 as a sum or difference.

8. $\sin 3x \cos x$

9. $\cos x \sin \frac{1}{2}x$

10. $\cos \frac{1}{3}\pi \sin \frac{2}{3}\pi$

11. $\cos 6x \cos 2x$

12. $\sin \frac{1}{4}\pi \sin \frac{1}{12}\pi$

In Exercises 13–18 verify the identities.

13. $\dfrac{\sin x + \sin y}{\cos x + \cos y} = \tan \frac{1}{2}(x + y)$

14. $\dfrac{\sin x + \sin y}{\sin x - \sin y} = \dfrac{\tan \frac{1}{2}(x + y)}{\tan \frac{1}{2}(x - y)}$

15. $\dfrac{\cos 3x + \cos x}{\sin 3x + \sin x} = \cot 2x$

16. $\cos 7x + \cos 5x + 2 \cos x \cos 2x = 4 \cos 4x \cos 2x \cos x$

17. $\dfrac{\sin 2x + \sin 2y}{\cos 2x + \cos 2y} = \tan (x + y)$

18. $\dfrac{\sin 9x - \sin 5x}{\sin 14x} = \dfrac{\sin 2x}{\sin 7x}$

19. Find the difference quotient for $\cos x$.

In Exercises 20–23 solve the following equations on $0 \le x \le \pi$.

20. $\sin 3x + \sin 5x = 0$

21. $\sin x - \sin 5x = 0$

22. $\cos 3x - \cos x = 0$

23. $\cos 2x - \cos 3x = 0$

24. Let $f(x) = \sin (2x + 1) + \sin (2x - 1)$. Make a sketch of the graph of the function. What is the period and amplitude?

25. Let $f(x) = \cos (3x + 1) + \cos (3x - 1)$. Make a sketch of the graph of this function and give the period and amplitude.

26. Make a sketch of the graph of the function $f(x) = \cos 99x - \cos 101x$.

In Exercises 27–30 use a calculator to show that the left member is equal to the right member.

27. $\sin 0.3 + \sin 1.2 = 2 \sin \frac{1}{2}(0.3 + 1.2) \cos \frac{1}{2}(0.3 - 1.2)$

28. $\cos 2.05 + \cos 0.72 = 2 \cos \frac{1}{2}(2.05 + 0.72) \cos \frac{1}{2}(2.05 - 0.72)$

29. $\cos 200° - \cos 76° = -2 \sin \frac{1}{2}(200° + 76°) \sin \frac{1}{2}(200° - 76°)$

30. $\sin 15° \cos 29° = \frac{1}{2}[\sin (15° + 29°) + \sin (15° - 29°)]$

Key Topics for Chapter 7

Define and/or discuss each of the following.

Cosine of Sum and Difference

Sine of Sum and Difference

Tangent of Sum and Difference

Double Angle Formulas

Half Angle Formulas

Product Formulas

Sum and Difference Formulas

Review Exercises for Chapter 7

1. Find $\sin (A + B)$ if $\cos A = \frac{3}{5}$ and $\sin B = \frac{2}{3}$, A in quadrant I and B in quadrant II.

2. Find $\cos (A + B)$ if $\sin A = \frac{1}{4}$, $\tan B = 1$, A and B in quadrant I.

3. Find $\tan 2A$ if $\cos A = -\frac{4}{5}$ in quadrant II.

4. Find $\sin 2A$ if $\tan A = \frac{1}{2}$ in quadrant III.

In Exercises 5–20 verify the identities.

5. $\cos \left(x - \dfrac{\pi}{6} \right) = \frac{1}{2}(\cos x + \sqrt{3} \sin x)$

6. $\sin (2x + 1) = \sin 2x \cos 1 + \cos 2x \sin 1$

7. $\tan (2x + \frac{1}{4}\pi) = \dfrac{1 + \tan 2x}{1 - \tan 2x}$

8. $2 \cot 2x = \cot x - \tan x$

9. $\csc x = \frac{1}{2} \csc \frac{1}{2}x \sec \frac{1}{2}x$

10. $\dfrac{\cos 2x}{\cos x} = \cos x - \tan x \sin x$

11. $\cos (1 + h) - \cos 1 = -2 \sin \frac{1}{2}(2 + h) \sin \frac{1}{2}h$

12. $\sin 4x - \sin 2x = 2 \sin x \cos 3x$

13. $\dfrac{\cos 2x - \cos 4x}{\sin 2x + \sin 4x} = \tan x$

14. $\dfrac{\cos 3t}{\cos t} = 1 - 4 \sin^2 t$

15. $\dfrac{\sin 3\theta}{\sin \theta} = 3 - 4 \sin^2 \theta$

16. $\dfrac{\sin x + \sin 3x}{\cos x + \cos 3x} = \tan 2x$

17. $\sin 4a \sin 2a + \sin^2 a = \frac{1}{2}(1 - \cos 6a)$

18. $\dfrac{\sin^3 x - \cos^3 x}{\sin x - \cos x} = 1 + \frac{1}{2} \sin 2x$

19. $\cos 4x = 4 \cos 2x - 3 + 8 \sin^4 x$

20. $\sin \frac{1}{2}A = \dfrac{\sec A - 1}{2 \sin \frac{1}{2}A \sec A}$

In Exercises 21–25 write each of the functions as a cosine and indicate the amplitude, period and phase shift.

21. $f(x) = 5 \cos 2x - 12 \sin 2x$

22. $f(x) = 3 \sin x - 4 \cos x$

23. $f(x) = \cos 3x - \sin 3x$

24. $f(x) = \sin 3x \cos 3x$

25. $f(x) = \sin^2 \frac{3}{2}x - \cos^2 \frac{3}{2}x$

Test 1 for Chapter 7

In Exercises 1–10 answer *true* or *false*.

1. $\cos 2x = 2 \cos x$

2. $\sin(x+y) = \sin x \cos y - \sin y \cos x$

3. $\tan(x+y) = \tan x + \tan y$

4. $\cos^2 5 - \sin^2 5 = \cos 10$

5. $|\sin 2x \cos 2x| \leq \frac{1}{2}$

6. $\sin x = \cos(x + \frac{1}{2}\pi)$

7. $\tan 30° = \frac{1}{2} \tan 60°$

8. $\sin xy = (\sin x)(\sin y)$

9. The graph of $y = (\cos x)(\sin x)$ is bounded by $-\frac{1}{2}$ and $\frac{1}{2}$.

10. $\cos 2\theta = 2 \cos^2 \theta - 1$.

11. Verify the identity $\sin 3x = \sin x \, (3 \cos^2 x - \sin^2 x)$.

12. Express as a sine function with a phase shift: $y = \sin \frac{1}{2}x + 2 \cos \frac{1}{2}x$.

13. If $\sin 2\theta = \frac{3}{5}$ and θ is in quadrant I, find $\sin \theta$.

14. Determine the exact value of $\cos \frac{1}{8}\pi$.

15. Verify the identity $\cos 2x = \dfrac{1 - \tan^2 x}{1 + \tan^2 x}$

Test 2 for Chapter 7

1. Compute $\sin 195°$ from the functions of $60°$ and $135°$.

2. Solve $\sin 2x = -2 \sin x, 0 \leq x < 2\pi$.

3. Solve $4 \sin x - \cos 2x - 5 = 0, 0 \leq x < 2\pi$.

In Exercises 4–7 verify the identities.

4. $\tan 3x \csc 3x = \sec 3x$

5. $\sec 2x = \dfrac{\sec^2 x}{1 - \tan^2 x}$

6. $\cot 2\phi + \csc 2\phi = \cot \phi$

7. $\tan \frac{1}{2}\theta = \csc \theta - \cot \theta$

8. Sketch the graph of $y = \sqrt{3} \cos 2x + \sin 2x$. What is the amplitude, period and phase shift?

9. Find $\sin(A+B)$ if $\cos A = \dfrac{5}{13}$, $\sin B = \dfrac{-\sqrt{3}}{2}$, A and B in quadrant IV.

10. Find $\sin 2\theta$ if $\tan \theta = -\frac{3}{4}$ in quadrant II.

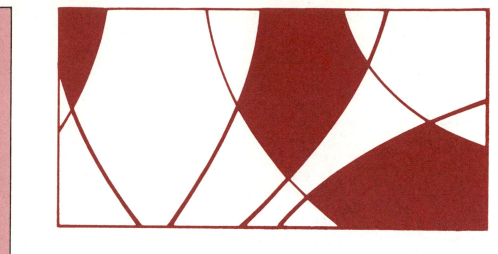

8

The Inverse Trigonometric Functions

8.1   Relations, Functions, and Inverses

The concept of a function, as explained in Chapter 1, means that for a given domain element there is one and only one range element. In many instances you were given a function in the form $y = f(x)$ and asked to determine the value of y corresponding to a given x. Sometimes, instead of computing values of the function for various domain elements, we are interested in computing the value of the domain element that corresponds to a given value of the function. For example, suppose the velocity (ft/sec) of a car varies with time (sec) according to $v = 10 + 4t$ and we would like to know how long it will take the car to reach a velocity of 70 ft/sec. Solving the given formula for t, we get $t = (v - 10)/4$. Substituting $v = 70$ into this expression yields $t = 15$. Therefore, it takes the car 15 seconds to reach a velocity of 70 ft/sec. Notice that the two formulas $v = 10 + 4t$ and $t = (v - 10)/4$ represent the same pairings of t and v, only in reverse order. Thus, the formula $v = 10 + 4t$ suggests the pairing $(15, 70)$ and $t = (v - 10)/4$ suggests the pairing $(70, 15)$. The process of interchanging the numbers in a functional pairing underlies an important mathematical concept.

Under certain conditions a new function may be obtained from a given function f by merely interchanging the numbers in the functional pairings $(x, f(x))$. A function obtained in this manner is called the **inverse function** of the given function and is denoted by f^{-1}. (The notation f^{-1} does not represent a negative exponent, it is simply a symbol for the inverse function of f.)

215

DEFINITION

> If f^{-1} is the inverse function of f, then by $y = f^{-1}(x)$ is meant $x = f(y)$.

EXAMPLE 1   The function $f = \{(2, 1), (-3, 2), (0, 5)\}$ has the inverse $f^{-1} = \{(1, 2), (2, -3), (5, 0)\}$.   ■

As the next example shows, interchanging the first and second elements in a functional pairing can lead to a set which does not represent a function.

EXAMPLE 2   If the function g is given by $g = \{(2, 7), (4, 6), (5, 7), (6, 9)\}$, then interchanging the first and second elements in each pair yields $\{(7, 2), (6, 4), (7, 5), (9, 6)\}$. This set of pairings is not a function since the domain element 7 has two different range elements, namely, 2 and 5.   ■

The algebraic procedure for finding the inverse function of a function given by the formula of $y = f(x)$ is summarized as follows:

(1)   Interchange the x and y variables.

(2)   Solve the new equation for the y variable (the resulting expression is $f^{-1}(x)$.)

EXAMPLE 3   If $f(x) = 2x + 4$, find $f^{-1}(x)$.

SOLUTION   Let $y = 2x + 4$ and then interchange x and y to get

$$x = 2y + 4$$

Solving this equation for y, we have

$$2y = x - 4$$
$$y = \tfrac{1}{2}x - 2$$

Substituting $f^{-1}(x)$ for y, the inverse is

$$f^{-1}(x) = \tfrac{1}{2}x - 2$$ ■

EXAMPLE 4   Let $f(x) = (2x + 1)/(x + 3)$. Find the inverse of f.

SOLUTION   Let $y = (2x + 1)/(x + 3)$. Then by interchanging the x and y variables,

$$x = \frac{2y + 1}{y + 3}$$

Solving for y,

$$xy + 3x = 2y + 1$$
$$xy - 2y = 1 - 3x$$

$$y(x-2) = 1 - 3x$$

$$y = \frac{1-3x}{x-2}$$

Hence, the inverse of f is

$$f^{-1}(x) = \frac{1-3x}{x-2}$$

∎

Notice that the domain of the inverse function in Example 4 is all real numbers except $x = 2$. Restrictions of this type should always be specified when computing inverse functions. Further, the domain of f is the range of f^{-1} and the range of f is the domain of f^{-1}. Thus the function and its inverse are obtained by interchanging the domain and range and keeping the same correspondence.

Scientific calculators have an [inv] *or* [arc] *button, but the operation of this button applies to only a few selected preprogrammed functions and will not perform the inverse operation in general.*

EXAMPLE 5   Consider the function defined by $y = x^2$. When interchanging the x and y variables we obtain $x = y^2$ and, when solving for y, $y = \pm\sqrt{x}$ which is not a function (Why not?)   ∎

If, as the preceding examples show, some functions have inverse functions while others do not, is it possible to determine this in advance? The answer to this is "Yes, we can." Specifically, a function has an inverse if its rule of correspondence does not assign the same range element to any two different domain elements. Functions that have this special property are called **one-to-one functions** and we make the following definition.

DEFINITION

> A function f is one-to-one if for a and b in the domain of f, $f(a) = f(b)$ implies that $a = b$.

Thus, the function

$$f = \{(2, 7), (4, 6), (5, 7), (6, 9)\}$$

is not one-to-one because $f(2) = 7 = f(5)$, but $2 \neq 5$.

Graphically, a one-to-one function is one for which *both* horizontal and vertical lines intersect the graph in, at most, one point. In Figure 8.1 the first function is one-to-one; the second is not.

As previously noted a function that is not one-to-one does not have an inverse function. In such cases we sometimes restrict the domain of the function to a specified interval in order to make a one-to-one function, thereby, allowing the possibility of an inverse function.

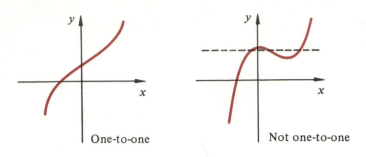

FIGURE 8.1   One-to-one   Not one-to-one

EXAMPLE 6   The function $y = x^2 - 1$ is not a one-to-one function because a horizontal line will intersect its graph in two places for all $y > -1$. See Figure 8.2(a). However, if we restrict the domain to the interval $x \geq 0$ the function is one-to-one on this interval and we compute its inverse function as follows: Interchanging x and y in the original function we get $x = y^2 - 1$. Solving for y, the inverse function is $y = \sqrt{x+1}$. Figure 8.2(b) shows the graph of the inverse function as a dashed line.

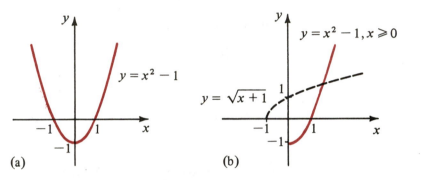

FIGURE 8.2   (a)   (b)

The graphs of f and f^{-1} have an interesting relationship. As the next two examples show, the graphs of f and f^{-1} are mirror reflections of each other in the line bisecting the first and third quadrants.

EXAMPLE 7   Draw the graphs of $f = \{(2, 1), (-3, 2), (0, 5)\}$ and $f^{-1} = \{(1, 2), (2, -3), (5, 0)\}$ on the same coordinate axes. (See Example 1.)

SOLUTION   By our definition of an inverse function, the second set of ordered pairs is the inverse of the first. The graphs of f and f^{-1} are shown in Figure 8.3. You can see that the points in the graphs are mirror reflections in the solid line.

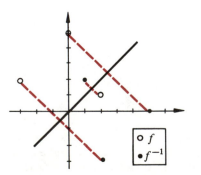

FIGURE 8.3

EXAMPLE 8   Draw the graphs of $y = 2x + 4$ and $y = \frac{1}{2}x - 2$ on the same coordinate axes.

SOLUTION   Several solution pairs for each equation are given in the respective tables and the graphs plotted in Figure 8.4. Notice that each graph is the mirror reflection in the dashed line of the other. The inverse nature of these functions was established in Example 3.

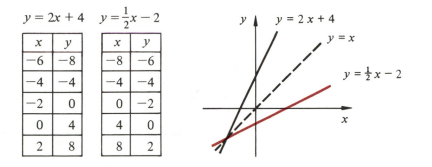

$y = 2x + 4$		$y = \frac{1}{2}x - 2$	
x	y	x	y
-6	-8	-8	-6
-4	-4	-4	-4
-2	0	0	-2
0	4	4	0
2	8	8	2

FIGURE 8.4

Exercises for Section 8.1

1. If $f(x) = 2x + 3$, for which x is $f(x) = 6$?   2. If $f(x) = 4/x$, for which x is $f(x) = 3$?

3. If $v = 10 + 4t$, when is $v = 20$?

4. If $f = \{(3, 0), (2, 6), (1, 5)\}$, for which x is $f(x) = 5$? For which x is $f(x) = 0$?

5. If $f = \{(0, 1), (2, 5), (3, 8), (5, 6)\}$, for which x is $f(x) = 8$? For which x is $f(x) = 3$?

Determine the inverse function in Exercises 6–21 *if it exists.*

6. $\{(2, 6), (3, 5), (0, 4)\}$ 7. $\{(3, 7), (5, 9), (7, 3), (9, 5)\}$

8. $\{(-1, 2), (2, 3), (6, -2)\}$ 9. $\{(1, 2), (2, 2)\}$

10. $\{(3, 2), (5, 4), (7, 2), (9, 8)\}$ 11. $\{(-2, 3), (-1, 4), (0, 0)\}$

12. $\{(0, 1), (1, 0)\}$ 13. $\{(0, 3), (1, 5), (2, 3), (6, 7)\}$

14. $f(x) = 3x + 2$ 15. $f(x) = x - 3$

16. $f(x) = \frac{1}{2}x + 5$ 17. $f(x) = \dfrac{1}{x + 1}$

18. $f(x) = x^2 + 2x$ 19. $f(x) = 2 - 3x^2$

20. $f(x) = \dfrac{2x - 1}{3x + 5}$ 21. $f(x) = \dfrac{x - 1}{x + 1}$

22. The function $f(x) = x^2 - 4x$ is not one-to-one and hence does not have an inverse. Show how to define an inverse on a restricted domain.

23. Which of the functions whose graphs are shown in the following figures have inverses? Sketch the graph of the inverse for those that do.

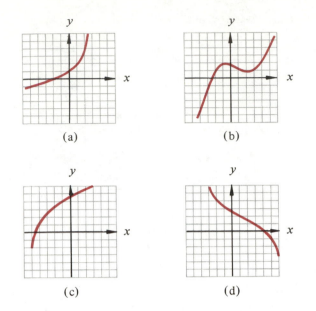

(a)   (b)

(c)   (d)

Determine graphically which of the pairs of functions in Exercises 24–31 are inverses of one another.

24. $y = 3x, y = \frac{1}{3}x$

25. $y = -2x, y = -\frac{1}{2}x$

26. $y = 6 - 3x, y = -\frac{1}{3}x + 2$

27. $y = x + 1, y = x - 2$

28. $y = \frac{1}{2}x + 1, y = 2x - 1$

29. $y = 5(x + 1), y = \frac{1}{5}x - 1$

30. $y = x^2, y = \sqrt{x}$

31. $y = x^3, y = \sqrt[3]{x}$

8.2   **Inverse Trigonometric Functions**

In Chapter 5 you learned that the function $y = \sin x$ can be considered as a function whose domain is the set of all real numbers and whose range is the interval $[-1, 1]$. By its functional nature, each value of x will give only one value of y. If, however, we wish to determine a domain element for a given functional value, the problem becomes ambiguous. For example, we might want the value of x for which $\sin x = \frac{1}{2}$. Unfortunately, there are infinitely many values, some of which are shown in Figure 8.5.

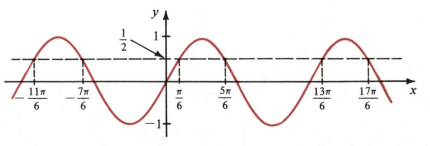

FIGURE 8.5

Solutions to

$\sin x = \frac{1}{2}$

While it is important to understand that $-\frac{7}{6}\pi$, $\frac{1}{6}\pi$, $\frac{5}{6}\pi$, $\frac{13}{6}\pi$ etc. all satisfy $\sin x = \frac{1}{2}$, you should be aware of the inherent difficulties that this presents. If you were asked to design a calculator to solve the equation $\sin x = \frac{1}{2}$, what value would you have the calculator display as the solution? (It takes a long time to display infinitely many solutions!) To avoid this ambiguity we restrict the domain of $\sin x$ to the interval $[-\frac{1}{2}\pi, \frac{1}{2}\pi]$. Then $x = \frac{1}{6}\pi$ is the only value of x for which $\sin x = \frac{1}{2}$. Note that on the interval $[-\frac{1}{2}\pi, \frac{1}{2}\pi]$ $\sin x$ is a one-to-one function.

The situation described for the sine function is true for each trigonometric function. We restrict the domain so that the domain and range values are paired in a one-to-one manner. The domain values to which a trigonometric function is so limited are called the **principal values** of the function. They are chosen near $x = 0$ to give a one-to-one pairing and are given in Table 8.1.

TABLE 8.1

Table of Principal Values of the Trigonometric Functions

$\sin x$,	$-\frac{1}{2}\pi \leq x \leq \frac{1}{2}\pi$	$\cot x$,	$0 < x < \pi$
$\cos x$,	$0 \leq x \leq \pi$	$\csc x$,	$-\frac{1}{2}\pi \leq x \leq \frac{1}{2}\pi, x \neq 0$
$\tan x$,	$-\frac{1}{2}\pi < x < \frac{1}{2}\pi$	$\sec x$,	$0 \leq x \leq \pi, x \neq \frac{1}{2}\pi$

On the domain of principal values, equations such as $\cos x = -\frac{1}{2}$, or $\sin x = -\frac{1}{2}$ have only one possible solution.

EXAMPLE 1   Find the principal value of x for which   (a) $\cos x = -\frac{1}{2}$;   (b) $\sin x = -\frac{1}{2}$.

SOLUTION   Figure 8.6(a) shows the cosine function, and 8.6(b) shows the sine function intersecting the line $y = -\frac{1}{2}$. In each case, the interval of principal values is in color so that you can see that only one value of x on that interval satisfies the equation. Thus, the principal value of x for which $\cos x = -\frac{1}{2}$ is $x = \frac{2}{3}\pi$. The principal value for x for which $\sin x = -\frac{1}{2}$ is for $x = -\frac{1}{6}\pi$.

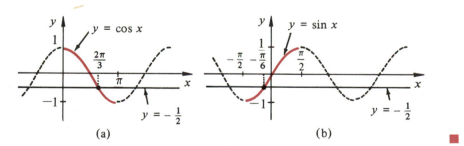

FIGURE 8.6   (a)   (b)

The restriction to the principal values permits the definition of inverse trigonometric, functions, that is functions obtained by pairwise interchanging the domain and range elements in the original function. In this text **we denote the inverse sine function by arcsin** and give the following definition.

DEFINITION

$y = \arcsin x$ if and only if $\sin y = x$,

$-1 \leq x \leq 1$, $-\frac{1}{2}\pi \leq y \leq \frac{1}{2}\pi$

The notation $y = \arcsin x$* is particularly descriptive of the inverse sine function since you can think of it as meaning "y is the arclength whose sine is x."

EXAMPLE 2   Evaluate $y = \arcsin (\sqrt{3}/2)$.

SOLUTION   By this is meant the principal value of the angle y for which $\sin y = \sqrt{3}/2$. Recalling the special angle values, we see that $y = \arcsin (\sqrt{3}/2) = \frac{1}{3}\pi$. Keep in mind the importance of the principal values. Thus, we can only say that $y = \arcsin (\sqrt{3}/2)$ is equivalent to $\sin y = \sqrt{3}/2$ if y is restricted to its principal values.   ■

The definition of each of the other five inverse trigonometric functions parallels that of the inverse sine function. Table 8.2 lists the domain and principal values for each of the inverse trigonometric functions.

TABLE 8.2
The Inverse Trigonometric Functions

Function	Domain	Principal Values
$y = \arcsin x$ $(x = \sin y)$	$-1 \leq x \leq 1$	$\frac{1}{2}\pi \leq \arcsin x \leq \frac{1}{2}\pi$
$y = \arccos x$ $(x = \cos y)$	$-1 \leq x \leq 1$	$0 \leq \arccos x \leq \pi$
$y = \arctan x$ $(x = \tan y)$	$-\infty < x < \infty$	$-\frac{1}{2}\pi \leq \arctan x \leq \frac{1}{2}\pi$
$y = \text{arccot } x$ $(x = \cot y)$	$-\infty < x < \infty$	$0 \leq \text{arccot } x \leq \pi$
$y = \text{arcsec } x$ $(x = \sec y)$	$x \leq -1$ or $x \geq 1$	$0 \leq \text{arcsec } x \leq \pi, \text{arcsec } x \neq \frac{1}{2}\pi$
$y = \text{arccsc } x$ $(x = \csc y)$	$x \leq -1$ or $x \geq 1$	$-\frac{1}{2}\pi \leq \text{arccsc } x \leq \frac{1}{2}\pi, \text{arccsc } x \neq 0$

Study Table 8.2 carefully so that you will know what limitations apply to the various inverse trigonometric functions. The information in this table is fundamental to the examples and exercises that follow.

Most scientific calculators have the capability of displaying the values of the inverse trigonometric functions, although not necessarily directly. For example, one model requires that you insert the number and then press the button $\boxed{\text{inv}}$ *and the desired trigonometric function in order. Some other models have a separate button for* $\sin^{-1}$, $\cos^{-1}$ *and* $\tan^{-1}$. *At this stage, if you haven't already, you should familiarize yourself with how your specific calculator functions.*

*Another notation that is used for the inverse to $y = \sin x$ is $y = \sin^{-1} x$. (read "the inverse sine of x"). This notation is somewhat confusing since it suggests taking the reciprocal of $\sin x$. For this reason, we will use arcsin to designate the inverse sine function.

EXAMPLE 3   Find arccos $(\frac{1}{3})$.

SOLUTION   We let $y = \arccos(\frac{1}{3})$. Then,

$$\frac{1}{3} = \cos y \quad \text{where } 0 \le y \le \pi$$

By use of Table C we find that $y = 1.23$.   ■

EXAMPLE 4   Find arctan (-1) and arccot (-1).

SOLUTION   Let $y = \arctan(-1)$ and $u = \text{arccot}(-1)$. Then,

$$-1 = \tan y, \qquad -\tfrac{1}{2}\pi < y < \tfrac{1}{2}\pi$$

and

$$-1 = \cot u, \quad 0 < u < \pi$$

Thus,

$$y = \arctan(-1) = -\tfrac{1}{4}\pi \quad \text{and} \quad u = \text{arccot}(-1) = \tfrac{3}{4}\pi$$ ■

Since the tangent and cotangent are reciprocal functions, you might have expected that arctan (-1) and arccot (-1) yield the same value. This example shows the necessity of adhering strictly to the definitions of the inverse functions, giving close attention to the principal values.

The following examples are concerned with taking a trigonometric function of some inverse trigonometric function. In these circumstances it helps to show the functional value by drawing a right triangle, always keeping track of the principal values.

EXAMPLE 5   Find sin $(\arccos \frac{1}{2})$.

SOLUTION   We first let $\theta = \arccos \frac{1}{2}$. Then, θ is the angle as shown in Figure 8.7 from which it is easy to see that

$$\sin(\arccos \tfrac{1}{2}) = \sin \theta = \sqrt{3}/2$$

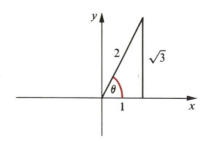

FIGURE 8.7   ■

EXAMPLE 6   Find cos(arcsin x)

SOLUTION   Letting $\theta = \arcsin x$, we want to find $\cos \theta$. Since the range values of arcsin x are $[-\tfrac{1}{2}\pi, \tfrac{1}{2}\pi]$ the angle θ will be one of the angles shown in Figure 8.8. In either case, $\cos \theta = \sqrt{1 - x^2}$.

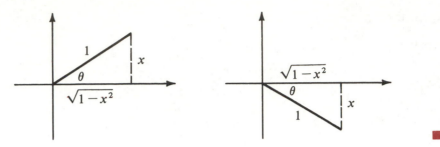

FIGURE 8.8

EXAMPLE 7   Find $\sin (\arccos \frac{1}{3} + \arctan (-2))$

SOLUTION   We let $\theta = \arccos \frac{1}{3}$ and $\phi = \arctan (-2)$. See Figure 8.9. Can you tell why θ is drawn as a first quadrant angle and ϕ as a fourth quadrant angle?

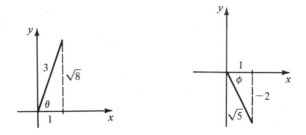

FIGURE 8.9

Then, since $\sin (\theta + \phi) = \sin \theta \cos \phi + \sin \phi \cos \theta$ we have

$$\sin (\arccos \tfrac{1}{3} + \arctan (-2)) = \frac{\sqrt{8}}{3}\left(\frac{1}{\sqrt{5}}\right) + \frac{-2}{\sqrt{5}}\left(\frac{1}{3}\right)$$

$$= \frac{-2 + \sqrt{8}}{3\sqrt{5}}$$

EXAMPLE 8   Verify the identity $\cos (2 \arccos x) = 2x^2 - 1$

SOLUTION   By letting $\theta = \arccos x$, we have

$$\cos (2 \arccos x) = \cos 2\theta = \cos^2 \theta - \sin^2 \theta$$

From Figure 8.10, we read off the values of $\cos \theta$ to be x and $\sin \theta$ to be $\sqrt{1 - x^2}$. Hence,

$$\cos (2 \arccos x) = x^2 - (1 - x^2) = 2x^2 - 1$$

which is what we wanted to show.

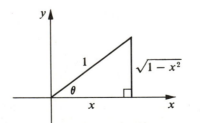

FIGURE 8.10

EXAMPLE 9 Find x if   (a) arccos $x = 0.578$

(b) arccos $x = 2\pi$

SOLUTION (a) If arccos $x = 0.578$, then $x = \cos 0.578$ if 0.578 is one of the principal values, that is, if 0.578 is between 0 and π. Since this is so,

$$x = \cos 0.578 = 0.8376.$$

(b) Observe that 2π is not in the set of principal values for the cosine function so that the equation

$$\text{arccos } x = 2\pi$$

has no solution. If you were not on the lookout for the principal values you might conclude that since arccos $x = 2\pi$, then $x = \cos 2\pi = 1$ which is incorrect.   ■

In passing, we note that

$$\sin (\arcsin x) = x$$

$$\cos (\arccos x) = x$$

and so forth, for all six trigonometric functions. This result is a direct consequence of the inverse nature of the functions involved. Also **if x is limited to the principal values** of the function,

$$\arcsin (\sin x) = x$$

$$\arccos (\cos x) = x$$

and so forth. For example, arcsin $(\sin \frac{1}{4}\pi) = \frac{1}{4}\pi$ since $\frac{1}{4}\pi$ is a principal value of the arcsin function. But, arcsin $(\sin \frac{5}{6}\pi) = \frac{1}{6}\pi$, since $\frac{5}{6}\pi$ is not a principal value.

Exercises for Section 8.2

Find the exact values of Exercises 1–25 without the use of tables or a calculator.

1. arcsin $\frac{1}{2}$

2. arcsin 1

3. arctan 1

4. arccos $(\sqrt{3}/2)$

5. arccos $(-\sqrt{3}/2)$

6. arcsin $(-\sqrt{2}/2)$

7. arcsec (-2)

8. arccot $(-\sqrt{3})$

9. arcsec 1

10. arccsc $\sqrt{2}$

11. arctan $(-\sqrt{3})$

12. sin $(\arccos (-\frac{3}{5}))$

13. cos $(\arcsin (-\frac{5}{13}))$

14. sin $(\arcsin 1)$

15. sin $(\arctan 2)$

16. sec $(\arccos \frac{1}{3})$

17. cos $(\arcsin \frac{1}{4})$

18. sin $(2 \arcsin \frac{1}{3})$

19. cos $(2 \arcsin \frac{1}{4})$

20. cos$(\arccos (-\frac{1}{3}) - \arcsin (-\frac{1}{3}))$

21. sin $(\arctan 1 - \arctan 0.8)$

22. cos $(\arcsin \frac{1}{4} + \arccos (-\frac{1}{3}))$

23. tan $(\arcsin (-\frac{1}{2}) - \arctan (-2))$

24. cos $(\arctan 2 - \arccos \frac{1}{2})$

25. tan $(2 \arctan 2)$

In Exercises 26–30 simplify the given expression.

26. $\sin (\arccos x^2)$

27. $\tan (\arcsin x)$

28. $\tan (2 \arccos x)$

29. $\sin (\arcsin y + \arcsin x)$

30. $\cos (\arccos x + \arcsin y)$

In Exercises 31–40, solve for x or show that there is no solution.

31. $\arccos x = 0.241$

32. $\arcsin x = -0.314$

33. $\arccos x = -0.5$

34. $\arcsin x = -\pi$

35. $\arctan x = 1.2$

36. $\arctan x = 2.43$

37. $\arctan x = -1.34$

38. $\arccos x = \frac{3}{4}\pi$

39. $\arcsin x = 0.8947$

40. $\arccos x = 2.815$

In Exercises 41–45, verify the given identity.

41. $\cos (2 \arcsin x) = 1 - 2x^2$

42. $\tan (\arctan x + \arctan 1) = \dfrac{1 + x}{1 - x}$

43. $\tan (2 \arctan x) = \dfrac{2x}{1 - x^2}$

44. $\sin (3 \arcsin \theta) = 3\theta - 4\theta^3$

45. $\cos (2 \arccos y) = 2y^2 - 1$

46. Show that the inverse sine function does not have the linearity property by showing that $\arcsin 2x \neq 2 \arcsin x$.

47. Show that the inverse cosine function does not have the linearity property by showing that $\arccos x + \arccos y \neq \arccos (x + y)$.

Sketch the graph of:

48. $\arcsin (\sin x)$

49. $\arcsin (\cos x)$

50. A picture u ft high is placed on a wall with its base v ft above the level of the observer's eye. If the observer stands x ft from the wall, show that the angle of vision α subtended by the picture is given by

$$\alpha = \text{arccot} \frac{x}{u + v} - \text{arccot} \frac{x}{v}$$

51. Explain carefully how to use your calculator to find the values of $\text{arccot } x$, $\text{arcsec } x$, and $\text{arccsc } x$ assuming that your calculator can find $\arcsin x$, $\arccos x$, and $\arctan x$.

Use your calculator or Table B to determine the values in Exercises 52–61. Give answer in radians or real numbers.

52. $\arcsin 0.7863$

53. $\arctan 2.659$

54. arccos 0.3547   **55.** arcsec 5.78

56. arccot 2.76   **57.** arcsin 0.9866

58. arccos 0.9034   **59.** arccot 0.8966

8.3   Graphs of the Inverse Trigonometric Functions

The graphs of the six inverse trigonometric functions are found by direct appeal to the definition and by a knowledge of the graphs of the trigonometric functions. For example: $y = \arcsin x$ if, and only if, $x = \sin y$ where $-\frac{1}{2}\pi \le y \le \frac{1}{2}\pi$. It follows that $y = \arcsin x$ looks like a piece of the relation $x = \sin y$. In Figure 8.11, we see such a graph.

The other parts of Figure 8.11 show the graphs of the remaining inverse functions. In each case, you can think of the graph of the original function wrapped around the y axis and then consider the portion that corresponds to the principal values.

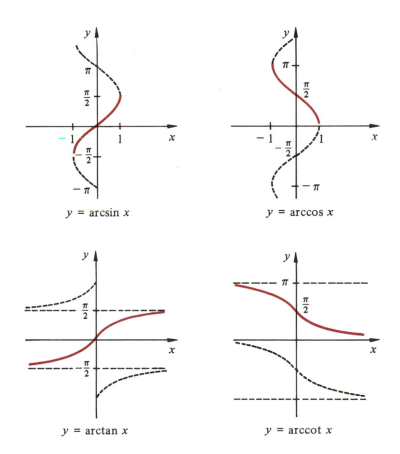

$y = \arcsin x$ $y = \arccos x$

$y = \arctan x$ $y = \text{arccot } x$

FIGURE 8.11

The inverse trigonometric functions

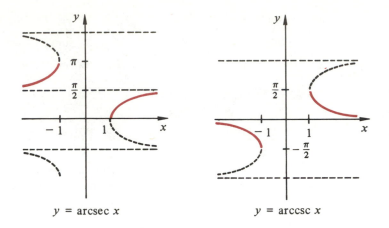

**FIGURE 8.11
(continued)**

$y = \text{arcsec } x$ $y = \text{arccsc } x$

Modifying the Inverse Trigonometric Functions

The graphs of the inverse trigonometric functions can be modified by adding or multiplying by certain constants much in the same way the trigonometric functions are modified. The process of graphing the inverse trigonometric functions can be facilitated by observing the affects of constants on the shape and location of the graphs. We will explain the modifications in terms of arcsin x with the understanding that the results apply to the other inverse trigonometric functions.

(1) Multiplication of the Argument by a Constant

The function $y = \text{arcsin } Ax$ is equivalent to $\sin y = Ax$ or $(1/A) \sin y = x$. We interpret this to mean that the domain of $y = \text{arcsin } Ax$ is $1/A$ times the domain of $y = \text{arcsin } x$. Thus, **the constant A expands or contracts the domain of the inverse function**. The principal values remain unaltered.

EXAMPLE 1   Sketch the graph of arcsin $2x$.

SOLUTION   In this case, $A = 2$, so we multiply the domain elements by $\frac{1}{2}$. Thus, the domain of arcsin $2x$ is $-\frac{1}{2} \leq x \leq \frac{1}{2}$. Letting $y = \text{arcsin } 2x$ the graph is shown in Figure 8.12

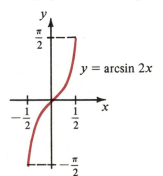

$y = \text{arcsin } 2x$

FIGURE 8.12

(2) Multiplication of the Function by a Constant

Consider the multiplication of arcsin x by a constant B. In this case, the function $y = B \arcsin x$ is equivalent to $\sin y/B = x$. The principal values for $\sin y/B$ can be written $-\frac{1}{2}\pi \leq y/B \leq \frac{1}{2}\pi$ or multiplying each member of the inequality by B as $-\frac{1}{2}\pi B \leq y \leq \frac{1}{2}\pi B$. We conclude from this that **the constant B alters the principal value interval**, leaving the domain unchanged. To find the principal value interval of arcsin Bx simply multiply the endpoints by B.

EXAMPLE 2   Sketch the graph of $2 \arccos x$.

SOLUTION   Let $y = 2 \arccos x$. Since the principal value interval for $y = \arccos x$ is $0 \leq y \leq \pi$, it follows that the principal value interval for $y = 2 \arccos x$ is $0 \leq y \leq 2\pi$. See Figure 8.13.

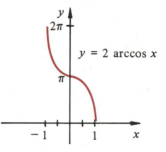

FIGURE 8.13

(3) The Addition of a Constant to the Function

The addition of a constant C to arcsin x can be represented by $y = C + \arcsin x$. Thus, the graph of $y = C + \arcsin x$ is the same as $y = \arcsin x$ only translated C units up or down. **If C is positive the graph moves up C units, if it is negative the graph moves down C units**.

EXAMPLE 3   Sketch the graph of $y = \frac{1}{4}\pi + \arcsin x$.

SOLUTION   The graph of this function is just the graph of $y = \arcsin x$ translated up $\frac{1}{4}\pi$ units. Therefore, the principal value interval for $y = \frac{1}{4}\pi + \arcsin x$ is $-\frac{1}{4}\pi \leq y \leq \frac{3}{4}\pi$. See Figure 8.14.

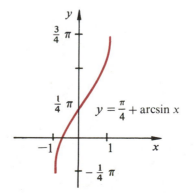

FIGURE 8.14

(4) The Addition of a Constant to the Argument

The function $y = \arcsin(x + D)$ is equivalent to $\sin y = x + D$ which means that the graph will be moved left or right D units. The translation will be D units to the left if D is positive and D units to the right if D is negative. More generally the graph of $y = \arcsin(Ax + D)$ is translated D/A units left or right.

EXAMPLE 4   Sketch the graph of $y = \arccos(x - 1)$.

SOLUTION   Note that the graph is the same as $y = \arccos x$ only translated 1 unit to the right. The graph is shown in Figure 8.15.

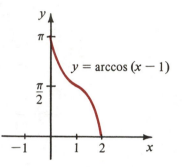

FIGURE 8.15

■

The following example considers the different translations, contractions, and expansions acting simultaneously.

EXAMPLE 5   Sketch the graph of $y = \frac{1}{2}\pi + 3 \arcsin(2x - 1)$.

SOLUTION   The constant $\frac{1}{2}\pi$ causes a vertical upward translation and the constant -1 causes a translation $\frac{1}{2}$ unit to the right of the function $f(x) = 3 \arcsin 2x$. The range of $3 \arcsin 2x$ is

$$3(-\tfrac{1}{2}\pi) \leq y \leq 3(\tfrac{1}{2}\pi)$$

and its domain is

$$-1 \leq 2x \leq 1, \text{ that is } -\tfrac{1}{2} \leq x \leq \tfrac{1}{2}.$$

Figure 8.16 shows the sketch of $f(x)$ and then of the given function as the translated form of $f(x)$.
Notice that the graph crosses the x-axis where $y = 0$, that is, where $\frac{1}{2}\pi + 3 \arcsin(2x - 1) = 0$. Solving this equation, we have

$$\arcsin(2x - 1) = -\tfrac{1}{6}\pi$$

from which

$$2x - 1 = \sin(-\tfrac{1}{6}\pi) = -0.5$$

$$x = 0.25$$

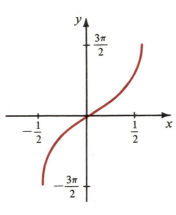 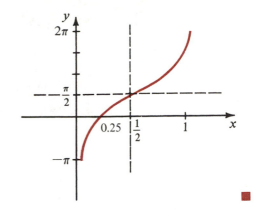

FIGURE 8.16

Exercises for Section 8.3

Sketch the graphs in Exercises 1–22.

1. $y = \arcsin 2x$ **2.** $y = \arccos 2x$ **3.** $y = 2 \arccos \frac{1}{2} x$

4. $y = \frac{1}{2} \arcsin \frac{1}{2} x$ **5.** $y = \arctan 3x$ **6.** $y = \arctan x + \frac{1}{2}\pi$

7. $y = 2 \arcsin x$ **8.** $y = \frac{1}{4} \arccos x$ **9.** $y = \arcsin x + \pi$

10. $y = 3 \arctan 2x$ **11.** $y = \frac{1}{2} \arcsin x + \frac{1}{2}\pi$ **12.** $y = \pi + 2 \arcsin \frac{1}{2} x$

13. $y = -\arcsin x$ **14.** $y = -2 \arccos x$ **15.** $y = \arcsin (x + 1)$

16. $y = \arccos (x + 1)$ **17.** $y = \arctan (x - 1)$ **18.** $y = \arctan (x + 1)$

19. $y = \arcsin (x + \frac{1}{2})$ **20.** $y = \arccos (x - \frac{1}{2}\sqrt{3})$ **21.** $y = \arcsin (2x - 1)$

22. $y = \arccos (2x - \frac{1}{2}\sqrt{3})$

Key Topics for Chapter 8

Define and/or discuss each of the following.

Inverse Functions in General

Principal Values of Trigonometric Functions

Definitions of Inverse Trigonometric Functions

Graphs of Inverse Trigonometric Functions

Trigonometric Functions of Sum of Inverse Functions

Graphs of Inverse Trigonometric Functions translated, expanded and contracted

Review Exercises for Chapter 8

1. If $y = 3x - 2$, for which value of x is $y = 10$?

2. If $y = \dfrac{2x+1}{x-1}$, for which value of x is $y = 3$?

3. Determine the function inverse to $y = 2x + 5$.

4. Determine the function inverse to $y = \dfrac{2x+5}{3x-1}$.

5. Determine the function inverse to $y = \sin(2x - 5) + 3$. What restrictions must you place on x?

In Exercises 6–15, find the exact value of the given expression.

6. $\arcsin(-1/2)$

7. $\arctan\sqrt{3}$

8. $\sin(\arccos(1/3))$

9. $\tan(\arcsin(-0.4))$

10. $\cos(2\arccos(0.3))$

11. $\sin(2\arcsin(-0.6))$

12. $\sin(\arctan(2) - \arcsin(1/3))$

13. $\cos(\arccos(1/4) + \arccos(-1/3))$

14. $\tan(\arctan(1) - \text{arccot}(-2))$

15. $\sin(\arccos(-1/3) + \arcsin(-1/3))$

Sketch the graphs in Exercises 16–25.

16. $y = \arcsin 3x$

17. $y = \frac{1}{2}\pi + \arcsin x$

18. $y = \arccos 2x - \frac{1}{4}\pi$

19. $y = \arccos(2x - 1)$

20. $y = 2\arctan x$

21. $y = \arctan 2x + \pi$

22. $y = \arcsin(x - \frac{1}{2}\sqrt{2})$

23. $y = -\arctan 2x$

24. $y = 2\arctan(-x + 1)$

25. $y = -3\arccos(-2x + 1)$

In Exercises 26–27 verify the given equality, for x and y in the domain of the arcsin function.

26. $\sin(2\arcsin x) = 2x(1 - x^2)^{1/2}$

27. $\sin(\arcsin x + \arcsin y) = x(1 - y^2)^{1/2} + y(1 - x^2)^{1/2}$

Test 1 for Chapter 8

Answer *true* or *false* in Exercises 1–10.

1. The inverse of $\arcsin x$ is $\dfrac{1}{\sin x}$.

2. $\tan x (\arctan x) = x$.

3. $y = 3\arccos x$ is equivalent to $\cos\frac{1}{3}y = x$.

4. The principal value interval for $\arctan x$ is $[-1, 1]$.

5. The domain of $\text{arcsec } x$ is $[-1, 1]$.

6. If $\tan x = 1$, the principal value of x is $\frac{1}{4}\pi$.

7. If $\cos x = 0$, the principal value of x is $-\frac{1}{2}\pi$.

8. $\sin (\arcsin x) = x$

9. $\arcsin (\sin x) = x$

10. The inverse sine function and inverse cosine function have the same domain.

11. Evaluate $\cos (2 \arctan 2)$.

12. Evaluate $\sin (\arctan 3 - \arccos \frac{1}{3})$.

13. Evaluate $\cos (\arctan \frac{1}{2})$.

14. Evaluate $\arcsin (\sin \frac{5}{4}\pi)$.

15. Make a sketch of the graph of the function $y = \arcsin 2x$.

16. Sketch the graph of $y = 3 \arccos (x + 2)$.

Test 2 for Chapter 8

1. What is the domain of $y = 3 \arcsin 2x$.

2. Evaluate $\sin (\arccos \sqrt{3}/2)$.

3. Evaluate $\sin (2 \arcsin \frac{1}{2})$.

4. Evaluate $\sin (\arctan 2 - \arcsin \frac{1}{3})$.

5. Determine the function inverse to $y = \dfrac{2x - 1}{x + 5}$.

6. Sketch the graph of $y = \arctan 2x$.

7. Sketch the graph of $y = \frac{1}{2}\pi + \arcsin x$.

8. Sketch the graph of $y = \frac{1}{3} \arccos 2x$.

Complex Numbers

The Need for Complex Numbers

We sometimes think of numbers as the invention of the human mind because they may be developed on the basis of obtaining solutions to certain types of equations. For example, beginning with the set of positive integers (counting numbers), the negative integers were invented in order to be able to solve an equation like $x + 7 = 4$. Likewise, the set of rational numbers was invented so that linear equations such as $2x = 3$ would have a solution. In order to solve $x^2 = 2$, it was necessary to invent the irrational numbers $\pm \sqrt{2}$. The irrational numbers together with the rational numbers comprise the set of real numbers.

For most applications, the set of real numbers is sufficient, but there are instances in which this set is inadequate. For instance, in solving the equation $x^2 + 1 = 0$, we obtain the root $x = \sqrt{-1}$. Since the square of every real number is non-negative, it is apparent that $\sqrt{-1}$ is not a real number. If we use i to represent $\sqrt{-1}$, then i is a number that has the property that $i^2 = -1$. With this understanding, we can write the roots of $x^2 + 1 = 0$ as $x = \pm i$. The number i is called a **pure imaginary*** number and, in general, is the square root of -1. Thus,

$$\sqrt{-4} = 2i$$
$$\sqrt{-7} = i\sqrt{7}$$

Numbers of the form bi, where b is a real number, make up the set of imaginary numbers.

---

*The word "imaginary" is, in a sense, an unfortunate choice of words since it may lead you to believe that they have a more "fictitious" character than the so-called "real" numbers.

EXAMPLE 1   Solve the equation $x^2 + 9 = 0$.

SOLUTION $x^2 + 9 = 0$

$$x^2 = -9$$

$$x = \pm\sqrt{-9} = \pm 3i$$ ■

In solving the equation $x^2 - 2x + 5 = 0$, we find the solution to be $x = 1 \pm \sqrt{-4}$. Using the concept of an imaginary number, this can be written $x = 1 \pm 2i$. Thus, we have a number that is a combination of a real number and an imaginary number; such numbers are called **complex numbers**. For convenience, we make the following definition:

DEFINITION 9.1

> A complex number z is any number of the form $z = a + bi$, where a and b are real numbers and $i = \sqrt{-1}$.

The real number a is called the **real part** of z while the real number b is called the **imaginary part** of z. By convention, if $b = 1$, the number is written $a + i$. Further, if $b = 0$, the imaginary part is customarily omitted and the number is said to be pure real. If $a = 0$ and $b \neq 0$, the real part is omitted and the number is said to be pure imaginary.

Two complex numbers are equal if, and only if, their real parts are equal and their imaginary parts are equal. Thus, $a + bi$ and $c + di$ are equal if, and only if, $a = c$ and $b = d$.

Combinations of complex numbers may be assumed to obey the ordinary algebraic rules for real numbers. Thus the sum, difference, product, and quotient of two complex numbers are found in the same manner as the sum, difference, product, and quotient of two real binomials — bearing in mind that $i^2 = -1$.

EXAMPLE 2   Find the sum and difference of $3 + 5i$ and $-9 + 2i$.

SOLUTION   (a) $(3 + 5i) + (-9 + 2i) = (3 - 9) + (5 + 2)i = -6 + 7i$

(b) $(3 + 5i) - (-9 + 2i) = (3 + 9) + (5 - 2)i = 12 + 3i$ ■

EXAMPLE 3   Find the product $(3 - 2i)(4 + i)$.

SOLUTION $(3 - 2i)(4 + i) = 12 + 3i - 8i - 2i^2$

$$= 12 - 5i + 2$$

$$= 14 - 5i$$ ■

The number $a - bi$ is called the **conjugate** of $a + bi$. To find the quotient of the two complex numbers, we use the following scheme: **multiply numerator and** denominator of the given quotient by the conjugate of the denominator. The technique is illustrated in the next example.

EXAMPLE 4   Find the quotient $\dfrac{2 + 3i}{4 - 5i}$.

SOLUTION

$$\frac{2 + 3i}{4 - 5i} = \frac{(2 + 3i)(4 + 5i)}{(4 - 5i)(4 + 5i)}$$

$$= \frac{8 + (12 + 10)i + 15i^2}{16 - 25i^2}$$

$$= \frac{-7 + 22i}{16 + 25}$$

$$= \frac{-7 + 22i}{41}$$ ∎

Exercises for Section 9.1

Perform the operations indicated in Exercises 1–24, expressing all answers in the form $a + bi$.

1. $(3 + 2i) + (4 + 3i)$

2. $(6 + 3i) + (5 - i)$

3. $(5 - 2i) + (-7 + 5i)$

4. $(-1 + i) + (2 - i)$

5. $(1 + i) + (3 - i)$

6. $7 - (5 + 3i)$

7. $(3 + 5i) - 4i$

8. $(3 + 2i) + (3 - 2i)$

9. $(2 + 3i)(4 + 5i)$

10. $(7 + 2i)(-1 - i)$

11. $(5 - i)(5 + i)$

12. $(6 - 3i)(6 + 3i)$

13. $(4 + \sqrt{3}i)^2$

14. $(5 - 2i)^2$

15. $6i(4 - 3i)$

16. $3i(-2 - i)$

17. $\dfrac{3 + 2i}{1 + i}$

18. $\dfrac{4i}{2 + i}$

19. $\dfrac{3}{2 - 3i}$

20. $\dfrac{7 - 2i}{6 - 5i}$

21. $\dfrac{1}{5i}$

22. $\dfrac{-3 + i}{-2 - i}$

23. $\dfrac{-1 - 3i}{4 - \sqrt{2}i}$

24. $\dfrac{i}{2 + \sqrt{5}i}$

25. Show that the sum of a complex number and its conjugate is a real number.

26. Show that the product of a complex number and its conjugate is a real number.

9.2   Graphical Representation of Complex Numbers

Since complex numbers are ordered pairs of real numbers, some two-dimensional configuration is necessary to represent them graphically. The Cartesian coordinate system is often used for this purpose, in which case, it is called the **complex plane**. The x-axis is used to represent the real part of the complex number, and the y-axis to represent the imaginary part. Hence, they are called the real and imaginary axes respectively. Thus, the complex number $x + iy$ is represented by the point whose coordinates are (x, y) as shown in Figure 9.1. For this reason, the complex number $z = x + iy$ is said to be written in **rectangular form**.

It is often convenient to think of a complex number $x + iy$ as representing a vector. With reference to Figure 9.1, the complex number $x + iy$ can be represented by the vector drawn from the origin to the point $x + iy$ with coordinates (x, y).

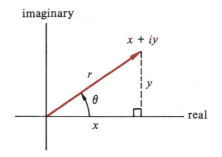

FIGURE 9.1

EXAMPLE 1   Represent $5 + 3i$, $-2 + 4i$, $-1 - 3i$, and $5 - i$ in the complex plane. (See Figure 9.2.)

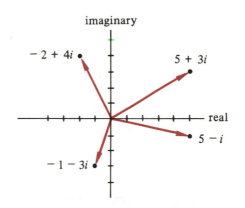

FIGURE 9.2

It is instructional to show the graphical representation of the sum of two complex numbers. Recalling that the sum of $a + bi$ and $c + di$ is given by

$$(a + bi) + (c + di) = (a + c) + (b + d)i$$

we have represented $a + bi$, $c + di$, and $(a + c) + (b + d)i$ in Figure 9.3. The result is the same as if we had applied the parallelogram law to the vectors representing $a + bi$ and

$c + di$. Note that $c + di$ is subtracted from $a + bi$ by plotting $a + bi$ and $-c - di$ and then using the parallelogram law.

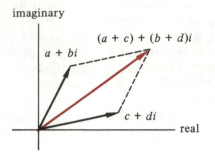

FIGURE 9.3
Addition of
Complex
Numbers

Exercises for Section 9.2

Perform the indicated operations in Exercises 1–16 graphically and check the results algebraically.

1. $(4 + i) + (3 + 5i)$

2. $(3 + 2i) + (1 - 3i)$

3. $(4 + 3i) + (-2 + i)$

4. $(-5 - 7i) + (-1 + 3i)$

5. $(2 - 4i) + (-3 + i)$

6. $i + (3 + 4i)$

7. $(5 + 3i) - 6$

8. $(3 + 2i) + (3 - 2i)$

9. $(5 - 3i) + (5 + 3i)$

10. $(6 + 4i) - 2i$

11. $(1 + 3i) - (2 - 5i)$

12. $(2 - i) - i$

13. $(2 + \sqrt{3}i) - (-1 - i)$

14. $(\sqrt{5} - i) - (\sqrt{5} + 3i)$

15. $(-3 + 2i) - (-3 - 2i)$

16. $(10 - 3i) - (10 + 3i)$

On the same coordinate system plot the number, its negative, and its conjugate. (Exercises 17–22.)

17. $-3 + 2i$

18. $4 - 3i$

19. $-2i$

20. $5 + i$

21. $-1 - i$

22. $3 + 5i$

9.3   Polar Coordinates

The rectangular coordinate system was used exclusively in the first eight chapters of this book. Another coordinate system widely used in science and mathematics is the so-called **polar** coordinate system. In this system, the position of a point is determined by specifying a distance from a given point and the direction from a given line. Actually, this is not a new concept; we frequently use this system to describe the relative location of geographic points. Thus, when we say that Cincinnati is about 300 miles southeast of Chicago, we are, in fact, using polar coordinates.

To establish a frame of reference for the polar coordinate system, we begin by choosing a point O and extending a line from this point. The point O is called the **pole**, and the extended line is called the **polar axis**. The position of any point P in the plane is then determined if we know the distance OP, and the angle AOP as indicated in Figure 9.4. The directed distance OP is called the **radius vector** of P and is denoted by r. The angle AOP is called the **vectorial angle** and is denoted by θ. The coordinates of a point P are then written as the ordered pair (r, θ). Notice that the radius vector is the first element and the vectorial angle is the second.

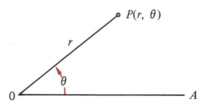

FIGURE 9.4

Polar coordinates, like rectangular coordinates, are regarded as signed quantities. When stating the polar coordinates of a point, it is customary to use the following sign conventions.

1.   The radius vector is positive when measured on the terminal side of the vectorial angle and is negative when measured in the opposite direction.

2.   The vectorial angle is positive when generated by a counter-clockwise rotation from the polar axis and negative when generated by a clockwise rotation.

The polar coordinates of a point determine the location of the point uniquely. However, the converse is not true, as we can see from Figure 9.5. Ignoring vectorial angles that are numerically greater than $360°$, we have four pairs of coordinates that yield the same point. Thus, the pairs $(5, 60°)$, $(5, -300°)$, $(-5, 240°)$, and $(-5, -120°)$ represent the same point.

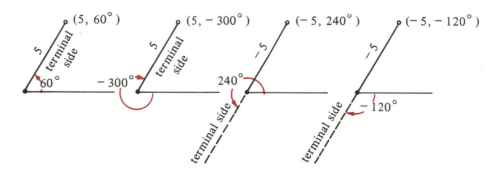

FIGURE 9.5

Polar coordinate paper is commercially available. As you can see in Figure 9.6, this paper consists of equally spaced concentric circles with radial lines extending at equal angles through the pole. While the use of polar coordinate paper is not mandatory, you will find that it is very helpful in plotting polar curves. Several points are plotted in Figure 9.6 for illustrative purposes.

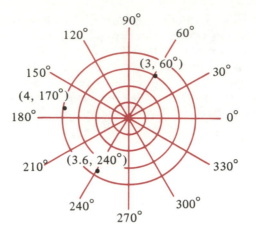

FIGURE 9.6

The relationship between the polar coordinates of a point and the rectangular coordinates of the same point can be found by superimposing the rectangular coordinate system on the polar coordinate system so that the origin corresponds to the pole and the positive x-axis to the polar axis. Under these circumstances, the point P shown in Figure 9.7 has both (x, y) and (r, θ) as coordinates. The desired relationship is then an immediate consequence of triangle OMP. Hence, the equations

(9.1) $x = r \cos \theta$

and

(9.2) $y = r \sin \theta$

can be used to transform a rectangular equation into a polar equation.

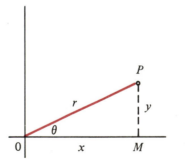

FIGURE 9.7

EXAMPLE 1   Find the polar equation of the circle whose rectangular equation is $x^2 + y^2 = a^2$.

SOLUTION   Substituting Equations (9.1) and (9.2) into the given equation, we have

$$r^2 \cos^2 \theta + r^2 \sin^2 \theta = a^2$$

$$r^2 (\cos^2 \theta + \sin^2 \theta) = a^2$$

$$r^2 = a^2$$

$$r = a$$

To make the transformation from polar coordinates into rectangular coordinates we use the following equations.

(9.3) $\quad r = \sqrt{x^2 + y^2}$

(9.4) $\quad \sin \theta = \dfrac{y}{\sqrt{x^2 + y^2}}$

(9.5) $\quad \cos \theta = \dfrac{x}{\sqrt{x^2 + y^2}}$

These equations are also derived from Figure 9.7.

EXAMPLE 2   Transform the following polar equation into a rectangular equation:

$$r = 1 - \cos \theta$$

SOLUTION   Substituting Equations (9.3) and (9.5) into the given equation, we have

$$\sqrt{x^2 + y^2} = 1 - \frac{x}{\sqrt{x^2 + y^2}}$$

$$x^2 + y^2 = \sqrt{x^2 + y^2} - x$$

$$x^2 + y^2 + x = \sqrt{x^2 + y^2}$$ ∎

EXAMPLE 3   Show that $r = 1/(1 - \cos \theta)$ is the polar form of a parabola.

SOLUTION   Here our work is simplified if we multiply both sides of the given equation by $1 - \cos \theta$ before making the substitution. Thus

$$r - r \cos \theta = 1$$

Substituting Equation (9.3) and (9.5), we get

$$\sqrt{x^2 + y^2} - x = 1$$

Transposing x to the right and squaring both sides of the resulting equation, we get

$$x^2 + y^2 = x^2 + 2x + 1$$

$$y^2 = 2x + 1$$

$$y^2 = 2(x + \tfrac{1}{2})$$

We recognize this as the standard form of a parabola having its vertex at $(-\tfrac{1}{2}, 0)$ and a horizontal axis. ∎

A polar equation has a graph in the polar coordinate plane just as a rectangular equation has a graph in the rectangular coordinate plane. To draw the graph of a polar equation, we start by assigning values to θ and finding the corresponding values of r. The desired graph is then generated by plotting the ordered pairs (r, θ) and connecting them with a smooth curve.

EXAMPLE 4   Sketch the graph of the equation $r = 1 + \cos \theta$.

SOLUTION   Using increments of $45°$ for θ, we obtain the following table.

θ	r
0	2.00
45°	1.71
90°	1.00
135°	0.29
180°	0.00
225	0.29
270°	1.00
315°	1.71
360°	2.00

The curve obtained by connecting these points with a smooth curve is called a *cardioid* (Figure 9.8).

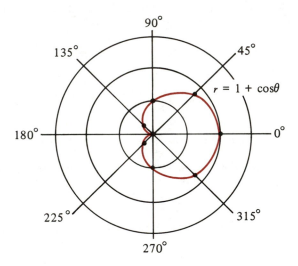

FIGURE 9.8

EXAMPLE 5   Sketch the graph of the equation $r = 4 \sin \theta$.

SOLUTION   We find the following table for values of r corresponding to the indicated values of θ. Drawing a smooth curve through the plotted points, we have the *circle* shown in Figure 9.9.

θ	r
0	0
$\pi/6$	2
$\pi/4$	$2\sqrt{2}$
$\pi/3$	$2\sqrt{3}$
$\pi/2$	4
$2\pi/3$	$2\sqrt{3}$
$3\pi/4$	$2\sqrt{2}$
$5\pi/6$	2
π	0

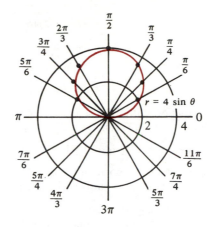

FIGURE 9.9

Notice that θ varies only from 0 to π radians. If we allow θ to vary from 0 to 2π radians, the graph will be traced out twice; once for $0 \le \theta \le \pi$ and again for $\pi < \theta \le 2\pi$. The student should demonstrate this by plotting points in the interval $\pi < \theta \le 2\pi$. ■

Exercises for Section 9.3

Plot the points in Exercises 1–10 on polar coordinate paper.

1. $(5, 30°)$ **2.** $(3.6, -45°)$

3. $(12, 2\pi/3)$ **4.** $(0.5, 220°)$

5. $(-7.1, 14°)$ **6.** $(-2, 7\pi/3)$

7. $(1.75, -200°)$ **8.** $(\sqrt{2}, -311°)$

9. $(-5, -30°)$ **10.** $(5, 150°)$

Convert the rectangular equations in Exercises 11–18 into equations in polar coordinates.

11. $2x + 3y = 6$ **12.** $y = x$

13. $x^2 + y^2 - 4x = 0$ **14.** $x^2 - y^2 = 4$

15. $x^2 + 4y^2 = 4$ **16.** $xy = 1$

17. $x^2 = 4y$ **18.** $y^2 = 16x$

Convert the polar equations in Exercises 19–26 into equations in rectangular coordinates.

19. $r = 5$ **20.** $r = \cos \theta$

21. $r = 10 \sin \theta$ **22.** $r = 2 (\sin \theta - \cos \theta)$

23. $r = 1 + 2 \sin \theta$ **24.** $r \sin \theta = 10$

25. $r = 5/(1 + \cos \theta)$ **26.** $r(1 - 2 \cos \theta) = 1$

Sketch the graph of the equations in Exercises 27–40.

27. $r = 5.6$ **28.** $r = \sqrt{2}$ **29.** $\theta = \frac{1}{3}\pi$

30. $\theta = 170°$ **31.** $r = 2 \sin \theta$ **32.** $r = 0.5 \cos \theta$

33. $r \sin \theta = 1$ **34.** $r \cos \theta = -10$ **35.** $r = 1 + \sin \theta$

36. $r = 1 - \cos \theta$ **37.** $r = \sec \theta$ **38.** $r = -\sin \theta$

39. $r = 4 \sin 3\theta$ **40.** $r = \sin 2\theta$

41. The radiation pattern of a particular two-element antenna is a cardioid of the form $r = 100 (1 + \cos \theta)$. Sketch the radiation pattern of this antenna.

42. The radiation pattern of a certain antenna is given by $r = 1/(2 - \cos \theta)$. Plot this pattern.

43. By transforming the polar equation in Exercise 42 into rectangular coordinates, show that the indicated radiation pattern is elliptical.

44. The feedback diagram of a certain electronic tachometer can be approximated by the curve $r = \frac{1}{2}\theta$. Sketch the feedback diagram of this tachometer from $\theta = 0$ to $\theta = 7\pi/6$.

9.4   Polar Representation of a Complex Number

The graphical representation of a complex number as a point in the complex plane is perhaps best conceptualized as a vector drawn from the origin to the point. We may thus use polar coordinates to describe a complex number. Referring to Figure 9.10 we see that a complex number $a + bi$ can be located with the polar coordinates (r, θ) where

(9.6) $r = \sqrt{a^2 + b^2}$ and $\tan \theta = \dfrac{b}{a}$

Therefore,

(9.7) $z = a + bi = r (\cos \theta + i \sin \theta)$

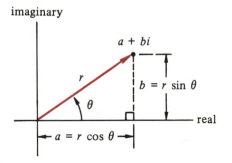

FIGURE 9.10
Polar form

The right hand side of Equation (9.7) is called the **polar form** of the complex number $a + bi$. The quantity $\cos \theta + i \sin \theta$ is sometimes written cis θ, in which case, we write

(9.8) $z = r$ cis θ

as the polar form of the complex number. The number r is called the **modulus** or magnitude of z and θ is called the **argument**. Note that a given complex number has many arguments, all differing by multiples of 2π. Sometimes we limit the argument to some interval of length 2π and thus obtain the **principal value**. In this book, unless we say otherwise, the principal values will be between $-\pi$ and π; that is, between $-180°$ and $180°$.

EXAMPLE 1   Represent $z = 1 + \sqrt{3}i$ in polar form. (See Figure 9.11.)

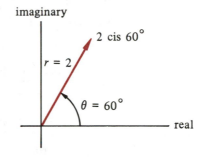

FIGURE 9.11

SOLUTION   Since

$$r = \sqrt{1^2 + (\sqrt{3})^2} = \sqrt{4} = 2$$

and

$$\theta = \arctan \sqrt{3} = 60°$$

we have

$$1 + \sqrt{3}i = 2(\cos 60° + i \sin 60°) = 2 \text{ cis } 60°$$

EXAMPLE 2   Express $z = 6(\cos 120° + i \sin 120°)$ in rectangular form. (See Figure 9.12.)

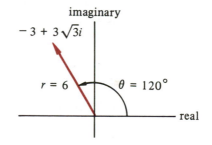

FIGURE 9.12

SOLUTION   Using the fact that $a = r \cos \theta$ and $b = r \sin \theta$, we have

$$a = 6 \cos 120° = 6\left(-\frac{1}{2}\right) = -3$$

$$b = 6 \sin 120° = 6\left(\frac{\sqrt{3}}{2}\right) = 3\sqrt{3}$$

Therefore,

$$z = a + bi = -3 + 3\sqrt{3}i$$

The polar form of complex numbers makes it easy to give a geometric interpretation to the product of two complex numbers. Thus, if $z_1 = r_1 \text{ cis } \theta_1$ and $z_2 = r_2 \text{ cis } \theta_2$, the product $z_1 z_2$ may be written

$$z_1 z_2 = r_1(\cos \theta_1 + i \sin \theta_1) \cdot r_2(\cos \theta_2 + i \sin \theta_2)$$

$$= r_1 r_2 [\cos \theta_1 \cos \theta_2 + i \cos \theta_1 \sin \theta_2$$

$$+ i \sin \theta_1 \cos \theta_2 + i^2 \sin \theta_1 \sin \theta_2]$$

$$= r_1 r_2 [(\cos \theta_1 \cos \theta_2 - \sin \theta_1 \sin \theta_2)$$

$$+ i(\cos \theta_1 \sin \theta_2 + \sin \theta_1 \cos \theta_2)]$$

Now, by using the identities for the sine and cosine of the sum of two angles, we have

(9.9) $z_1 z_2 = r_1 r_2 [\cos (\theta_1 + \theta_2) + i \sin (\theta_1 + \theta_2)] = r_1 r_2 \text{ cis } (\theta_1 + \theta_2)$

Therefore, the modulus of the product of two complex numbers is the product of the individual moduli and the argument of the product is the sum of the individual arguments. Graphically, multiplication of z_1 by z_2 results in a rotation of the vector through z_1 by an angle equal to the argument of z_2 and an expansion or contraction of the modulus depending on whether $|z_2| > 1$ or $|z_2| < 1$. (See Figure 9.13.)

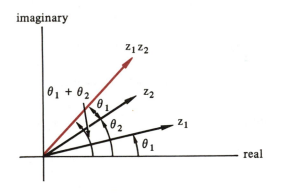

FIGURE 9.13

Multiplication of
complex numbers

EXAMPLE 3   Multiply $z_1 = -1 + \sqrt{3}i$ and $z_2 = 1 + i$, using the polar form of each.

SOLUTION   Computing the modulus and argument of each complex number yields

$$r_1 = \sqrt{(-1)^2 + (\sqrt{3})^2} = 2 \qquad \tan \theta_1 = \frac{\sqrt{3}}{-1}, \theta_1 = 120°$$

$$r_2 = \sqrt{1^2 + 1^2} = \sqrt{2} \qquad \tan \theta_2 = \frac{1}{1}, \theta_2 = 45°$$

Therefore,

$$z_1 z_2 = (2 \text{ cis } 120°)(\sqrt{2} \text{ cis } 45°) = 2\sqrt{2} \text{ cis } (120° + 45°)$$

$$= 2\sqrt{2} \text{ cis } 165°$$

In the same manner as in the above discussion, we may show that if $z_1 = r_1 \operatorname{cis} \theta_1$ and $z_2 = r_2 \operatorname{cis} \theta_2$, then

(9.10) $$\frac{z_1}{z_2} = \frac{r_1}{r_2} [\cos(\theta_1 - \theta_2) + i \sin(\theta_1 - \theta_2)] = \frac{r_1}{r_2} \operatorname{cis}(\theta_1 - \theta_2)$$

In words, the modulus of the quotient of two complex numbers is the quotient of the individual moduli, and the argument is the difference of the individual arguments.

EXAMPLE 4   Divide $z_1 = 2 \operatorname{cis} 120°$ by $z_2 = \sqrt{2} \operatorname{cis} 45°$.

SOLUTION

$$\frac{z_1}{z_2} = \frac{2 \operatorname{cis} 120°}{\sqrt{2} \operatorname{cis} 45°} = \frac{2}{\sqrt{2}} \operatorname{cis}(120° - 45°) = \sqrt{2} \operatorname{cis} 75°$$

■

Exercises for Section 9.4

Plot the complex numbers in Exercises 1–10 and then express each complex number in polar form.

1. $1 - \sqrt{3}i$ **2.** $3 + 4i$

3. $\sqrt{5} + 2i$ **4.** $\sqrt{3} - i$

5. 9 **6.** $5i$

7. $3 - 4i$ **8.** $-1 + i$

9. $5 - 6i$ **10.** $-3 - 4i$

Plot the complex numbers in Exercises 11–20 and then express each complex number in rectangular form.

11. $2 \operatorname{cis} 30°$ **12.** $4 \operatorname{cis} 60°$

13. $5 \operatorname{cis} 135°$ **14.** $10 \operatorname{cis} 90°$

15. $\sqrt{3} \operatorname{cis} 210°$ **16.** $\sqrt{5} \operatorname{cis} 180°$

17. $3 \operatorname{cis} 300°$ **18.** $7 \operatorname{cis} 0°$

19. $10 \operatorname{cis} 20°$ **20.** $2 \operatorname{cis} 100°$

Perform the operations indicated in Exercises 21–36. If the complex numbers are not already in polar form, put them in that form before proceeding.

21. $(4 \operatorname{cis} 30°)(3 \operatorname{cis} 60°)$ **22.** $(2 \operatorname{cis} 120°)(\sqrt{5} \operatorname{cis} 180°)$

23. $(\sqrt{2} \operatorname{cis} 90°)(\sqrt{2} \operatorname{cis} 240°)$ **24.** $(5 \operatorname{cis} 180°)(3 \operatorname{cis} 90°)$

25. $(10 \operatorname{cis} 35°)(2 \operatorname{cis} 100°)$ **26.** $(3 \operatorname{cis} 45°)(2 \operatorname{cis} 120°)$

27. $(3 + 4i)(\sqrt{3} - i)$ **28.** $3i(2 - i)$

29. $\dfrac{10 \operatorname{cis} 30°}{2 \operatorname{cis} 90°}$ **30.** $\dfrac{5 \operatorname{cis} 29°}{3 \operatorname{cis} 4°}$

31. $\dfrac{4 \text{ cis } 26°40'}{2 \text{ cis } 19°10'}$ **32.** $\dfrac{12 \text{ cis } 100°}{3 \text{ cis } 23°}$

33. $\dfrac{1-i}{\sqrt{3}+i}$ **34.** $\dfrac{\sqrt{3}+i}{\sqrt{3}-i}$

35. $\dfrac{4i}{-1+i}$ **36.** $\dfrac{5}{1+i}$

37. Prove *Euler's Identities:*

$$\cos \theta = \frac{1}{2}[\text{cis } \theta + \text{cis } (-\theta)]$$

$$\sin \theta = \frac{1}{2i}[\text{cis } \theta - \text{cis } (-\theta)]$$

9.5   DeMoivre's Theorem

The square of the complex number $z = r \text{ cis } \theta$ is given by

$$z^2 = (r \text{ cis } \theta)(r \text{ cis } \theta)$$
$$= r^2 \text{ cis } 2\theta$$

Likewise,

$$z^3 = z^2 \cdot z = (r^2 \text{ cis } 2\theta) \cdot (r \text{ cis } \theta)$$
$$= r^3 \text{ cis } 3\theta$$

We expect the pattern exhibited for z^2 and z^3 to apply as well to z^4, z^5, z^6, and so on. As a matter of fact if $z = r \text{ cis } \theta$, then

(9.11) $z^n = r^n \text{ cis } n\theta$

This result is known as **DeMoivre's Theorem**. The theorem is true for all real values of n, a fact that we shall accept without proof.

EXAMPLE 1   Use DeMoivre's Theorem to find $(-2 + 2i)^4$.

SOLUTION   Here we have

$$r = \sqrt{2^2 + (-2)^2} = \sqrt{8} \qquad \theta = 135°$$

Therefore,

$$(-2 + 2i)^4 = (\sqrt{8})^4 (\cos 135° + i \sin 135°)^4$$
$$= (\sqrt{8})^4 [\cos 4(135°) + i \sin 4(135°)]$$
$$= 64 [\cos 540° + i \sin 540°]$$
$$= 64 [\cos 180° + i \sin 180°]$$
$$= -64 \qquad \blacksquare$$

In the system of real numbers, there is no square root of -1, no fourth root of -81, and so on. However, if we use complex numbers, we can find the nth root of any number by using DeMoivre's Theorem.

Recalling that DeMoivre's Theorem is valid for all real n, it is possible to evaluate $[r \text{ cis } \theta]^{1/n}$ as

(9.12) $[r \text{ cis } \theta]^{1/n} = r^{1/n} \text{ cis } \dfrac{\theta}{n} = \sqrt[n]{r} \text{ cis } \dfrac{\theta}{n}$

Since $\cos \theta$ and $\sin \theta$ are periodic functions with a period of $360°$, we can write $\cos \theta = \cos (\theta + k \cdot 360°)$ and $\sin \theta = \sin (\theta + k \cdot 360°)$ where k is an integer. Hence,

(9.13) $[r \text{ cis } \theta]^{1/n} = \sqrt[n]{r} \text{ cis } \left(\dfrac{\theta + k \cdot 360°}{n} \right)$

For a given number n, the right side of this equation takes on n distinct values corresponding to $k = 0, 1, 2, ..., n - 1$. For $k > n - 1$, the result is merely a duplication of the first n values.

EXAMPLE 2   Find the square roots of $4i$.

SOLUTION   We first express $4i$ in polar form using

$$r = \sqrt{0^2 + 4^2} = 4 \text{ and } \theta = 90°$$

Thus,

$$4i = 4 \text{ cis } 90°$$

and, the square roots of $4i$ are given by

$$2 \text{ cis } \left(\frac{90° + k \cdot 360°}{2} \right)$$

Therefore, we have for $k = 0$,

$$2 \text{ cis } 45° = \sqrt{2} + \sqrt{2}i$$

and for $k = 1$,

$$2 \text{ cis } 225° = -\sqrt{2} - \sqrt{2}i$$

It is convenient and informative to plot these values in the complex plane as shown in Figure 9.14. Notice that both roots are located on a circle of radius 2, but $180°$ apart.

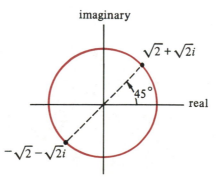

FIGURE 9.14

EXAMPLE 3   Find the three cube roots of unity.

SOLUTION   In polar form, the number 1 may be written 1 cis 0°. Thus,

$$\sqrt[3]{1 \text{ cis } 0°} = 1 \text{ cis } \left(\frac{0° + k \cdot 360°}{3} \right)$$

For $k = 0$,

$$1 \text{ cis } 0° = 1$$

For $k = 1$,

$$1 \text{ cis } 120° = \frac{-1 + \sqrt{3}i}{2}$$

For $k = 2$,

$$1 \text{ cis } 240° = \frac{-1 - \sqrt{3}i}{2}$$

These roots are displayed in Figure 9.15. Notice that they are located on a circle of radius 1 at equally spaced intervals of 120°.

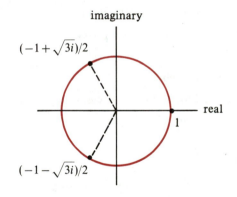

FIGURE 9.15

EXAMPLE 4   Find the fourth roots of $-1 + \sqrt{3}i$.

SOLUTION   Writing $-1 + \sqrt{3}i$ in polar form, we have

$$-1 + \sqrt{3}i = 2 \text{ cis } 120°$$

Therefore,

$$[-1 + \sqrt{3}i]^{1/4} = \sqrt[4]{2} \text{ cis } \frac{120° + k \cdot 360°}{4}$$

The four roots correspond to $k = 0, 1, 2, 3$; that is,

$$\text{for } k = 0, \sqrt[4]{2} \text{ cis } 30° = \sqrt[4]{2} \left(\frac{\sqrt{3}}{2} + \frac{1}{2}i \right)$$

$$\text{for } k = 1, \sqrt[4]{2} \text{ cis } 120° = \sqrt[4]{2} \left(-\frac{1}{2} + \frac{\sqrt{3}}{2}i \right)$$

$$\text{for } k = 2, \sqrt[4]{2} \text{ cis } 210° = \sqrt[4]{2}\left(-\frac{\sqrt{3}}{2} - \frac{1}{2}i\right)$$

$$\text{for } k = 3, \sqrt[4]{2} \text{ cis } 300° = \sqrt[4]{2}\left(\frac{1}{2} - \frac{\sqrt{3}}{2}i\right)$$ ∎

Exercises for Section 9.5

Use DeMoivre's Theorem to evaluate the powers in Exercises 1–10. Leave the answer in polar form.

1. $(-1 + \sqrt{3}i)^3$ 2. $(1 + i)^4$

3. $(\sqrt{3} \text{ cis } 60°)^4$ 4. $(\sqrt{3} - i)^6$

5. $(-2 + 2i)^5$ 6. $(-1 + 3i)^3$

7. $(-\sqrt{3} + i)^7$ 8. $(2 \text{ cis } 20°)^5$

9. $(2 + 5i)^4$ 10. $(3 + 2i)^{10}$

Find the roots indicated in Exercises 11–20 and sketch their location in the complex plane.

11. Fifth roots of 1   12. Cube roots of 64

13. Fourth roots of i 14. Fourth roots of -16

15. Square roots of $1 + i$ 16. Fifth roots of $\sqrt{3} + i$

17. Sixth roots of $-\sqrt{3} + i$ 18. Square roots of $-1 + i$

19. Fourth roots of $-1 + \sqrt{3}i$ 20. Sixth roots of $-i$

21. Obtain an expression for $\cos 2\theta$ and $\sin 2\theta$ in terms of trigonometric functions of θ by making use of DeMoivre's Theorem.

22. Find all roots of the equation $x^4 + 81 = 0$.

23. Find all roots of $x^3 + 64 = 0$.

Key Topics for Chapter 9

Define and/or discuss each of the following.

Complex numbers

Complex conjugate

Rectangular form of a complex number

Polar coordinates

Polar form of a complex number

Graphical representation of a complex number

DeMoivre's Theorem

Review Exercises for Chapter 9

In Exercises 1–15 perform the indicated operations and express in the form $a + bi$. (Notice that $\overline{a + bi} = a - bi$.)

1. $(3 - 2i) + (6 - i)$ **2.** $(3 - 2i) - (i + 2)$

3. $(i + 7) - (i + 2)$ **4.** $(2 + i)(2 - i)$

5. $(3 - i)\overline{(3 + i)}$ **6.** $(i - 2)(2 + i)i$

7. $i(i^2 - 1)(i^2 + 1)$ **8.** $(6 - i)^2 \overline{(6 + i)}$

9. $(3 + 2i)(2 - i) + \overline{(3 + 2i)(2 - i)}$ **10.** $i/(2 + i)$

11. $(2i + 1)/(i - 1)$ **12.** $\overline{(6 - i)}/i^2$

13. $(i - 1)/\overline{(i - 1)}$ **14.** $(9 - i)(9 - 2i)/(i + 2)$

15. $(4 - i)(3 + 2i)/(i + 1)$

In Exercises 16–20 plot and express in polar form.

16. $1 + i\sqrt{3}$ **17.** i **18.** $i - 1$ **19.** $1 + i$ **20.** 4

In Exercises 21–25 plot and express in rectangular form.

21. $2 \operatorname{cis} 45°$ **22.** $-3 \operatorname{cis} 75°$ **23.** $4 \operatorname{cis} (-20°)$

24. $(-2 \operatorname{cis} 30°)^2$ **25.** $(3 \operatorname{cis} 10°)^3$

In Exercises 26–30 convert to polar coordinates. Sketch.

26. $x^2 + y^2 + y = 0$ **27.** $y = 2x$

28. $y^2 + x^2 - 3x = 1$ **29.** $4x^2 + y^2 = 1$

30. $x = 4y^2$

In Exercises 31–40 convert to rectangular coordinates. Sketch.

31. $r = 2$ **32.** $r = 2 + 3 \cos \theta$

33. $r = \dfrac{2}{1 + \sin \theta}$ **34.** $r \cos \theta = 3$

35. $r = 3 \cos \theta$ **36.** $r = \theta$

37. $r = \sin 2\theta$ **38.** $r^2 = \sin \theta$

39. $r^2 - r = 0$ **40.** $r^2 - 3r + 2 = 0$

In Exercises 41–45 evaluate. Leave answers in polar form.

41. $(1 + i)^3$ **42.** $(-2 + 2i)^6$

43. $(-\sqrt{3} + i)^5$ **44.** $(\sqrt{3} \operatorname{cis} 60°)^4$

45. $(3 - i)^{10}$

Find the roots indicated in Exercises 46–50 and sketch their location in the complex plane. Leave answers in polar form.

46. Square roots of i

47. Fifth roots of -1

48. Cube roots of $\cdot i$

49. Cube roots of $1+i$

50. Eighth roots of 16 cis 24°

Test 1 for Chapter 9

In Exercises 1–10, answer *true* or *false*.

1. $2 + 5i$ is an imaginary number.

2. $3 - i$ is the complex conjugate of $3 + i$.

3. The sum of a complex number and its complex conjugate is a real number.

4. $x + iy$ is the polar form of a complex number.

5. Two complex numbers are multiplied by multiplying their magnitudes and adding their arguments.

6. Two complex numbers are added by adding the real part to the imaginary part.

7. Complex numbers are represented graphically as points on the real line.

8. The complex number $z = 3$ cis 35° has a modulus of 3.

9. DeMoivre's Theorem is valid for all real values of n.

10. DeMoivre's Theorem states that $x + iy = x^n$ cis ny.

11. Find the sum $(2 - 4i) + 2i - (1 + 8i)$.

12. Find the quotient $(2 + 3i)/(4 + i)$ and express it in the form $a + bi$.

13. Multiply $(2 + i) \cdot (6 - \sqrt{2}i)$ by first expressing the given numbers in polar form.

14. Use DeMoivre's Theorem to find $(1 + 3i)^5$.

15. Find the three cube roots of -1.

Test 2 for Chapter 9

1. Find the sum and the difference of $4 - i$ and $5 + 2i$.

2. Find the product $(-2 - i)(-7 + 2i)$.

3. Draw the graph of $z = -3 + 4i$. Draw the graph of the complex conjugate on the same axes.

4. Find the quotient $(4 - i)/(i - 3)$. Express your answer in the form $a + bi$.

5. Find the modulus of z, if $z = 1 - 2i$.

6. Plot $2 - \sqrt{3}i$ and then express it in polar form.

7. Plot 7 cis 315° and then express it in rectangular form.

8. Evaluate $(\sqrt{5}$ cis 120°$)(7$ cis 80°$)$.

9. Find the cube roots of $-1 + i$.

10. Find all the roots of $x^5 + 32 = 0$.

10

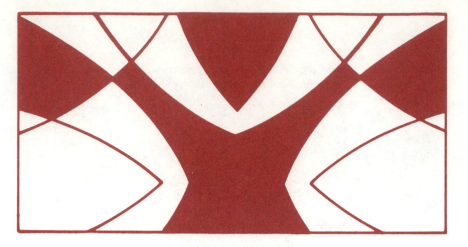

Logarithms

10.1   Definition of the Logarithm

In this chapter you will learn something of the usefulness of logarithms.

We begin with a statement of the definition of the logarithm and its relation to the exponent.

DEFINITION 10.1

> The logarithm of a positive number x to the base $b(\neq 1)$ is the power to which b must be raised to give x. That is,
>
> $y = \log_b x$ if and only if $x = b^y$

The equation $y = \log_b x$ is read, "y is the logarithm of x to the base b." The following examples illustrate this definition.

EXAMPLE 1
    (a) $\log_2 8 = 3$, since $2^3 = 8$.
    (b) $\log_3 \frac{1}{9} = -2$, since $3^{-2} = \frac{1}{9}$.
    (c) $\log_{10} 10{,}000 = 4$, since $10^4 = 10{,}000$.   ■

EXAMPLE 2
    Find the base a if $\log_a 16 = 4$.

SOLUTION
    Since $2^4 = 16$, we conclude that the desired base is $a = 2$.   ■

EXAMPLE 3
    Find the number x if $\log_3 x = -3$.

SOLUTION
    Since $3^{-3} = \frac{1}{27}$, we conclude that $x = \frac{1}{27}$.   ■

Since $y = \log_b x$ means that $x = b^y$, we may sketch the graph of the logarithm function by first constructing a table of values. The tables for $y = \log_2 x$ and $y = \log_{10} x$ are included below and the corresponding graphs are presented in Figure 10.1.

x	$\frac{1}{16}$	$\frac{1}{8}$	$\frac{1}{4}$	$\frac{1}{2}$	1	2	4	8	16
$\log_2 x$	-4	-3	-2	-1	0	1	2	3	4

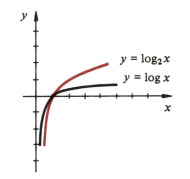

FIGURE 10.1

x	$\frac{1}{1000}$	$\frac{1}{100}$	$\frac{1}{10}$	1	10	100
$\log_{10} x$	-3	-2	-1	0	1	2

Each of the curves of Figure 10.1 is characteristic of what is called **logarithmic shape**. This figure clearly demonstrates the following functional characteristics of the logarithm function. Any function that obeys these properties is said to *behave logarithmically*.

(1) $\log_a x$ is not defined for $x \leq 0$.

(2) $\log_a 1 = 0$.

(3) $\log_a a = 1$.

(4) $\log_a x$ is negative for $0 < x < 1$ and positive for $x > 1$.

(5)   As x approaches 0, $\log_a x$ decreases without bound.

(6)   As x increases without bound, $\log_a x$ increases without bound.

Exercises for Section 10.1

Write a logarithmic equation equivalent to the given exponential equations in Exercises 1–5.

1. $x = 2^3$ **2.** $y = 3^8$ **3.** $M = 5^{-3}$

4. $N = 10^{-2}$ **5.** $L = 7^2$

Find the base of the logarithm function such that:

6. $\log_b 8 = 1$ **7.** $\log_b 4 = 2$ **8.** $\log_b (1/4) = -2$

9. $\log_b 100 = 2$ **10.** $\log_b (1/3) = -1$

Solve the equations in Exercises 11–28 for the unknown.

11. $\log_{10} x = 4$ **12.** $\log_5 N = 2$ **13.** $\log_x 10 = 1$

14. $\log_x 25 = 2$ **15.** $\log_x 64 = 3$ **16.** $\log_{16} x = 2$

17. $\log_{27} x = \frac{2}{3}$ **18.** $\log_2 \frac{1}{8} = x$ **19.** $\log_3 9 = x$

20. $\log_{10} 10^7 = x$ **21.** $\log_b b^a = x$ **22.** $\log_b x = b$

23. $\log_x 2 = \frac{1}{3}$ **24.** $\log_x 0.0001 = -2$ **25.** $\log_x 6 = \frac{1}{2}$

26. $\log_6 x = 0$ **27.** $6^{\log_6 x} = 6$ **28.** $x^{\log_x x} = 3$

29. Let $f(x) = \log_3 x$. Find $f(9)(\frac{1}{27})$, $f(81)$.

30. Let $f(x) = \log_2 x$. By example, show that:

     (a) $f(x + y) \neq f(x) + f(y)$ (b) $f(ax) \neq af(x)$

31. A power supply has a power output in watts approximated by the equation

$$P = 64(2)^{-3t}$$

where t is in days. How many days does it take for the power supply to reduce to a power output of 1 watt?

32. A certain radioactive material decays exponentially by the equation

$$A(t) = A_0 2^{-t/5}$$

Find the half-life of the material.

10.2   Basic Properties of the Logarithm

Logarithmic expressions must often be rearranged or simplified. These simplifications are accomplished by three basic Rules of Logarithms which correspond precisely to the three fundamental rules for exponents and are necessary consequences of them.

RULE 1

$$\log_a MN = \log_a M + \log_a N$$

PROOF

Let $u = \log_a M$ and $v = \log_a N$. Then,

$$a^u = M \text{ and } a^v = N$$

from which

$$MN = a^u a^v = a^{u+v}$$

Expressing, again, in terms of logarithms,

$$\log_a MN = u + v = \log_a M + \log_a N$$

RULE 2

$$\log_a M^c = c \log_a M, \text{ where } c \text{ is any real number.}$$

PROOF

Let $u = \log_a M$. Then,

$$a^u = M \text{ and } (a^u)^c = a^{uc} = M^c$$

In terms of logarithms, this may be expressed as

$$\log M^c = uc = c \log_a M$$

RULE 3

$$\log_a (M/N) = \log_a M - \log_a N$$

PROOF $M/N = M(N)^{-1}$. Now apply the previous two rules.

In words, Rule 1 states that the logarithm of a product is equal to the sum of the logarithms of the individual terms; Rule 2 states that the logarithm of a number to a power is the power times the logarithm of the number; and Rule 3 states that the logarithm of a quotient is the difference of the logarithms of the individual terms. Examine these rules carefully and notice where they apply as well as where they do *not* apply. For example, there is *no* rule for simplifying expressions of the form $\log_a (x + y)$ or $\log_a (x - y)$.

EXAMPLE 1   (a) $\log_2 (8)(64) = \log_2 8 + \log_2 64 = \log_2 2^3 + \log_2 2^6 = 3 + 6 = 9$.

(b) $\log_3 \sqrt{243} = \log_3 243^{1/2} = \frac{1}{2} \log_3 243 = \frac{1}{2}(5) = 2.5$.

(c) $\log_2 (\frac{3}{5}) = \log_2 3 - \log_2 5$.

(d) $\log (4 \cdot 29/5) = \log 4 + \log 29 - \log 5$.   (log x means $\log_{10} x$.)   ■

EXAMPLE 2   Write the expression $\log x - 2 \log x + 3 \log (x + 1) - \log (x^2 - 1)$ as a single term.

SOLUTION   Proceed as follows:

$$\log x - 2 \log x + 3 \log (x + 1) - \log (x^2 - 1)$$

$$= \log x - \log x^2 + \log (x + 1)^3 - \log (x^2 - 1)$$

$$= \log \frac{x(x + 1)^3}{x^2(x^2 - 1)} = \log \frac{(x + 1)^2}{x(x - 1)}$$ ■

EXAMPLE 3   Given $\log_a x = 3$, find $\log_a (1/x)$.

SOLUTION   By definition $\log_a x = 3$ means

$$x = a^3$$

Also, letting $\log_a (1/x) = y$, we can write

$$\frac{1}{x} = a^y \quad \text{or} \quad x = a^{-y}$$

Since a^3 and a^{-y} are both equal to x, we have

$$a^3 = a^{-y}$$

Therefore, $y = \log_a (1/x) = -3$.   ■

Exercises for Section 10.2

Evaluate the logarithms in Exercises 1–8.

1. $\log_2 32 \cdot 16$ **2.** $\log_2 16^5$ **3.** $\log_5 25^{1/4}$

4. $\log_3 27$ **5.** $\log_3 27 \cdot 9 \cdot 3$ **6.** $\log_2 64 \cdot 32 \cdot 8$

7. $\log_2 (8 \cdot 32)^3$ **8.** $\log_3 (9 \cdot 81)^8$

Given that $\log 2 = 0.3010$, $\log 3 = 0.4771$, and $\log 7 = 0.8451$, find the logarithms in Exercises 9–20.

9. $\log \frac{3}{2}$ **10.** $\log 4$ **11.** $\log 12$

12. $\log 30$ **13.** $\log 90$ **14.** $\log \sqrt{2}$

15. $\log \sqrt{5}$ **16.** $\log 21^{1/3}$ **17.** $\log 2400$

18. $\log 0.00018$ **19.** $\log 0.0014$ **20.** $\log 42000$

In Exercises 21–28 write the given expression as a single logarithmic term.

21. $\log_2 x^2 - \log_2 x$ **22.** $\log_2 (x^2 - 1) - \log_2 (x - 1)$

23. $\log x + \log \dfrac{1}{x}$ **24.** $\log 3x + 3 \log (x + 2) - \log (x^2 - 4)$

25. $\log 5t + 2 \log (t^2 - 4) - \frac{1}{2} \log (t + 3)$ **26.** $\log z - 3 \log 3z - \log (2z - 9)$

27. $3 \log u - 2 \log (u + 1) - 5 \log (u - 1)$ **28.** $\log t + 7 \log (2t - 8)$

29. Let $\log_e y = x + \log_e c$. Show that $y = ce^x$.

30. If y is directly proportional to x^p, what relation exists between $\log y$ and $\log x$?

31. Compare the functions $f(x) = \log x^2$ and $g(x) = 2 \log x$. In what way are they the same? In what way different?

32. Let $f(x) = \log_a x$ and $g(x) = \log_{1/a} x$. Show that $g(x) = f(1/x)$.

33. If $\log_a x = 2$, find $\log_{1/a} x$ and $\log_a (1/x)$.

34. Compare the graphs of the functions
$$f(x) = \log_2 2x,\ g(x) = \log_2 x,\ h(x) = \log_2 \sqrt{x},\ \text{and}\ m(x) = \log_2 x^2.$$

35. Given the graph of $y = \log x$, explain a convenient way to obtain the following graphs.

(a) $\log x^p$ (b) $\log px$

(c) $\log (x + p)$ (d) $\log (x/p)$

36. If $f(x) = \log_b x$, is $f(x + y) = f(x) + f(y)$?

In Exercises 37–40, use a calculator to verify the given statement.

37. $\log [25 \cdot 34] = \log 25 + \log 34$ **38.** $\log [16 \cdot \pi] = \log 16 + \log \pi$

39. $\log \frac{125}{73} = \log 125 - \log 73$ **40.** $\log \frac{17}{35} = \log 17 - \log 35$

10.3   **Common Logarithms**

Historically, logarithms to the base 10 have been used most frequently — especially in simplifying numerical computation. These logarithms have come to be known as **common logarithms**. We customarily omit the numerical subscript when discussing common logarithms. Thus, in this book, log x means $\log_{10} x$.

Table D in the Appendix is a listing of common logarithms of numbers between 1 and 10 in steps of 0.01. This table is a "four place" table which means that it gives values accurate to four decimal places. To find the logarithm of a number between 1 and 10, locate the first two digits of the number in the left column. The columns at the top are headed by the third digit of the number. Thus, to find the logarithm of 5.31, look down the left-hand column until you come to 5.3. Then move over to the column headed by 1 to find that log 5.31 = 0.7251.

To find the number whose logarithm is given, we use Table D "in reverse." Such a procedure is often called finding **antilogarithms**. Thus, since log 2 = 0.3010, antilog 0.3010 = 2.

Table D lists logarithms for numbers m such that $1 \leq m \leq 10$. The corresponding range values are $0 \leq \log m \leq 1$; that is, all the values in the table are fractions between 0 and 1.

To find the logarithm of any number we use the fact that any positive number M may be written as the product of a number m (where m is between 1 and 10) and 10^c (where c is an integer). That is,

$$M = m \cdot 10^c, \qquad 1 \leq m < 10.$$

A number so written is said to be in **scientific form**. Several examples are given in the next example.

EXAMPLE 1
$$53.1 = 5.31 \times 10^1$$
$$0.00531 = 5.31 \times 10^{-3}$$
$$5310000 = 5.31 \times 10^6$$
■

By using the scientific form of a number M, we can write

$$\log M = \log (m \cdot 10^c)$$

Using Rule 1 for logarithms, we get

$$\log M = \log 10^c + \log m$$

Finally, using Rule 2,

$$\log M = c \log 10 + \log m$$
$$= c + \log m$$

This is the standard or uniform way to express logarithms — as the sum of an integer and a positive number between 0 and 1. The quantity log m is called the **mantissa** of log M and is always a number between 0 and 1. The quantity c is called the **characteristic** of log M and is always an integer.

EXAMPLE 2   Find (a) log 5.31   (b) log 5,310,000   (c) log 0.00531.

SOLUTION

(a)   Since $53.1 = 5.31 \times 10^1$, we have

$$\log 53.1 = 1 + \log 5.31 = 1 + 0.7251 = 1.7251$$

(b)   Since $5{,}310{,}000 = 5.31 \times 10^6$, we have

$$\log 5{,}310{,}000 = 6 + \log 5.31 = 6 + 0.7251 = 6.7251$$

(c)   Since $0.00531 = 5.31 \times 10^{-3}$, we have

$$\log 0.00531 = -3 + \log 5.31 = -3 + 0.7251$$

We usually do not combine a negative characteristic with the mantissa because this tends to obscure both the mantissa and the characteristic; hence, it would be difficult to recover them when finding antilogarithms. Sometimes, we express the negative characteristic as a positive number minus ten. Thus, since $-3 = 7 - 10$, we can write

$$\log 0.00531 = -3 + 0.7251 = 7.7251 - 10 \qquad \blacksquare$$

The separation of the characteristic and the mantissa is convenient for computations using Tables of Logarithms but is unnecessary for calculator logic. Calculators (and computers) carry the characteristic and mantissa in combined form. For numbers greater than one the two forms are the same, but for numbers less than one the forms are different. For example, log 35.7 is displayed as `1.5527`, *which is the same value obtained from a Table. However, log 0.00357 is displayed as* `-2.4473`, *which does not appear to be the same as the table value of $7.5527 - 10$. However, the table value form can be converted to the calculator form by simply performing the indicated subtraction. To convert the calculator form to that of the table, write*

$$-2.4473 = -2.4473 + 10 - 10 = 7.5527 - 10$$

EXAMPLE 3   Use the fact than antilog $(0.4099) = 2.57$ to find (a) antilog (2.4099); (b) antilog $(7.4099 - 10)$; (c) antilog (-6.5901).

SOLUTION

(a)   antilog $(2.4099) = 2.57 \times 10^2 = 257$

(b)   antilog $(7.4099 - 10) = 2.57 \times 10^{-3} = 0.00257$

(c)   antilog $(-6.5901) = $ antilog $(3.4099 - 10) = 2.57 \times 10^{-7}$

In part (c) write the number -6.5901 as the sum of a number between 0 and 1 and an integer before consulting a table of logarithms. Note, for example that -6.5901 is *not* equal to $-6 + 0.5901$, but rather $-7 + 0.4099 = 3.4099 - 10$. $\qquad \blacksquare$

There are two popular methods for obtaining antilogs with calculators. Some models of calculators use an `inv` *key with the* `log` *key to compute antilogs. Under this scheme, if log $N = 3.7971$, , the value of N is given by the following key strokes.*

$$= 3.7971 \boxed{\text{inv}} \boxed{\text{log}} = 6267.58$$

Other models use a $\boxed{10^x}$ *key for antilogs. The use of these keys follows from the fact that* $\log_b N = x$ *means* $N = b^x$. *Thus, if* $\log N = 3.7971$, *the value of* N *is given by* $10^{3.7971}$, *that is*

$$= 3.7971 \boxed{10^x} = 6267.58$$

Of course, expressions like $\log N = 7.0915 - 10$ *must be written in the form* $\log N = -2.9085$ *before a calculator can be used.*

EXAMPLE 4   Find the value of $10^{-4.0970}$

SOLUTION   Let $N = 10^{-4.0970}$, then

$$\log N = -4.0970 = -5 + 0.9030 = 5.9030 - 10$$

Since antilog $0.9030 \approx 8$,

$$N = 8 \times 10^{-5} = 0.00008$$

EXAMPLE 5   Solve for x: $10^x = 4000$.

SOLUTION   The necessary power of 10 is nothing more than $\log 4000$. But $4000 = 4 \times 10^3$. Hence,

$$x = \log 4000 = 3 + \log 4 = 3 + 0.6021 = 3.6021$$

Exercises for Section 10.3

Evaluate Exercises 1–15.

1. (a) log 5.41   (b) log 5410   (c) log 0.00541

2. (a) log 1.25   (b) log 125   (c) log 0.125

3. (a) log 9.03   (b) log 0.000903   (c) log 903,000

4. (a) log 8.89   (b) log 88.9   (c) log 0.0889

5. (a) log (5.25)(3.65)   (b) log (5.25/3.65)

6. (a) log (8.03)(7.54)   (b) log (8.03/7.54)

7. (a) log (0.255)(85.6)   (b) log (0.255/85.6)

8. (a) log (0.295)(3.11)   (b) log (0.295/3.11)

9. (a) log (258)(3670)   (b) log (258/3670)

10. (a) log (1110)(56300)   (b) log (1110/56300)

11. (a) antilog 3.2601   (b) antilog 0.2601
    (c) antilog 8.2601 − 10

12. (a) antilog 0.9258   (b) antilog 5.9258
    (c) antilog 9.9258 − 10

13. (a) antilog 1.1818   (b) antilog 3.1818 − 10
      (c) antilog 6.1818

14. (a) antilog 2.8692   (b) antilog 7.8692
      (c) antilog 3.8692 − 10

15. (a) antilog 0.6053   (b) antilog −2.3947
      (c) antilog −0.3947

Solve for x in Exercises 16–20.

16. $10^x = 20$ **17.** $10^x = 25$ **18.** $10^x = (1.54)(674)$

19. $10^x = \sqrt{20}$ **20.** $10^{x+2} = 17^4$

Use a calculator to evaluate each of the common logarithms in Exercises 21–28. In each case indicate the mantissa and the characteristic.
Note: For numbers less than one, calculators carry the characteristic and the mantissa in combined form.

21. log 55.8   **22.** log 677,400   **23.** log 134

24. log 4,690   **25.** log 2.11   **26.** log 0.998

27. log 0.00391   **28.** log 0.000025

In Exercises 29–36 use a calculator to evaluate the given antilog.

29. antilog 3.1160   **30.** antilog 1.0555   **31.** antilog 0.7623

32. antilog (8.3324 − 10)   **33.** antilog (9.1414 − 10)   **34.** antilog (7.4444 − 10)

35. antilog (−5.8709)   **36.** antilog (−2.1602)

10.4   Interpolation

Table D does not include the logarithm of every number between 1 and 10, only those in steps of 0.01. To approximate logarithms of numbers written with one more digit accuracy, you may proceed in a variety of ways. Perhaps the most common is the linear interpolation method based on proportional parts introduced in Section 2.5. This type of estimation assumes that for a small increase in the number, the increase in the logarithm is proportional. Rather than completely describe the process (which is detailed in 2.5), we will be content with a few examples.

EXAMPLE 1   Use linear interpolation to approximate log 2573.

SOLUTION   Since $2573 = 2.573 \times 10^3$, the characteristic is 3. The mantissa lies between the mantissa for 2.570 and 2.580. From Table D, the entries for 2.570 and 2.580 are found to be 0.4099 and 0.4116, respectively.

Number			Mantissa		
10	3	$\begin{bmatrix} 2.570 \\ 2.573 \\ 2.580 \end{bmatrix}$	$\begin{bmatrix} .4099 \\ \ldots \\ .4116 \end{bmatrix}$	c	17

We can establish the following proportion

$$\frac{c}{17} = \frac{3}{10} \quad \text{or} \quad c = \frac{3}{10}(17) = 5.1 \approx 5$$

The required mantissa of log 2.573 is then $0.4099 + 0.0005 = 0.4104$. Therefore, log $2573 = 3.4104$.

∎

Note that in the preceding example we rounded off the number c to four digits after the decimal point. Otherwise, it would seem that interpolation was increasing the acccuracy of the table which, of course, is impossible.

EXAMPLE 2   Find antilog of 2.4059.

SOLUTION   Since the mantissa is 0.4059 and the characteristic is 2, we must find the number corresponding to the mantissa 0.4059 and then multiply by 10^2. Using Table D, we arrange the work as follows:

Number			Mantissa		
10	n	$\begin{bmatrix} 2.540 \\ \ldots \\ 2.550 \end{bmatrix}$	$\begin{bmatrix} .4048 \\ .4059 \\ .4065 \end{bmatrix}$	11	17

$$\frac{n}{10} = \frac{11}{17}$$

$$n = \frac{11}{17}(10) = 6.4 \approx 6$$

Therefore, the desired number is $2.546 \times 10^2 = 254.6$.

∎

Exercises for Section 10.4

Use the method of linear interpolation to approximate the common logarithm of the numbers in Exercises 1–10.

1. 2.361	**2.** 5842	**3.** .009573	**4.** 3.142	**5.** 2.718
6. 49,990	**7.** 642,300	**8.** 5.011	**9.** 1005	**10.** 62.45

Find the number x to 4 significant digits by using linear interpolation for Exercises 11–20.

11. $\log x = 2.1110$ **12.** $\log x = 8.1284 - 10$

13. $\log x = 7.814$ **14.** $\log x = 3.4141$

15. $\log x = 7.7228 - 10$

16. $\log x = \frac{1}{2}$

17. $\log x = 0.25$

18. $\log x = \frac{1}{3}$

19. $\log x = \pi$

20. $\log x = -\pi$

Find the number x in Exercises 21–25. (The number e is approximately equal to 2.718.)

21. $10^x = e$ **22.** $10^x = \pi$ **23.** $10^x = e^2$ **24.** $10^x = 0.1441$

25. $10^x = \sqrt{2.169}$

Write Exercises 26–30 as an approximate decimal.

26. $\sqrt[4]{20}$ **27.** 10^π **28.** π^e **29.** $\sqrt{10}$ **30.** $10^{0.4368}$

31. Using $\log 3 = 0.4771$ and $\log 4 = 0.6020$, approximate $\log 3.5$ by linear interpolation and compare to the value in Table D.

32. Use the fact that $\sqrt{4} = 2$ and $\sqrt{9} = 3$ and linear interpolation to approximate $\sqrt{7}$. Is the approximation high or low?

33. Let $f(x) = 2x - 3$. Use the values of $f(2)$ and $f(3)$ along with linear interpolation to approximate $f(2.5)$. How accurate is your result?

10.5   Computations with Logarithms

With the advent of high speed mechanical and electronic computing devices, the use of common logarithms for computational purposes has been relegated to a minor role. Even so, basic logarithmic computation remains important. Further, the arithmetic of logarithmic computation should increase your appreciation for some of the background theory. We choose not to explore all the "short cuts" and conventions related to logarithmic computation; a few examples should suffice.

Some of the following examples could as easily have been worked by conventional arithmetic, but the aim here is to illustrate and not to be too concerned with whether the use of logarithms is the preferred technique.

EXAMPLE 1   Use logarithms to evaluate

$$M = \frac{2158 \times 0.512}{0.00042}$$

SOLUTION   We find $\log M$ and use the rules of logarithms to write

$$\log M = \log 2158 + \log 0.0512 - \log 0.00042$$

$\log 2158$ is found by interpolation to be 3.3341.
$\log 0.512$ is found directly to be $-1 + 0.7093$.
$\log 0.00042$ is found directly to be $-4 + 0.6232$.

Hence,

$$\log M = 3.3341 + (-1 + 0.7093) - (-4 + 0.6232)$$
$$= 3 + 0.3341 - 1 + 0.7093 + 4 - 0.6232$$

Adding the characteristics, yields

$$\log M = 6 + (0.3341 + 0.7093 - 0.6232)$$
$$= 6 + 0.4202$$
$$= 6.4202$$

By interpolation, antilog $0.4202 = 2.631$, and hence,

$$M = 2.631 \times 10^6$$

EXAMPLE 2   Use logarithms to approximate $(25.4)^{1/4}$.

SOLUTION   Let $M = (25.4)^{1/4}$. Then $\log M = \log (25.4)^{1/4} = (1/4) \log 25.4$. From Table D, $\log 25.4 = 1.4048$ and hence,

$$\log M = 0.3512,$$
$$M = \text{antilog } 0.3512$$

By interpolation

$$M = 2.245$$

EXAMPLE 3   Find the power to which 3 must be raised to give 5.

SOLUTION   We are looking for a number x such that

$$3^x = 5$$

Taking the common logarithm of both sides, we have

$$\log 3^x = \log 5$$

or, using Rule 2 for logarithms,

$$x \log 3 = \log 5$$
$$x = \frac{\log 5}{\log 3}$$
$$= \frac{0.6990}{0.4771}$$

Using logarithms to evaluate x, we have

$$\log x = \log \frac{0.6990}{0.4771} = \log 0.6990 - \log 0.4771$$
$$= (9.8445 - 10) - (9.6786 - 10) = 0.1659$$
$$x = \text{antilog } 0.1659 = 1.465$$

EXAMPLE 4   An approximate rule for atmospheric pressure at altitudes less than 50 miles can be shown to be

$$P = 14.7(0.5)^{h/3.25}$$

where P is in psi when h is altitude in miles. What altitude is just above 99 percent of the atmosphere?

SOLUTION   Because density and pressure are proportional, the desired altitude is the point at which the pressure is 1 percent of standard atmospheric pressure. Hence

$$(0.01)(14.7) = 14.7(\tfrac{1}{2})^{h/3.25},$$

$$0.01 = (\tfrac{1}{2})^{h/3.25}$$

Taking the logarithm of both sides, we obtain,

$$\log 0.01 = \frac{h}{3.25} \log \tfrac{1}{2}$$

Solving for h, we have

$$h = \frac{(3.25)(\log 0.01)}{\log \tfrac{1}{2}}$$

$$= \frac{(3.25)(-2)}{-0.3010}$$

$$= 21.6 \text{ miles}$$

Exercises for Section 10.5

Use logarithms to approximate the computations in Exercises 1–15 to three significant digits.

1. $(65.7)(0.00411)$

2. $\dfrac{365 \times 1.423}{120 \times 0.00133}$

3. $2563^{1/3}$

4. $\dfrac{(0.39)(23.7)}{0.00505}$

5. $\sqrt[5]{8}$

6. $(0.16)^3$

7. $3.1^{3.1}$

8. $\dfrac{21.5 + 46.3}{\sqrt{5}}$

9. $\dfrac{123.8 + 95.3}{543.2}$

10. $\dfrac{76.1 + \log(8 \times 10^{40})}{\log 5.3}$

11. $\dfrac{-52.3 \times 64}{\log 3.83}$

12. $10^{-4.213}$

13. $2^{-1.431}$

14. $3^{1.212}$

15. $(\sqrt{2})^{\sqrt{2}}$

In Exercises 16–20 solve for x:

16. $10^x = \pi$ **17.** $2^x = \pi$ **18.** $2^x = 3$

19. $3^x = \dfrac{201}{314}$ **20.** $10^x = 5.555$

21. The period T, in seconds, of a simple pendulum of length L, in feet, is given by $T = 2\pi\sqrt{L/32.2}$. Approximate the period of a pendulum that is a yard long.

22. The difference in intensity level of two sounds with intensities I and I_0 is defined to be $10 \log (I/I_0)$ decibels. Find the intensity level in decibels of the sound produced by an electric motor that is 175.6 times greater than I_0.

Key Topics for Chapter 10

Define and/or discuss each of the following.

Definition of a logarithm

Properties of logarithms

Common logarithms

Interpolation

Computation with logarithms

Review Exercises for Chapter 10

In Exercises 1–8, solve for x.

1. $\log x = 3$ **2.** $\log_2 x = 5$

3. $\log_3 81 = x$ **4.** $\log_5 \frac{1}{25} = x$

5. $\log_x 64 = 6$ **6.** $\log_x 0.027 = -3$

7. $\log_x 0.01 = -2$ **8.** $\log x = -8$

Evaluate the logarithms in Exercises 9–14 without a table or a calculator.

9. $\log_2 (8 \cdot 2)$ **10.** $\log_2 \frac{1}{16}$

11. $\log_3 \sqrt{27}$ **12.** $\log_5 (25)^{1/3}$

13. $\log_5 (625 \cdot 125)^3$ **14.** $\log_2 (128 \cdot 32)^4$

In Exercises 15–20 write the given expression as a single logarithmic term.

15. $\log x + \log x^2$ **16.** $\log 2x - \log x$

17. $\log_2 2x + 3 \log_2 x$ **18.** $3 \log_2 x - \log_2 x$

19. $3 \log_5 x - \log_2 (2x - 3)$ **20.** $5 \log (x + 2) - 2 \log x$

Use logarithms to evaluate each of the expressions in Exercises 21–30.

21. $29.6 \, (397.0)$

22. $3970 \, (2.79)$

23. $0.760 \, (25.97)$

24. $0.00301 \, (17.42)$

25. $\dfrac{7.16 \, (311.2)}{90.34}$

26. $\dfrac{80.02 \, (3.92)}{88.6}$

27. $\sqrt[3]{0.9773}$

28. $\sqrt{1.209}$

29. $\dfrac{15.6\sqrt{11.3}}{\pi}$

30. $\dfrac{208.5 \, (0.514)^3}{\sqrt{2}}$

Test 1 for Chapter 10

In Exercises 1–10, answer *true* or *false*.

1. $\log_b 0 = 1$

2. If $x_1 < x_2$, then $2^{x_1} < 2^{x_2}$

3. The domain of $\log (-x)$ is the empty set

4. The domain of $\log (-x^2)$ is the empty set

5. $\log_b(x + y) = \log_b x + \log_b y$

6. $\log_b x^n = n \log_b x$

7. If $\log_b x = y$, then $x = b^y$

8. $2^{\log_2 7} = 7$

9. The value of $\log x$ is negative for all values of x greater than 0 and less than 1.

10. The characteristic of a logarithm is always a number greater than 0 and less than or equal to 1.

11. Make a careful sketch of $y = \log_3 x$ on the interval $0 \le x \le 3$.

12. If $\log y = 5.4371$, find y using Table D. Show the interpolation.

13. Find x, if $10^{x^2 - 4} = 7$.

14. Use logarithms to evaluate $\dfrac{0.00035 \, (4912)}{17.9 \, (0.0803)}$.

15. Use logarithms to evaluate $\sqrt[3]{0.762}$.

Test 2 for Chapter 10

1. Sketch the graph of $y = \log_2 (x - 5)$.

2. Solve for x if   (a) $\log_x 8 = -3$ (b) $\log_3 x = 4$.

3. Express as a single logarithmic term: $3 \log x - \log (x^2 - 2) + 2 \log (x + 1)$.

4. Evaluate   (a) $\log_5 (125 \cdot 625)$ (b) $\log_2 (128 \cdot 64)$.

5. Use Table D to evaluate (a) log 3956   (b) log 0.00075 and (c) Antilog 6.7838 − 10.

6. Using the results in Problem 5, evaluate $N = \dfrac{3956}{0.00075}$.

7. Solve for x if $3^x = 2^{x+1}$

8. Solve for x if $\log x + \log (x - 3) = 1$.

9. Use logarithms to evaluate $(0.137)^{0.2}$.

TABLE A

Values of the
Trigonometric
Functions for
Degrees

x	sin x	cos x	tan x	cot x	sec x	csc x	
0° 0′	.00000	1.0000	.00000		1.0000		90° 0′
10′	.00291	1.0000	.00291	343.77	1.0000	343.78	50′
20′	.00582	1.0000	.00582	171.88	1.0000	171.89	40′
30′	.00873	1.0000	.00873	114.59	1.0000	114.59	30′
40′	.01164	.9999	.01164	85.940	1.0001	85.946	20′
50′	.01454	.9999	.01455	68.750	1.0001	68.757	10′
1° 0′	.01745	.9998	.01746	57.290	1.0002	57.299	89° 0′
10′	.02036	.9998	.02036	49.104	1.0002	49.114	50′
20′	.02327	.9997	.02328	42.964	1.0003	42.976	40′
30′	.02618	.9997	.02619	38.188	1.0003	38.202	30′
40′	.02908	.9996	.02910	34.368	1.0004	34.382	20′
50′	.03199	.9995	.03201	31.242	1.0005	31.258	10′
2° 0′	.03490	.9994	.03492	28.6363	1.0006	28.654	88° 0′
10′	.03781	.9993	.03783	26.4316	1.0007	26.451	50′
20′	.04071	.9992	.04075	24.5418	1.0008	24.562	40′
30′	.04362	.9990	.04366	22.9038	1.0010	22.926	30′
40′	.04653	.9989	.04658	21.4704	1.0011	21.494	20′
50′	.04943	.9988	.04949	20.2056	1.0012	20.230	10′
3° 0′	.05234	.9986	.05241	19.0811	1.0014	19.107	87° 0′
10′	.05524	.9985	.05533	18.0750	1.0015	18.103	50′
20′	.05814	.9983	.05824	17.1693	1.0017	17.198	40′
30′	.06105	.9981	.06116	16.3499	1.0019	16.380	30′
40′	.06395	.9980	.06408	15.6048	1.0021	15.637	20′
50′	.06685	.9978	.06700	14.9244	1.0022	14.958	10′
4° 0′	.06976	.9976	.06993	14.3007	1.0024	14.336	86° 0′
10′	.07266	.9974	.07285	13.7267	1.0027	13.763	50′
20′	.07556	.9971	.07578	13.1969	1.0029	13.235	40′
30′	.07846	.9969	.07870	12.7062	1.0031	12.746	30′
40′	.08136	.9967	.08163	12.2505	1.0033	12.291	20′
50′	.08426	.9964	.08456	11.8262	1.0036	11.868	10′
5° 0′	.08716	.9962	.08749	11.4301	1.0038	11.474	85° 0′
10′	.09005	.9959	.09042	11.0594	1.0041	11.105	50′
20′	.09295	.9957	.09335	10.7119	1.0044	10.758	40′
30′	.09585	.9954	.09629	10.3854	1.0046	10.433	30′
40′	.09874	.9951	.09923	10.0780	1.0049	10.128	20′
50′	.10164	.9948	.10216	9.7882	1.0052	9.839	10′
6° 0′	.10453	.9945	.10510	9.5144	1.0055	9.5668	84° 0′
	cos x	sin x	cot x	tan x	csc x	sec x	x

TABLE A

Trigonometric
Functions for
Degrees
(continued)

x	sin x	cos x	tan x	cot x	sec x	csc x	
6° 0′	.1045	.9945	.10510	9.5144	1.0055	9.5668	84° 0′
10′	.1074	.9942	.10805	9.2553	1.0058	9.3092	50′
20′	.1103	.9939	.11099	9.0098	1.0061	9.0652	40′
30′	.1132	.9936	.11394	8.7769	1.0065	8.8337	30′
40′	.1161	.9932	.11688	8.5555	1.0068	8.6138	20′
50′	.1190	.9929	.11983	8.3450	1.0072	8.4647	10′
7° 0′	.1219	.9925	.12278	8.1443	1.0075	8.2055	83° 0′
10′	.1248	.9922	.12574	7.9530	1.0079	8.0157	50′
20′	.1276	.9918	.12869	7.7704	1.0083	7.8344	40′
30′	.1305	.9914	.13165	7.5958	1.0086	7.6613	30′
40′	.1334	.9911	.1346	7.4287	1.0090	7.4957	20′
50′	.1363	.9907	.1376	7.2687	1.0094	7.3372	10′
8° 0′	.1392	.9903	.1405	7.1154	1.0098	7.1853	82° 0′
10′	.1421	.9899	.1435	6.9682	1.0102	7.0396	50′
20′	.1449	.9894	.1465	6.8269	1.0107	6.8998	40′
30′	.1478	.9890	.1495	6.6912	1.0111	6.7655	30′
40′	.1507	.9886	.1524	6.5606	1.0116	6.6363	20′
50′	.1536	.9881	.1554	6.4348	1.0120	6.5121	10′
9° 0′	.1564	.9877	.1584	6.3138	1.0125	6.3925	81° 0′
10′	.1593	.9872	.1614	6.1970	1.0129	6.2772	50′
20′	.1622	.9868	.1644	6.0844	1.0134	6.1661	40′
30′	.1650	.9863	.1673	5.9758	1.0139	6.0589	30′
40′	.1679	.9858	.1703	5.8708	1.0144	5.9554	20′
50′	.1708	.9853	.1733	5.7694	1.0149	5.8554	10′
10° 0′	.1736	.9848	.1763	5.6713	1.0154	5.7588	80° 0′
10′	.1765	.9843	.1793	5.5764	1.0160	5.6653	50′
20′	.1794	.9838	.1823	5.4845	1.0165	5.5749	40′
30′	.1822	.9833	.1853	5.3955	1.0170	5.4874	30′
40′	.1851	.9827	.1883	5.3093	1.0176	5.4026	20′
50′	.1880	.9822	.1914	5.2257	1.0182	5.3205	10′
11° 0′	.1908	.9816	.1944	5.1446	1.0187	5.2408	79° 0′
10′	.1937	.9811	.1974	5.0658	1.0193	5.1636	50′
20′	.1965	.9805	.2004	4.9894	1.0199	5.0886	40′
30′	.1994	.9799	.2035	4.9152	1.0205	5.0159	30′
40′	.2022	.9793	.2065	4.8430	1.0211	4.9452	20′
50′	.2051	.9787	.2095	4.7729	1.0217	4.8765	10′
12° 0′	.2079	.9781	.2126	4.7046	1.0223	4.8097	78° 0′
	cos x	sin x	cot x	tan x	csc x	sec x	x

TABLE A

Trigonometric
Functions for
Degrees
(continued)

x	sin x	cos x	tan x	cot x	sec x	csc x	
12° 0′	.2079	.9781	.2126	4.7046	1.0223	4.8097	78° 0′
10′	.2108	.9775	.2156	4.6382	1.0230	4.7448	50′
20′	.2136	.9769	.2186	4.5736	1.0236	4.6817	40′
30′	.2164	.9763	.2217	4.5107	1.0243	4.6202	30′
40′	.2193	.9757	.2247	4.4494	1.0249	4.5604	20′
50′	.2221	.9750	.2278	4.3897	1.0256	4.5022	10′
13° 0′	.2250	.9744	.2309	4.3315	1.0263	4.4454	77° 0′
10′	.2278	.9737	.2339	4.2747	1.0270	4.3901	50′
20′	.2306	.9730	.2370	4.2193	1.0277	4.3362	40′
30′	.2334	.9724	.2401	4.1653	1.0284	4.2837	30′
40′	.2363	.9717	.2432	4.1126	1.0291	4.2324	20′
50′	.2391	.9710	.2462	4.0611	1.0299	4.1824	10′
14° 0′	.2419	.9703	.2493	4.0108	1.0306	4.1336	76° 0′
10′	.2447	.9696	.2524	3.9617	1.0314	4.0859	50′
20′	.2476	.9689	.2555	3.9136	1.0321	4.0394	40′
30′	.2504	.9681	.2586	3.8667	1.0329	3.9939	30′
40′	.2532	.9674	.2617	3.8208	1.0337	3.9495	20′
50′	.2560	.9667	.2648	3.7760	1.0345	3.9061	10′
15° 0′	.2588	.9659	.2679	3.7321	1.0353	3.8637	75° 0′
10′	.2616	.9652	.2711	3.6891	1.0361	3.8222	50′
20′	.2644	.9644	.2742	3.6470	1.0369	3.7817	40′
30′	.2672	.9636	.2773	3.6059	1.0377	3.7420	30′
40′	.2700	.9628	.2805	3.5656	1.0386	3.7032	20′
50′	.2728	.9621	.2836	3.5261	1.0394	3.6652	10′
16° 0′	.2756	.9613	.2867	3.4874	1.0403	3.6280	74° 0′
10′	.2784	.9605	.2899	3.4495	1.0412	3.5915	50′
20′	.2812	.9596	.2931	3.4124	1.0421	3.5559	40′
30′	.2840	.9588	.2962	3.3759	1.0430	3.5209	30′
40′	.2868	.9580	.2994	3.3402	1.0439	3.4867	20′
50′	.2896	.9572	.3026	3.3052	1.0448	3.4532	10′
17° 0′	.2924	.9563	.3057	3.2709	1.0457	3.4203	73° 0′
10′	.2952	.9555	.3089	3.2371	1.0466	3.3881	50′
20′	.2979	.9546	.3121	3.2041	1.0476	3.3565	40′
30′	.3007	.9537	.3153	3.1716	1.0485	3.3255	30′
40′	.3035	.9528	.3185	3.1397	1.0495	3.2951	20′
50′	.3062	.9520	.3217	3.1084	1.0505	3.2653	10′
18° 0′	.3090	.9511	.3249	3.0777	1.0515	3.2361	72° 0′
	cos x	sin x	cot x	tan x	csc x	sec x	x

TABLE A

Trigonometric
Functions for
Degrees
(continued)

x	sin x	cos x	tan x	cot x	sec x	csc x	
18° 0′	.3090	.9511	.3249	3.0777	1.0515	3.2361	72° 0′
10′	.3118	.9502	.3281	3.0475	1.0525	3.2074	50′
20′	.3145	.9492	.3314	3.0178	1.0535	3.1792	40′
30′	.3173	.9483	.3346	2.9887	1.0545	3.1516	30′
40′	.3201	.9474	.3378	2.9600	1.0555	3.1244	20′
50′	.3228	.9465	.3411	2.9319	1.0566	3.0977	10′
19° 0′	.3256	.9455	.3443	2.9042	1.0576	3.0716	71° 0′
10′	.3283	.9446	.3476	2.8770	1.0587	3.0458	50′
20′	.3311	.9436	.3508	2.8502	1.0598	3.0206	40′
30′	.3338	.9426	.3541	2.8239	1.0609	2.9957	30′
40′	.3365	.9417	.3574	2.7980	1.0620	2.9714	20′
50′	.3393	.9407	.3607	2.7725	1.0631	2.9474	10′
20° 0′	.3420	.9397	.3640	2.7475	1.0642	2.9238	70° 0′
10′	.3448	.9387	.3673	2.7228	1.0653	2.9006	50′
20′	.3475	.9377	.3706	2.6985	1.0665	2.8879	40′
30′	.3502	.9367	.3739	2.6746	1.0676	2.8555	30′
40′	.3529	.9356	.3772	2.6511	1.0688	2.8334	20′
50′	.3557	.9346	.3805	2.6279	1.0700	2.8118	10′
21° 0′	.3584	.9336	.3839	2.6051	1.0712	2.7904	69° 0′
10′	.3611	.9325	.3872	2.5826	1.0724	2.7695	50′
20′	.3638	.9315	.3906	2.5605	1.0736	2.7488	40′
30′	.3665	.9304	.3939	2.5386	1.0748	2.7285	30′
40′	.3692	.9293	.3973	2.5172	1.0760	2.7085	20′
50′	.3719	.9283	.4006	2.4960	1.0773	2.6888	10′
22° 0′	.3746	.9272	.4040	2.4751	1.0785	2.6695	68° 0′
10′	.3773	.9261	.4074	2.4545	1.0798	2.6504	50′
20′	.3800	.9250	.4108	2.4342	1.0811	2.6316	40′
30′	.3827	.9239	.4142	2.4142	1.0824	2.6131	30′
40′	.3854	.9228	.4176	2.3945	1.0837	2.5949	20′
50′	.3881	.9216	.4210	2.3750	1.0850	2.5770	10′
23° 0′	.3907	.9205	.4245	2.3559	1.0864	2.5593	67° 0′
10′	.3934	.9194	.4279	2.3369	1.0877	2.5419	50′
20′	.3961	.9182	.4314	2.3183	1.0891	2.5247	40′
30′	.3987	.9171	.4348	2.2998	1.0904	2.5078	30′
40′	.4014	.9159	.4383	2.2817	1.0918	2.4912	20′
50′	.4041	.9147	.4417	2.2637	1.0932	2.4748	10′
24° 0′	.4067	.9135	.4452	2.2460	1.0946	2.4586	66° 0′
	cos x	sin x	cot x	tan x	csc x	sec x	x

TABLE A

Trigonometric
Functions for
Degrees
(continued)

x	sin x	cos x	tan x	cot x	sec x	csc x	
24° 0′	.4067	.9135	.4452	2.2460	1.0946	2.4586	66° 0′
10′	.4094	.9124	.4487	2.2286	1.0961	2.4426	50′
20′	.4120	.9112	.4522	2.2113	1.0975	2.4269	40′
30′	.4147	.9100	.4557	2.1943	1.0990	2.4114	30′
40′	.4173	.9088	.4592	2.1775	1.1004	2.3961	20′
50′	.4200	.9075	.4628	2.1609	1.1019	2.3811	10′
25° 0′	.4226	.9063	.4663	2.1445	1.1034	2.3662	65° 0′
10′	.4253	.9051	.4699	2.1283	1.1049	2.3515	50′
20′	.4279	.9038	.4734	2.1123	1.1064	2.3371	40′
30′	.4305	.9026	.4770	2.0965	1.1079	2.3228	30′
40′	.4331	.9013	.4806	2.0809	1.1095	2.3088	20′
50′	.4358	.9001	.4841	2.0655	1.1110	2.2949	10′
26° 0′	.4384	.8988	.4877	2.0503	1.1126	2.2812	64° 0′
10′	.4410	.8975	.4913	2.0353	1.1142	2.2677	50′
20′	.4436	.8962	.4950	2.0204	1.1158	2.2543	40′
30′	.4462	.8949	.4986	2.0057	1.1174	2.2412	30′
40′	.4488	.8936	.5022	1.9912	1.1190	2.2282	20′
50′	.4514	.8923	.5059	1.9768	1.1207	2.2154	10′
27° 0′	.4540	.8910	.5095	1.9626	1.1223	2.2027	63° 0′
10′	.4566	.8897	.5132	1.9486	1.1240	2.1902	50′
20′	.4592	.8884	.5169	1.9347	1.1257	2.1779	40′
30′	.4617	.8870	.5206	1.9210	1.1274	2.1657	30′
40′	.4643	.8857	.5243	1.9074	1.1291	2.1537	20′
50′	.4669	.8843	.5280	1.8940	1.1308	2.1418	10′
28° 0′	.4695	.8829	.5317	1.8807	1.1326	2.1301	62° 0′
10′	.4720	.8816	.5354	1.8676	1.1343	2.1185	50′
20′	.4746	.8802	.5392	1.8546	1.1361	2.1070	40′
30′	.4772	.8788	.5430	1.8418	1.1379	2.0957	30′
40′	.4797	.8774	.5467	1.8291	1.1397	2.0846	20′
50′	.4823	.8760	.5505	1.8165	1.1415	2.0736	10′
29° 0′	.4848	.8746	.5543	1.8040	1.1434	2.0627	61° 0′
10′	.4874	.8732	.5581	1.7917	1.1452	2.0519	50′
20′	.4899	.8718	.5619	1.7796	1.1471	2.0413	40′
30′	.4924	.8704	.5658	1.7675	1.1490	2.0308	30′
40′	.4950	.8689	.5696	1.7556	1.1509	2.0204	20′
50′	.4975	.8675	.5735	1.7437	1.1528	2.0101	10′
30° 0′	.5000	.8660	.5774	1.7321	1.1547	2.0000	60° 0′
	cos x	sin x	cot x	tan x	csc x	sec x	x

TABLE A

Trigonometric
Functions for
Degrees
(continued)

x	sin x	cos x	tan x	cot x	sec x	csc x	
30° 0′	.5000	.8660	.5774	1.7321	1.1547	2.0000	60° 0′
10′	.5025	.8646	.5812	1.7205	1.1567	1.9900	50′
20′	.5050	.8631	.5851	1.7090	1.1586	1.9801	40′
30′	.5075	.8616	.5890	1.6977	1.1606	1.9703	30′
40′	.5100	.8601	.5930	1.6864	1.1626	1.9606	20′
50′	.5125	.8587	.5969	1.6753	1.1646	1.9511	10′
31° 0′	.5150	.8572	.6009	1.6643	1.1666	1.9416	59° 0′
10′	.5175	.8557	.6048	1.6534	1.1687	1.9323	50′
20′	.5200	.8542	.6088	1.6426	1.1708	1.9230	40′
30′	.5225	.8526	.6128	1.6319	1.1728	1.9139	30′
40′	.5250	.8511	.6168	1.6212	1.1749	1.9049	20′
50′	.5275	.8496	.6208	1.6107	1.1770	1.8959	10′
32° 0′	.5299	.8480	.6249	1.6003	1.1792	1.8871	58° 0′
10′	.5324	.8465	.6289	1.5900	1.1813	1.8783	50′
20′	.5348	.8450	.6330	1.5798	1.1835	1.8699	40′
30′	.5373	.8434	.6371	1.5697	1.1857	1.8612	30′
40′	.5398	.8418	.6412	1.5597	1.1879	1.8527	20′
50′	.5422	.8403	.6453	1.5497	1.1901	1.8444	10′
33° 0′	.5446	.8387	.6494	1.5399	1.1924	1.8361	57° 0′
10′	.5471	.8371	.6536	1.5301	1.1946	1.8279	50′
20′	.5495	.8355	.6577	1.5204	1.1969	1.8198	40′
30′	.5519	.8339	.6619	1.5108	1.1992	1.8118	30′
40′	.5544	.8323	.6661	1.5013	1.2015	1.8039	20′
50′	.5568	.8307	.6703	1.4919	1.2039	1.7960	10′
34° 0′	.5592	.8290	.6745	1.4826	1.2062	1.7883	56° 0′
10′	.5616	.8274	.6787	1.4733	1.2086	1.7806	50′
20′	.5640	.8258	.6830	1.4641	1.2110	1.7730	40′
30′	.5664	.8241	.6873	1.4550	1.2134	1.7655	30′
40′	.5688	.8225	.6916	1.4460	1.2158	1.7581	20′
50′	.5712	.8208	.6959	1.4370	1.2183	1.7507	10′
35° 0′	.5736	.8192	.7002	1.4281	1.2208	1.7435	55° 0′
10′	.5760	.8175	.7046	1.4193	1.2233	1.7362	50′
20′	.5783	.8158	.7089	1.4106	1.2258	1.7291	40′
30′	.5807	.8141	.7133	1.4019	1.2283	1.7221	30′
40′	.5831	.8124	.7177	1.3934	1.2309	1.7151	20′
50′	.5854	.8107	.7221	1.3848	1.2335	1.7082	10′
36° 0′	.5878	.8090	.7265	1.3764	1.2361	1.7013	54° 0′
	cos x	sin x	cot x	tan x	csc x	sec x	x

TABLE A

Trigonometric
Functions for
Degrees
(continued)

x	sin x	cos x	tan x	cot x	sec x	csc x	
36° 0′	.5878	.8090	.7265	1.3764	1.2361	1.7013	54° 0′
10′	.5901	.8073	.7310	1.3680	1.2387	1.6945	50′
20′	.5925	.8056	.7355	1.3597	1.2413	1.6878	40′
30′	.5948	.8039	.7400	1.3514	1.2440	1.6812	30′
40′	.5972	.8021	.7445	1.3432	1.2467	1.6746	20′
50′	.5995	.8004	.7490	1.3351	1.2494	1.6681	10′
37° 0′	.6018	.7986	.7536	1.3270	1.2521	1.6616	53° 0′
10′	.6041	.7969	.7581	1.3190	1.2549	1.6553	50′
20′	.6065	.7951	.7627	1.3111	1.2577	1.6489	40′
30′	.6088	.7934	.7673	1.3032	1.2605	1.6427	30′
40′	.6111	.7916	.7720	1.2954	1.2633	1.6365	20′
50′	.6134	.7898	.7766	1.2876	1.2662	1.6304	10′
38° 0′	.6157	.7880	.7813	1.2799	1.2690	1.6243	52° 0′
10′	.6180	.7862	.7860	1.2723	1.2719	1.6183	50′
20′	.6202	.7844	.7907	1.2647	1.2748	1.6123	40′
30′	.6225	.7826	.7954	1.2572	1.2779	1.6064	30′
40′	.6248	.7808	.8002	1.2497	1.2808	1.6005	20′
50′	.6271	.7790	.8050	1.2423	1.2837	1.5948	10′
39° 0′	.6293	.7771	.8098	1.2349	1.2868	1.5890	51° 0′
10′	.6316	.7753	.8146	1.2276	1.2898	1.5833	50′
20′	.6338	.7735	.8195	1.2203	1.2929	1.5777	40′
30′	.6361	.7716	.8243	1.2131	1.2960	1.5721	30′
40′	.6383	.7698	.8292	1.2059	1.2991	1.5666	20′
50′	.6406	.7679	.8342	1.1988	1.3022	1.5611	10′
40° 0′	.6428	.7660	.8391	1.1918	1.3054	1.5557	50° 0′
10′	.6450	.7642	.8441	1.1847	1.3086	1.5504	50′
20′	.6472	.7623	.8491	1.1778	1.3118	1.5450	40′
30′	.6494	.7604	.8541	1.1708	1.3151	1.5398	30′
40′	.6517	.7585	.8591	1.1640	1.3184	1.5346	20′
50′	.6539	.7566	.8642	1.1571	1.3217	1.5294	10′
41° 0′	.6561	.7547	.8693	1.1504	1.3250	1.5243	49° 0′
10′	.6583	.7528	.8744	1.1436	1.3284	1.5192	50′
20′	.6604	.7509	.8796	1.1369	1.3318	1.5142	40′
30′	.6626	.7490	.8847	1.1303	1.3352	1.5092	30′
40′	.6648	.7470	.8899	1.1237	1.3386	1.5042	20′
50′	.6670	.7451	.8952	1.1171	1.3421	1.4993	10′
42° 0′	.6691	.7431	.9004	1.1106	1.3456	1.4945	48° 0′
	cos x	sin x	cot x	tan x	csc x	sec x	x

TABLE A

Trigonometric
Functions for
Degrees
(continued)

x	sin x	cos x	tan x	cot x	sec x	csc x	
42° 0′	.6691	.7431	.9004	1.1106	1.3456	1.4945	48° 0′
10′	.6713	.7412	.9057	1.1041	1.3492	1.4897	50′
20′	.6734	.7392	.9110	1.0977	1.3527	1.4849	40′
30′	.6756	.7373	.9163	1.0913	1.3563	1.4802	30′
40′	.6777	.7353	.9217	1.0850	1.3600	1.4755	20′
50′	.6799	.7333	.9271	1.0786	1.3636	1.4709	10′
43° 0′	.6820	.7314	.9325	1.0724	1.3673	1.4663	47° 0′
10′	.6841	.7294	.9380	1.0661	1.3711	1.4617	50′
20′	.6862	.7274	.9435	1.0599	1.3748	1.4572	40′
30′	.6884	.7254	.9490	1.0538	1.3786	1.4527	30′
40′	.6905	.7234	.9545	1.0477	1.3824	1.4483	20′
50′	.6926	.7214	.9601	1.0416	1.3863	1.4439	10′
44° 0′	.6947	.7193	.9657	1.0355	1.3902	1.4396	46° 0′
10′	.6967	.7173	.9713	1.0295	1.3941	1.4352	50′
20′	.6988	.7153	.9770	1.0235	1.3980	1.4310	40′
30′	.7009	.7133	.9827	1.0176	1.4020	1.4267	30′
40′	.7030	.7112	.9884	1.0117	1.4061	1.4225	20′
50′	.7050	.7092	.9942	1.0058	1.4101	1.4184	10′
45° 0′	.7071	.7071	1.0000	1.0000	1.4142	1.4142	45° 0′
	cos x	sin x	cot x	tan x	csc x	sec x	x

TABLE B

Values of
Trigonometric
Functions —
Decimal
Subdivisions
(csc $\theta = 1/\sin \theta$;
sec $\theta = 1/\cos \theta$)

θ	$\sin \theta$	$\cos \theta$	$\tan \theta$	$\cot \theta$	
0.0	0.00000	1.0000	0.00000	∞	90.0
.1	.00175	1.0000	.00175	573.0	.9
.2	.00349	1.0000	.00349	286.5	.8
.3	.00524	1.0000	.00524	191.0	.7
.4	.00698	1.0000	.00698	143.24	.6
.5	.00873	1.0000	.00873	114.59	.5
.6	.01047	0.9999	.01047	95.49	.4
.7	.01222	.9999	.01222	81.85	.3
.8	.01396	.9999	.01396	71.62	.2
.9	.01571	.9999	.01571	63.66	.1
1.0	0.01745	0.9998	0.01746	57.29	89.0
.1	.01920	.9998	.01920	52.08	.9
.2	.02094	.9998	.02095	47.74	.8
.3	.02269	.9997	.02269	44.07	.7
.4	.02443	.9997	.02444	40.92	.6
.5	.02618	.9997	.02619	38.19	.5
.6	.02792	.9996	.02793	35.80	.4
.7	.02967	.9996	.02968	33.69	.3
.8	.03141	.9995	.03143	31.82	.2
.9	.03316	.9995	.03317	30.14	.1
2.0	0.03490	0.9994	0.03492	28.64	88.0
.1	.03664	.9993	.03667	27.27	.9
.2	.03839	.9993	.03842	26.03	.8
.3	.04013	.9992	.04016	24.90	.7
.4	.04188	.9991	.04191	23.86	.6
.5	.04362	.9990	.04366	22.90	.5
.6	.04536	.9990	.04541	22.02	.4
.7	.04711	.9989	.04716	21.20	.3
.8	.04885	.9988	.04891	20.45	.2
.9	.05059	.9987	.05066	19.74	.1
3.0	0.05234	0.9986	0.05241	19.081	87.0
.1	.05408	.9985	.05416	18.464	.9
.2	.05582	.9984	.05591	17.886	.8
.3	.05756	.9983	.05766	17.343	.7
.4	.05931	.9982	.05941	16.832	.6
.5	.06105	.9981	.06116	16.350	.5
.6	.06279	.9980	.06291	15.895	.4
.7	.06453	.9979	.06467	15.464	.3
.8	.06627	.9978	.06642	15.056	.2
.9	.06802	.9977	.06817	14.669	.1
4.0	0.06976	0.9976	0.06993	14.301	86.0
	$\cos \theta$	$\sin \theta$	$\cot \theta$	$\tan \theta$	θ

TABLE B

Values of
Trigonometric
Functions —
Decimal
Subdivisions
(csc $\theta = 1/\sin \theta$;
sec $\theta = 1/\cos \theta$)
(continued)

θ	$\sin \theta$	$\cos \theta$	$\tan \theta$	$\cot \theta$	
4.0	0.06976	0.9976	0.06993	14.301	86.0
.1	.07150	.9974	.07168	13.951	.9
.2	.07324	.9973	.07344	13.617	.8
.3	.07498	.9972	.07519	13.300	.7
.4	.07672	.9971	.07695	12.996	.6
.5	.07846	.9969	.07870	12.706	.5
.6	.08020	.9968	.08046	12.429	.4
.7	.08194	.9966	.08221	12.163	.3
.8	.08368	.9965	.08397	11.909	.2
.9	.08542	.9963	.08573	11.664	.1
5.0	0.08716	0.9962	0.08749	11.430	85.0
.1	.08889	.9960	.08925	11.205	.9
.2	.09063	.9959	.09101	10.988	.8
.3	.09237	.9957	.09277	10.780	.7
.4	.09411	.9956	.09453	10.579	.6
.5	.09585	.9954	.09629	10.385	.5
.6	.09758	.9952	.09805	10.199	.4
.7	.09932	.9951	.09981	10.019	.3
.8	.10106	.9949	.10158	9.845	.2
.9	.10279	.9947	.10334	9.677	.1
6.0	0.10453	0.9945	0.10510	9.514	84.0
.1	.10626	.9943	.10687	9.357	.9
.2	.10800	.9942	.10863	9.205	.8
.3	.10973	.9940	.11040	9.058	.7
.4	.11147	.9938	.11217	8.915	.6
.5	.11320	.9936	.11394	8.777	.5
.6	.11494	.9934	.11570	8.643	.4
.7	.11667	.9932	.11747	8.513	.3
.8	.11840	.9930	.11924	8.386	.2
.9	.12014	.9928	.12101	8.264	.1
7.0	0.12187	0.9925	0.12278	8.144	83.0
.1	.12360	.9923	.12456	8.028	.9
.2	.12533	.9921	.12633	7.916	.8
.3	.12706	.9919	.12810	7.806	.7
.4	.12880	.9917	.12988	7.700	.6
.5	.13053	.9914	.13165	7.596	.5
.6	.13226	.9912	.13343	7.495	.4
.7	.13399	.9910	.13521	7.396	.3
.8	.13572	.9907	.13698	7.300	.2
.9	.13744	.9905	.13876	7.207	.1
8.0	0.13917	0.9903	0.14054	7.115	82.0
	$\cos \theta$	$\sin \theta$	$\cot \theta$	$\tan \theta$	θ

TABLE B

Values of
Trigonometric
Functions —
Decimal
Subdivisions
($\csc \theta = 1/\sin \theta$;
$\sec \theta = 1/\cos \theta$)
(continued)

θ	$\sin \theta$	$\cos \theta$	$\tan \theta$	$\cot \theta$	
8.0	0.13917	0.9903	0.14054	7.115	82.0
.1	.14090	.9900	.14232	7.026	.9
.2	.14263	.9898	.14410	6.940	.8
.3	.14436	.9895	.14588	6.855	.7
.4	.14608	.9893	.14767	6.772	.6
.5	.14781	.9890	.14945	6.691	.5
.6	.14954	.9888	.15124	6.612	.4
.7	.15126	.9885	.15302	6.535	.3
.8	.15299	.9882	.15481	6.460	.2
.9	.15471	.9880	.15660	6.386	.1
9.0	0.15643	0.9877	0.15838	6.314	81.0
.1	.15816	.9874	.16017	6.243	.9
.2	.15988	.9871	.16196	6.174	.8
.3	.16160	.9869	.16376	6.107	.7
.4	.16333	.9866	.16555	6.041	.6
.5	.16505	.9863	.16734	5.976	.5
.6	.16677	.9860	.16914	5.912	.4
.7	.16849	.9857	.17093	5.850	.3
.8	.17021	.9854	.17273	5.789	.2
.9	.17193	.9851	.17453	5.730	.1
10.0	0.1736	0.9848	0.1763	5.671	80.0
.1	.1754	.9845	.1781	5.614	.9
.2	.1771	.9842	.1799	5.558	.8
.3	.1788	.9839	.1817	5.503	.7
.4	.1805	.9836	.1835	5.449	.6
.5	.1822	.9833	.1853	5.396	.5
.6	.1840	.9829	.1871	5.343	.4
.7	.1857	.9826	.1890	5.292	.3
.8	.1874	.9823	.1908	5.242	.2
.9	.1891	.9820	.1926	5.193	.1
11.0	0.1908	0.9816	0.1944	5.145	79.0
.1	.1925	.9813	.1962	5.097	.9
.2	.1942	.9810	.1980	5.050	.8
.3	.1959	.9806	.1998	5.005	.7
.4	.1977	.9803	.2016	4.959	.6
.5	.1994	.9799	.2035	4.915	.5
.6	.2011	.9796	.2053	4.872	.4
.7	.2028	.9792	.2071	4.829	.3
.8	.2045	.9789	.2089	4.787	.2
.9	.2062	.9785	.2107	4.745	.1
12.0	0.2079	0.9781	0.2126	4.705	78.0
	$\cos \theta$	$\sin \theta$	$\cot \theta$	$\tan \theta$	θ

TABLE B

Values of
Trigonometric
Functions —
Decimal
Subdivisions
(csc $\theta = 1/\sin \theta$;
sec $\theta = 1/\cos \theta$)
(continued)

θ	$\sin \theta$	$\cos \theta$	$\tan \theta$	$\cot \theta$	
12.0	0.2079	0.9781	0.2126	4.705	78.0
.1	.2096	.9778	.2144	4.665	.9
.2	.2113	.9774	.2162	4.625	.8
.3	.2130	.9770	.2180	4.586	.7
.4	.2147	.9767	.2199	4.548	.6
.5	.2164	.9763	.2217	4.511	.5
.6	.2181	.9759	.2235	4.474	.4
.7	.2198	.9755	.2254	4.437	.3
.8	.2215	.9751	.2272	4.402	.2
.9	.2233	.9748	.2290	4.366	.1
13.0	0.2250	0.9744	0.2309	4.331	77.0
.1	.2267	.9740	.2327	4.297	.9
.2	.2284	.9736	.2345	4.264	.8
.3	.2300	.9732	.2364	4.230	.7
.4	.2317	.9728	.2382	4.198	.6
.5	.2334	.9724	.2401	4.165	.5
.6	.2351	.9720	.2419	4.134	.4
.7	.2368	.9715	.2438	4.102	.3
.8	.2385	.9711	.2456	4.071	.2
.9	.2402	.9707	.2475	4.041	.1
14.0	0.2419	0.9703	0.2493	4.011	76.0
.1	.2436	.9699	.2512	3.981	.9
.2	.2453	.9694	.2530	3.952	.8
.3	.2470	.9690	.2549	3.923	.7
.4	.2487	.9686	.2568	3.895	.6
.5	.2504	.9681	.2586	3.867	.5
.6	.2521	.9677	.2605	3.839	.4
.7	.2538	.9673	.2623	3.812	.3
.8	.2554	.9668	.2642	3.785	.2
.9	.2571	.9664	.2661	3.758	.1
15.0	0.2588	0.9659	0.2679	3.732	75.0
.1	.2605	.9655	.2698	3.706	.9
.2	.2622	.9650	.2717	3.681	.8
.3	.2639	.9646	.2736	3.655	.7
.4	.2656	.9641	.2754	3.630	.6
.5	.2672	.9636	.2773	3.606	.5
.6	.2689	.9632	.2792	3.582	.4
.7	.2706	.9627	.2811	3.558	.3
.8	.2723	.9622	.2830	3.534	.2
.9	.2740	.9617	.2849	3.511	.1
16.0	0.2756	0.9613	0.2867	3.487	74.0
	$\cos \theta$	$\sin \theta$	$\cot \theta$	$\tan \theta$	θ

TABLE B

Values of
Trigonometric
Functions —
Decimal
Subdivisions
(csc $\theta = 1/\sin \theta$;
sec $\theta = 1/\cos \theta$)
(continued)

θ	$\sin \theta$	$\cos \theta$	$\tan \theta$	$\cot \theta$	
16.0	0.2756	0.9613	0.2867	3.487	74.0
.1	.2773	.9608	.2886	3.465	.9
.2	.2790	.9603	.2905	3.442	.8
.3	.2807	.9598	.2924	3.420	.7
.4	.2823	.9593	.2943	3.398	.6
.5	.2840	.9588	.2962	3.376	.5
.6	.2857	.9583	.2981	3.354	.4
.7	.2874	.9578	.3000	3.333	.3
.8	.2890	.9573	.3019	3.312	.2
.9	.2907	.9568	.3038	3.291	.1
17.0	0.2924	0.9563	0.3057	3.271	73.0
.1	.2940	.9558	.3076	3.251	.9
.2	.2957	.9553	.3096	3.230	.8
.3	.2974	.9548	.3115	3.211	.7
.4	.2990	.9542	.3134	3.191	.6
.5	.3007	.9537	.3153	3.172	.5
.6	.3024	.9532	.3172	3.152	.4
.7	.3040	.9527	.3191	3.133	.3
.8	.3057	.9521	.3211	3.115	.2
.9	.3074	.9516	.3230	3.096	.1
18.0	0.3090	0.9511	0.3249	3.078	72.0
.1	.3107	.9505	.3269	3.060	.9
.2	.3123	.9500	.3288	3.042	.8
.3	.3140	.9494	.3307	3.024	.7
.4	.3156	.9489	.3327	3.006	.6
.5	.3173	.9483	.3346	2.989	.5
.6	.3190	.9478	.3365	2.971	.4
.7	.3206	.9472	.3385	2.954	.3
.8	.3223	.9466	.3404	2.937	.2
.9	.3239	.9461	.3424	2.921	.1
19.0	0.3256	0.9455	0.3443	2.904	71.0
.1	.3272	.9449	.3463	2.888	.9
.2	.3289	.9444	.3482	2.872	.8
.3	.3305	.9438	.3502	2.856	.7
.4	.3322	.9432	.3522	2.840	.6
.5	.3338	.9426	.3541	2.824	.5
.6	.3355	.9421	.3561	2.808	.4
.7	.3371	.9415	.3581	2.793	.3
.8	.3387	.9409	.3600	2.778	.2
.9	.3404	.9403	.3620	2.762	.1
20.0	0.3420	0.9397	0.3640	2.747	70.0
	$\cos \theta$	$\sin \theta$	$\cot \theta$	$\tan \theta$	θ

TABLE B

Values of
Trigonometric
Functions —
Decimal
Subdivisions
(csc $\theta = 1/\sin \theta$;
sec $\theta = 1/\cos \theta$)
(continued)

θ	$\sin \theta$	$\cos \theta$	$\tan \theta$	$\cot \theta$	
20.0	0.3420	0.9397	0.3640	2.747	70.0
.1	.3437	.9391	.3659	2.733	.9
.2	.3453	.9385	.3679	2.718	.8
.3	.3469	.9379	.3699	2.703	.7
.4	.3486	.9373	.3719	2.689	.6
.5	.3502	.9367	.3739	2.675	.5
.6	.3518	.9361	.3759	2.660	.4
.7	.3535	.9354	.3779	2.646	.3
.8	.3551	.9348	.3799	2.633	.2
.9	.3567	.9342	.3819	2.619	.1
21.0	0.3584	0.9336	0.3839	2.605	69.0
.1	.3600	.9330	.3859	2.592	.9
.2	.3616	.9323	.3879	2.578	.8
.3	.3633	.9317	.3899	2.565	.7
.4	.3649	.9311	.3919	2.552	.6
.5	.3665	.9304	.3939	2.539	.5
.6	.3681	.9298	.3959	2.526	.4
.7	.3697	.9291	.3979	2.513	.3
.8	.3714	.9285	.4000	2.500	.2
.9	.3730	.9278	.4020	2.488	.1
22.0	0.3746	0.9272	0.4040	2.475	68.0
.1	.3762	.9265	.4061	2.463	.9
.2	.3778	.9259	.4081	2.450	.8
.3	.3795	.9252	4101	2.438	.7
.4	.3811	.9245	.4122	2.426	.6
.5	.3827	.9239	.4142	2.414	.5
.6	.3843	.9232	.4163	2.402	.4
.7	.3859	.9225	.4183	2.391	.3
.8	.3875	.9219	.4204	2.379	.2
.9	.3891	.9212	.4224	2.367	.1
23.0	0.3907	0.9205	0.4245	2.356	67.0
.1	.3923	.9198	.4265	2.344	.9
.2	.3939	.9191	.4286	2.333	.8
.3	.3955	.9184	.4307	2.322	.7
.4	.3971	.9178	.4327	2.311	.6
.5	.3987	.9171	.4348	2.300	.5
.6	.4003	.9164	.4369	2.289	.4
.7	.4019	.9157	.4390	2.278	.3
.8	.4035	.9150	.4411	2.267	.2
.9	.4051	.9143	.4431	2.257	.1
24.0	0.4067	0.9135	0.4452	2.246	66.0
	$\cos \theta$	$\sin \theta$	$\cot \theta$	$\tan \theta$	θ

TABLE B

Values of
Trigonometric
Functions —
Decimal
Subdivisions
(csc $\theta = 1/\sin \theta$;
sec $\theta = 1/\cos \theta$)
(continued)

θ	$\sin \theta$	$\cos \theta$	$\tan \theta$	$\cot \theta$	
24.0	0.4067	0.9135	0.4452	2.246	66.0
.1	.4083	.9128	.4473	2.236	.9
.2	.4099	.9121	.4494	2.225	.8
.3	.4115	.9114	.4515	2.215	.7
.4	.4131	.9107	.4536	2.204	.6
.5	.4147	.9100	.4557	2.194	.5
.6	.4163	.9092	.4578	2.184	.4
.7	.4179	.9085	.4599	2.174	.3
.8	.4195	.9078	.4621	2.164	.2
.9	.4210	.9070	.4642	2.154	.1
25.0	0.4226	0.9063	0.4663	2.145	65.0
.1	.4242	.9056	.4684	2.135	.9
.2	.4258	.9048	.4706	2.125	.8
.3	.4274	.9041	.4727	2.116	.7
.4	.4289	.9033	.4748	2.106	.6
.5	.4305	.9026	.4770	2.097	.5
.6	.4321	.9018	.4791	2.087	.4
.7	.4337	.9011	.4813	2.078	.3
.8	.4352	.9003	.4834	2.069	.2
.9	.4368	.8996	.4856	2.059	.1
26.0	0.4384	0.8988	0.4877	2.050	64.0
.1	.4399	.8980	.4899	2.041	.9
.2	.4415	.8973	.4921	2.032	.8
.3	.4431	.8965	.4942	2.023	.7
.4	.4446	.8957	.4964	2.014	.6
.5	.4462	.8949	.4986	2.006	.5
.6	.4478	.8942	.5008	1.997	.4
.7	.4493	.8934	.5029	1.988	.3
.8	.4509	.8926	.5051	1.980	.2
.9	.4524	.8918	.5073	1.971	.1
27.0	0.4540	0.8910	0.5095	1.963	63.0
.1	.4555	.8902	.5117	1.954	.9
.2	.4571	.8894	.5139	1.946	.8
.3	.4586	.8886	.5161	1.937	.7
.4	.4602	.8878	.5184	1.929	.6
.5	.4617	.8870	.5206	1.921	.5
.6	.4633	.8862	.5228	1.913	.4
.7	.4648	.8854	.5250	1.905	.3
.8	.4664	.8846	.5272	1.897	.2
.9	.4679	.8838	.5295	1.889	.1
28.0	0.4695	0.8829	0.5317	1.881	62.0
	$\cos \theta$	$\sin \theta$	$\cot \theta$	$\tan \theta$	θ

TABLE B

Values of
Trigonometric
Functions —
Decimal
Subdivisions
(csc $\theta = 1/\sin \theta$;
sec $\theta = 1/\cos \theta$)
(continued)

θ	$\sin \theta$	$\cos \theta$	$\tan \theta$	$\cot \theta$	
28.0	0.4695	0.8829	0.5317	1.881	62.0
.1	.4710	.8821	.5340	1.873	.9
.2	.4726	.8813	.5362	1.865	.8
.3	.4741	.8805	.5384	1.857	.7
.4	.4756	.8796	.5407	1.849	.6
.5	.4772	.8788	.5430	1.842	.5
.6	.4787	.8780	.5452	1.834	.4
.7	.4802	.8771	.5475	1.827	.3
.8	.4818	.8763	.5498	1.819	.2
.9	.4833	.8755	.5520	1.811	.1
29.0	0.4848	0.8746	0.5543	1.804	61.0
.1	.4863	.8738	.5566	1.797	.9
.2	.4879	.8729	.5589	1.789	.8
.3	.4894	.8721	.5612	1.782	.7
.4	.4909	.8712	.5635	1.775	.6
.5	.4924	.8704	.5658	1.767	.5
.6	.4939	.8695	.5681	1.760	.4
.7	.4955	.8686	.5704	1.753	.3
.8	.4970	.8678	.5727	1.746	.2
.9	.4985	.8669	.5750	1.739	.1
30.0	0.5000	0.8660	0.5774	1.7321	60.0
.1	.5015	.8652	.5797	1.7251	.9
.2	.5030	.8643	.5820	1.7182	.8
.3	.5045	.8634	.5844	1.7113	.7
.4	.5060	.8625	.5867	1.7045	.6
.5	.5075	.8616	.5890	1.6977	.5
.6	.5090	.8607	.5914	1.6909	.4
.7	.5105	.8599	.5938	1.6842	.3
.8	.5120	.8590	.5961	1.6775	.2
.9	.5135	.8581	.5985	1.6709	.1
31.0	0.5150	0.8572	0.6009	1.6643	59.0
.1	.5165	.8563	.6032	1.6577	.9
.2	.5180	.8554	.6056	1.6512	.8
.3	.5195	.8545	.6080	1.6447	.7
.4	.5210	.8536	.6104	1.6383	.6
.5	.5225	.8526	.6128	1.6319	.5
.6	.5240	.8517	.6152	1.6255	.4
.7	.5255	.8508	.6176	1.6191	.3
.8	.5270	.8499	.6200	1.6128	.2
.9	.5284	.8490	.6224	1.6066	.1
32.0	0.5299	0.8480	0.6249	1.6003	58.0
	$\cos \theta$	$\sin \theta$	$\cot \theta$	$\tan \theta$	θ

TABLE B

Values of
Trigonometric
Functions —
Decimal
Subdivisions
(csc $\theta = 1/\sin \theta$;
sec $\theta = 1/\cos \theta$)
(continued)

θ	$\sin \theta$	$\cos \theta$	$\tan \theta$	$\cot \theta$	
32.0	0.5299	0.8480	0.6249	1.6003	58.0
.1	.5314	.8471	.6273	1.5941	.9
.2	.5329	.8462	.6297	1.5880	.8
.3	.5344	.8453	.6322	1.5818	.7
.4	.5358	.8443	.6346	1.5757	.6
.5	.5373	.8434	.6371	1.5697	.5
.6	.5388	.8425	.6395	1.5637	.4
.7	.5402	.8415	.6420	1.5577	.3
.8	.5417	.8406	.6445	1.5517	.2
.9	.5432	.8396	.6469	1.5458	.1
33.0	0.5446	0.8387	0.6494	1.5399	57.0
.1	.5461	.8377	.6519	1.5340	.9
.2	.5476	.8368	.6544	1.5282	.8
.3	.5490	.8358	.6569	1.5224	.7
.4	.5505	.8348	.6594	1.5166	.6
.5	.5519	.8339	.6619	1.5108	.5
.6	.5534	.8329	.6644	1.5051	.4
.7	.5548	.8320	.6669	1.4994	.3
.8	.5563	.8310	.6694	1.4938	.2
.9	.5577	.8300	.6720	1.4882	.1
34.0	0.5592	0.8290	0.6745	1.4826	56.0
.1	.5606	.8281	.6771	1.4770	.9
.2	.5621	.8271	.6796	1.4715	.8
.3	.5635	.8261	.6822	1.4659	.7
.4	.5650	.8251	.6847	1.4605	.6
.5	.5664	.8241	.6873	1.4550	.5
.6	.5678	.8231	.6899	1.4496	.4
.7	.5693	.8221	.6924	1.4442	.3
.8	.5707	.8211	.6950	1.4388	.2
.9	.5721	.8202	.6976	1.4335	.1
35.0	0.5736	0.8192	0.7002	1.4281	55.0
.1	.5750	.8181	.7028	1.4229	.9
.2	.5764	.8171	.7054	1.4176	.8
.3	.5779	.8161	.7080	1.4124	.7
.4	.5793	.8151	.7107	1.4071	.6
.5	.5807	.8141	.7133	1.4019	.5
.6	.5821	.8131	.7159	1.3968	.4
.7	.5835	.8121	.7186	1.3916	.3
.8	.5850	.8111	.7212	1.3865	.2
.9	.5864	.8100	.7239	1.3814	.1
36.0	0.5878	0.8090	0.7265	1.3764	54.0
	$\cos \theta$	$\sin \theta$	$\cot \theta$	$\tan \theta$	θ

TABLE B

Values of
Trigonometric
Functions —
Decimal
Subdivisions
(csc $\theta = 1/\sin \theta$;
sec $\theta = 1/\cos \theta$)
(continued)

θ	sin θ	cos θ	tan θ	cot θ	
36.0	0.5878	0.8090	0.7265	1.3764	54.0
.1	.5892	.8080	.7292	1.3713	.9
.2	.5906	.8070	.7319	1.3663	.8
.3	.5920	.8059	.7346	1.3613	.7
.4	.5934	.8049	.7373	1.3564	.6
.5	.5948	.8039	.7400	1.3514	.5
.6	.5962	.8028	.7427	1.3465	.4
.7	.5976	.8018	.7454	1.3416	.3
.8	.5990	.8007	.7481	1.3367	.2
.9	.6004	.7997	.7508	1.3319	.1
37.0	0.6018	0.7986	0.7536	1.3270	53.0
.1	.6032	.7976	.7563	1.3222	.9
.2	.6046	.7965	.7590	1.3175	.8
.3	.6060	.7955	.7618	1.3127	.7
.4	.6074	.7944	.7646	1.3079	.6
.5	.6088	.7934	.7673	1.3032	.5
.6	.6101	.7923	.7701	1.2985	.4
.7	.6115	.7912	.7729	1.2938	.3
.8	.6129	.7902	.7757	1.2892	.2
.9	.6143	.7891	.7785	1.2846	.1
38.0	0.6157	0.7880	0.7813	1.2799	52.0
.1	.6170	.7869	.7841	1.2753	.9
.2	.6184	.7859	.7869	1.2708	.8
.3	.6198	.7848	.7898	1.2662	.7
.4	.6211	.7837	.7926	1.2617	.6
.5	.6225	.7826	.7954	1.2572	.5
.6	.6239	.7815	.7983	1.2527	.4
.7	.6252	.7804	.8012	1.2482	.3
.8	.6266	.7793	.8040	1.2437	.2
.9	.6280	.7782	.8069	1.2393	.1
39.0	0.6293	0.7771	0.8098	1.2349	51.0
.1	.6307	.7760	.8127	1.2305	.9
.2	.6320	.7749	.8156	1.2261	.8
.3	.6334	.7738	.8185	1.2218	.7
.4	.6347	.7727	.8214	1.2174	.6
.5	.6361	.7716	.8243	1.2131	.5
.6	.6374	.7705	.8273	1.2088	.4
.7	.6388	.7694	.8302	1.2045	.3
.8	.6401	.7683	.8332	1.2002	.2
.9	.6414	.7672	.8361	1.1960	.1
40.0	0.6428	0.7660	0.8391	1.1918	50.0
	cos θ	sin θ	cot θ	tan θ	θ

TABLE B

Values of
Trigonometric
Functions —
Decimal
Subdivisions
($\csc \theta = 1/\sin \theta$;
$\sec \theta = 1/\cos \theta$)
(continued)

θ	$\sin \theta$	$\cos \theta$	$\tan \theta$	$\cot \theta$	
40.0	0.6428	0.7660	0.8391	1.1918	50.0
.1	.6441	.7649	.8421	1.1875	.9
.2	.6455	.7638	.8451	1.1833	.8
.3	.6468	.7627	.8481	1.1792	.7
.4	.6481	.7615	.8511	1.1750	.6
.5	.6494	.7604	.8541	1.1708	.5
.6	.6508	.7593	.8571	1.1667	.4
.7	.6521	.7581	.8601	1.1626	.3
.8	.6534	.7570	.8632	1.1585	.2
.9	.6547	.7559	.8662	1.1544	.1
41.0	0.6561	0.7547	0.8693	1.1504	49.0
.1	.6574	.7536	.8724	1.1463	.9
.2	.6587	.7524	.8754	1.1423	.8
.3	.6600	.7513	.8785	1.1383	.7
.4	.6613	.7501	.8816	1.1343	.6
.5	.6626	.7490	.8847	1.1303	.5
.6	.6639	.7478	.8878	1.1263	.4
.7	.6652	.7466	.8910	1.1224	.3
.8	.6665	.7455	.8941	1.1184	.2
.9	.6678	.7443	.8972	1.1145	.1
42.0	0.6691	0.7431	0.9004	1.1106	48.0
.1	.6704	.7420	.9036	1.1067	.9
.2	.6717	.7408	.9067	1.1028	.8
.3	.6730	.7396	.9099	1.0990	.7
.4	.6743	.7385	.9131	1.0951	.6
.5	.6756	.7373	.9163	1.0913	.5
.6	.6769	.7361	.9195	1.0875	.4
.7	.6782	.7349	.9228	1.0837	.3
.8	.6794	.7337	.9260	1.0799	.2
.9	.6807	.7325	.9293	1.0761	.1
43.0	0.6820	0.7314	0.9325	1.0724	47.0
.1	.6833	.7302	.9358	1.0686	.9
.2	.6845	.7290	.9391	1.0649	.8
.3	.6858	.7278	.9424	1.0612	.7
.4	.6871	.7266	.9457	1.0575	.6
.5	.6884	.7254	.9490	1.0538	.5
.6	.6896	.7242	.9523	1.0501	.4
.7	.6909	.7230	.9556	1.0464	.3
.8	.6921	.7218	.9590	1.0428	.2
.9	.6934	.7206	.9623	1.0392	.1
44.0	0.6947	0.7193	0.9657	1.0355	46.0
	$\cos \theta$	$\sin \theta$	$\cot \theta$	$\tan \theta$	θ

TABLE B

Values of
Trigonometric
Functions —
Decimal
Subdivisions
(csc $\theta = 1/\sin\theta$;
sec $\theta = 1/\cos\theta$)
(continued)

θ	$\sin\theta$	$\cos\theta$	$\tan\theta$	$\cot\theta$	
44.0	0.6947	0.7193	0.9657	1.0355	46.0
.1	.6959	.7181	.9691	1.0319	.9
.2	.6972	.7169	.9725	1.0283	.8
.3	.6984	.7157	.9759	1.0247	.7
.4	.6997	.7145	.9793	1.0212	.6
.5	.7009	.7133	.9827	1.0176	.5
.6	.7022	.7120	.9861	1.0141	.4
.7	.7034	.7108	.9896	1.0105	.3
.8	.7046	.7096	.9930	1.0070	.2
.9	.7059	.7083	.9965	1.0035	.1
45.0	0.7071	0.7071	1.0000	1.0000	45.0
	$\cos\theta$	$\sin\theta$	$\cot\theta$	$\tan\theta$	θ

TABLE C

Values of the
Trigonometric
Functions for
Radians and Real
Numbers

t	$\sin t$	$\cos t$	$\tan t$	$\cot t$	$\sec t$	$\csc t$
.00	.0000	1.0000	.0000	.....	1.000	.....
.01	.0100	1.0000	.0100	99.997	1.000	100.00
.02	.0200	.9998	.0200	49.993	1.000	50.00
.03	.0300	.9996	.0300	33.323	1.000	33.34
.04	.0400	.9992	.0400	24.987	1.001	25.01
.05	.0500	.9988	.0500	19.983	1.001	20.01
.06	.0600	.9982	.0601	16.647	1.002	16.68
.07	.0699	.9976	.0701	14.262	1.002	14.30
.08	.0799	.9968	.0802	12.473	1.003	12.51
.09	.0899	.9960	.0902	11.081	1.004	11.13
.10	.0998	.9950	.1003	9.967	1.005	10.02
.11	.1098	.9940	.1104	9.054	1.006	9.109
.12	.1197	.9928	.1206	8.293	1.007	8.353
.13	.1296	.9916	.1307	7.649	1.009	7.714
.14	.1395	.9902	.1409	7.096	1.010	7.166
.15	.1494	.9888	.1511	6.617	1.011	6.692
.16	.1593	.9872	.1614	6.197	1.013	6.277
.17	.1692	.9856	.1717	5.826	1.015	5.911
.18	.1790	.9838	.1820	5.495	1.016	5.586
.19	.1889	.9820	.1923	5.200	1.018	5.295
.20	.1987	.9801	.2027	4.933	1.020	5.033
.21	.2085	.9780	.2131	4.692	1.022	4.797
.22	.2182	.9759	.2236	4.472	1.025	4.582
.23	.2280	.9737	.2341	4.271	1.027	4.386
.24	.2377	.9713	.2447	4.086	1.030	4.207
.25	.2474	.9689	.2553	3.916	1.032	4.042
.26	.2571	.9664	.2660	3.759	1.035	3.890
.27	.2667	.9638	.2768	3.613	1.038	3.749
.28	.2764	.9611	.2876	3.478	1.041	3.619
.29	.2860	.9582	.2984	3.351	1.044	3.497
.30	.2955	.9553	.3093	3.233	1.047	3.384
.31	.3051	.9523	.3203	3.122	1.050	3.278
.32	.3146	.9492	.3314	3.018	1.053	3.179
.33	.3240	.9460	.3425	2.920	1.057	3.086
.34	.3335	.9428	.3537	2.827	1.061	2.999
.35	.3429	.9394	.3650	2.740	1.065	2.916
.36	.3523	.9359	.3764	2.657	1.068	2.839
.37	.3616	.9323	.3879	2.578	1.073	2.765
.38	.3709	.9287	.3994	2.504	1.077	2.696
.39	.3802	.9249	.4111	2.433	1.081	2.630

TABLE C

Values of the
Trigonometric
Functions for
Radians and Real
Numbers
(continued)

t	sin t	cos t	tan t	cot t	sec t	csc t
.40	.3894	.9211	.4228	2.365	1.086	2.568
.41	.3986	.9171	.4346	2.301	1.090	2.509
.42	.4078	.9131	.4466	2.239	1.095	2.452
.43	.4169	.9090	.4586	2.180	1.100	2.399
.44	.4259	.9048	.4708	2.124	1.105	2.348
.45	.4350	.9004	.4831	2.070	1.111	2.299
.46	.4439	.8961	.4954	2.018	1.116	2.253
.47	.4529	.8916	.5080	1.969	1.122	2.208
.48	.4618	.8870	.5206	1.921	1.127	2.166
.49	.4706	.8823	.5334	1.875	1.133	2.125
.50	.4794	.8776	.5463	1.830	1.139	2.086
.51	.4882	.8727	.5594	1.788	1.146	2.048
.52	.4969	.8678	.5726	1.747	1.152	2.013
.53	.5055	.8628	.5859	1.707	1.159	1.978
.54	.5141	.8577	.5994	1.668	1.166	1.945
.55	.5227	.8525	.6131	1.631	1.173	1.913
.56	.5312	.8473	.6269	1.595	1.180	1.883
.57	.5396	.8419	.6410	1.560	1.188	1.853
.58	.5480	.8365	.6552	1.526	1.196	1.825
.59	.5564	.8309	.6696	1.494	1.203	1.797
.60	.5646	.8253	.6841	1.462	1.212	1.771
.61	.5729	.8196	.6989	1.431	1.220	1.746
.62	.5810	.8139	.7139	1.401	1.229	1.721
.63	.5891	.8080	.7291	1.372	1.238	1.697
.64	.5972	.8021	.7445	1.343	1.247	1.674
.65	.6052	.7961	.7602	1.315	1.256	1.652
.66	.6131	.7900	.7761	1.288	1.266	1.631
.67	.6210	.7838	.7923	1.262	1.276	1.610
.68	.6288	.7776	.8087	1.237	1.286	1.590
.69	.6365	.7712	.8253	1.212	1.297	1.571
.70	.6442	.7648	.8423	1.187	1.307	1.552
.71	.6518	.7584	.8595	1.163	1.319	1.534
.72	.6594	.7518	.8771	1.140	1.330	1.517
.73	.6669	.7452	.8949	1.117	1.342	1.500
.74	.6743	.7385	.9131	1.095	1.354	1.483
.75	.6816	.7317	.9316	1.073	1.367	1.467
.76	.6889	.7248	.9505	1.052	1.380	1.452
.77	.6961	.7179	.9697	1.031	1.393	1.437
.78	.7033	.7109	.9893	1.011	1.407	1.422
.79	.7104	.7038	1.009	.9908	1.421	1.408

TABLE C

Values of the
Trigonometric
Functions for
Radians and Real
Numbers
(continued)

t	$\sin t$	$\cos t$	$\tan t$	$\cot t$	$\sec t$	$\csc t$
.80	.7174	.6967	1.030	.9712	1.435	1.394
.81	.7243	.6895	1.050	.9520	1.450	1.381
.82	.7311	.6822	1.072	.9331	1.466	1.368
.83	.7379	.6749	1.093	.9146	1.482	1.355
.84	.7446	.6675	1.116	.8964	1.498	1.343
.85	.7513	.6600	1.138	.8785	1.515	1.331
.86	.7578	.6524	1.162	.8609	1.533	1.320
.87	.7643	.6448	1.185	.8437	1.551	1.308
.88	.7707	.6372	1.210	.8267	1.569	1.297
.89	.7771	.6294	1.235	.8100	1.589	1.287
.90	.7833	.6216	1.260	.7936	1.609	1.277
.91	.7895	.6137	1.286	.7774	1.629	1.267
.92	.7956	.6058	1.313	.7615	1.651	1.257
.93	.8016	.5978	1.341	.7458	1.673	1.247
.94	.8076	.5898	1.369	.7303	1.696	1.238
.95	.8134	.5817	1.398	.7151	1.719	1.229
.96	.8192	.5735	1.428	.7001	1.744	1.221
.97	.8249	.5653	1.459	.6853	1.769	1.212
.98	.8305	.5570	1.491	.6707	1.795	1.204
.99	.8360	.5487	1.524	.6563	1.823	1.196
1.00	.8415	.5403	1.557	.6421	1.851	1.188
1.01	.8468	.5319	1.592	.6281	1.880	1.181
1.02	.8521	.5234	1.628	.6142	1.911	1.174
1.03	.8573	.5148	1.665	.6005	1.942	1.166
1.04	.8624	.5062	1.704	.5870	1.975	1.160
1.05	.8674	.4976	1.743	.5736	2.010	1.153
1.06	.8724	.4889	1.784	.5604	2.046	1.146
1.07	.8772	.4801	1.827	.5473	2.083	1.140
1.08	.8820	.4713	1.871	.5344	2.122	1.134
1.09	.8866	.4625	1.917	.5216	2.162	1.128
1.10	.8912	.4536	1.965	.5090	2.205	1.122
1.11	.8957	.4447	2.014	.4964	2.249	1.116
1.12	.9001	.4357	2.066	.4840	2.295	1.111
1.13	.9044	.4267	2.120	.4718	2.344	1.106
1.14	.9086	.4176	2.176	.4596	2.395	1.101
1.15	.9128	.4085	2.234	.4475	2.448	1.096
1.16	.9168	.3993	2.296	.4356	2.504	1.091
1.17	.9208	.3902	2.360	.4237	2.563	1.086
1.18	.9246	.3809	2.427	.4120	2.625	1.082
1.19	.9284	.3717	2.498	.4003	2.691	1.077

TABLE C

Values of the
Trigonometric
Functions for
Radians and Real
Numbers
(continued)

t	$\sin t$	$\cos t$	$\tan t$	$\cot t$	$\sec t$	$\csc t$
1.20	.9320	.3624	2.572	.3888	2.760	1.073
1.21	.9356	.3530	2.650	.3773	2.833	1.069
1.22	.9391	.3436	2.733	.3659	2.910	1.065
1.23	.9425	.3342	2.820	.3546	2.992	1.061
1.24	.9458	.3248	2.912	.3434	3.079	1.057
1.25	.9490	.3153	3.010	.3323	3.171	1.054
1.26	.9521	.3058	3.113	.3212	3.270	1.050
1.27	.9551	.2963	3.224	.3102	3.375	1.047
1.28	.9580	.2867	3.341	.2993	3.488	1.044
1.29	.9608	.2771	3.467	.2884	3.609	1.041
1.30	.9636	.2675	3.602	.2776	3.738	1.038
1.31	.9662	.2579	3.747	.2669	3.878	1.035
1.32	.9687	.2482	3.903	.2562	4.029	1.032
1.33	.9711	.2385	4.072	.2456	4.193	1.030
1.34	.9735	.2288	4.256	.2350	4.372	1.027
1.35	.9757	.2190	4.455	.2245	4.566	1.025
1.36	.9779	.2092	4.673	.2140	4.779	1.023
1.37	.9799	.1994	4.913	.2035	5.014	1.021
1.38	.9819	.1896	5.177	.1931	5.273	1.018
1.39	.9837	.1798	5.471	.1828	5.561	1.017
1.40	.9854	.1700	5.798	.1725	5.883	1.015
1.41	.9871	.1601	6.165	.1622	6.246	1.013
1.42	.9887	.1502	6.581	.1519	6.657	1.011
1.43	.9901	.1403	7.055	.1417	7.126	1.010
1.44	.9915	.1304	7.602	.1315	7.667	1.009
1.45	.9927	.1205	8.238	.1214	8.299	1.007
1.46	.9939	.1106	8.989	.1113	9.044	1.006
1.47	.9949	.1006	9.887	.1011	9.938	1.005
1.48	.9959	.0907	10.983	.0910	11.029	1.004
1.49	.9967	.0807	12.350	.0810	12.390	1.003
1.50	.9975	.0707	14.101	.0709	14.137	1.003
1.51	.9982	.0608	16.428	.0609	16.458	1.002
1.52	.9987	.0508	19.670	.0508	19.695	1.001
1.53	.9992	.0408	24.498	.0408	24.519	1.001
1.54	.9995	.0308	32.461	.0308	32.476	1.000
1.55	.9998	.0208	48.078	.0208	48.089	1.000
1.56	.9999	.0108	92.620	.0108	92.626	1.000
1.57	1.0000	.0008	1255.8	.0008	1255.8	1.000

TABLE D

Four-Place Table
of Logarithms

n	0	1	2	3	4	5	6	7	8	9
1.0	+0.0000	0043	0086	0128	0170	0212	0253	0294	0334	0374
1.1	.0414	0453	0492	0531	0569	0607	0645	0682	0719	0755
1.2	.0792	0828	0864	0899	0934	0969	1004	1038	1072	1106
1.3	.1139	1173	1206	1239	1271	1303	1335	1367	1399	1430
1.4	.1461	1492	1523	1553	1584	1614	1644	1673	1703	1732
1.5	.1761	1790	1818	1847	1875	1903	1931	1959	1987	2014
1.6	.2041	2068	2095	2122	2148	2175	2201	2227	2253	2279
1.7	.2304	2330	2355	2380	2405	2430	2455	2480	2504	2529
1.8	.2553	2577	2601	2625	2648	2672	2695	2718	2742	2765
1.9	.2788	2810	2833	2856	2878	2900	2923	2945	2967	2989
2.0	.3010	3032	3054	3075	3096	3118	3139	3160	3181	3201
2.1	.3222	3243	3263	3284	3304	3324	3345	3365	3385	3404
2.2	.3424	3444	3464	3483	3502	3522	3541	3560	3579	3598
2.3	.3617	3636	3655	3674	3692	3711	3729	3747	3766	3784
2.4	.3802	3820	3838	3856	3874	3892	3909	3927	3945	3962
2.5	.3979	3997	4014	4031	4048	4065	4082	4099	4116	4133
2.6	.4150	4166	4183	4200	4216	4232	4249	4265	4281	4298
2.7	.4314	4330	4346	4362	4378	4393	4409	4425	4440	4456
2.8	.4472	4487	4502	4518	4533	4548	4564	4579	4594	4609
2.9	.4624	4639	4654	4669	4683	4698	4713	4728	4742	4757
3.0	.4771	4786	4800	4814	4829	4843	4857	4871	4886	4900
3.1	.4914	4928	4942	4955	4969	4983	4997	5011	5024	5038
3.2	.5051	5065	5079	5092	5105	5119	5132	5145	5159	5172
3.3	.5185	5198	5211	5224	5237	5250	5263	5276	5289	5302
3.4	.5315	5328	5340	5353	5366	5378	5391	5403	5416	5428
3.5	.5441	5453	5465	5478	5490	5502	5514	5527	5539	5551
3.6	.5563	5575	5587	5599	5611	5623	5635	5647	5658	5670
3.7	.5682	5694	5705	5717	5729	5740	5752	5763	5775	5786
3.8	.5798	5809	5821	5832	5843	5855	5866	5877	5888	5899
3.9	.5911	5922	5933	5944	5955	5966	5977	5988	5999	6010
4.0	.6021	6031	6042	6053	6064	6075	6085	6096	6107	6117
4.1	.6128	6138	6149	6160	6170	6180	6191	6201	6212	6222
4.2	.6232	6243	6253	6263	6274	6284	6294	6304	6314	6325
4.3	.6335	6345	6355	6365	6375	6385	6395	6405	6415	6425
4.4	.6435	6444	6454	6464	6474	6484	6493	6503	6513	6522
4.5	.6532	6542	6551	6561	6571	6580	6590	6599	6609	6618
4.6	.6628	6637	6645	6656	6665	6675	6684	6693	6702	6712
4.7	.6721	6730	6739	6749	6758	6767	6776	6785	6794	6803
4.8	.6812	6821	6830	6839	6848	6857	6866	6875	6884	6893
4.9	.6902	6911	6920	6928	6937	6946	6955	6964	6972	6981
5.0	+.6990	6998	7007	7016	7024	7033	7042	7050	7059	7067
5.1	.7076	7084	7093	7101	7110	7118	7126	7135	7143	7152
5.2	.7160	7168	7177	7185	7193	7202	7210	7218	7226	7235
5.3	.7243	7251	7259	7267	7275	7284	7292	7300	7308	7316
5.4	.7324	7332	7340	7348	7356	7364	7372	7380	7388	7396

TABLE D

Four-Place Table
of Logarithms
(continued)

n	0	1	2	3	4	5	6	7	8	9
5.5	.7404	7412	7419	7427	7435	7443	7451	7459	7466	7474
5.6	.7482	7490	7497	7505	7513	7520	7528	7536	7543	7551
5.7	.7559	7566	7574	7582	7589	7597	7604	7612	7619	7627
5.8	.7634	7642	7649	7657	7664	7672	7679	7686	7694	7701
5.9	.7709	7716	7723	7731	7738	7745	7752	7760	7767	7774
6.0	.7782	7789	7796	7803	7810	7818	7825	7832	7839	7846
6.1	.7853	7860	7868	7875	7882	7889	7896	7903	7910	7917
6.2	.7924	7931	7938	7945	7952	7959	7966	7973	7980	7987
6.3	.7993	8000	8007	8014	8021	8028	8035	8041	8048	8055
6.4	.8062	8069	8075	8082	8089	8096	8102	8109	8116	8122
6.5	.8129	8136	8142	8149	8156	8162	8169	8176	8182	8189
6.6	.8195	8202	8209	8215	8222	8228	8235	8241	8248	8254
6.7	.8261	8267	8274	8280	8287	8293	8299	8306	8312	8319
6.8	.8325	8331	8338	8344	8351	8357	8363	8370	8376	8382
6.9	.8388	8395	8401	8407	8414	8420	8426	8432	8439	8445
7.0	.8451	8457	8463	8470	8476	8482	8488	8494	8500	8506
7.1	.8513	8519	8525	8531	8537	8543	8549	8555	8561	8567
7.2	.8573	8579	8585	8591	8597	8603	8609	8615	8621	8627
7.3	.8633	8639	8645	8651	8657	8663	8669	8675	8681	8686
7.4	.8692	8698	8704	8710	8716	8722	8727	8733	8739	8745
7.5	.8751	8756	8762	8768	8774	8779	8785	8791	8797	8802
7.6	.8808	8814	8820	8825	8831	8837	8842	8848	8854	8859
7.7	.8865	8871	8876	8882	8887	8893	8899	8904	8910	8915
7.8	.8921	8927	8932	8938	8943	8949	8954	8960	8965	8971
7.9	.8976	8982	8987	8993	8998	9004	9009	9015	9020	9025
8.0	.9031	9036	9042	9047	9053	9058	9063	9069	9074	9079
8.1	.9085	9090	9096	9101	9106	9112	9117	9122	9128	9133
8.2	.9138	9143	9149	9154	9159	9165	9170	9175	9180	9186
8.3	.9191	9196	9201	9206	9212	9217	9222	9227	9232	9238
8.4	.9243	9248	9253	9258	9263	9269	9274	9279	9284	9289
8.5	.9294	9299	9304	9309	9315	9320	9325	9330	9335	9340
8.6	.9345	9350	9355	9360	9365	9370	9375	9380	9385	9390
8.7	.9395	9400	9405	9410	9415	9420	9425	9430	9435	9440
8.8	.9445	9450	9455	9460	9465	9469	9474	9479	9484	9489
8.9	.9494	9499	9504	9509	9513	9518	9523	9528	9533	9538
9.0	.9542	9547	9552	9557	9562	9566	9571	9576	9581	9586
9.1	.9590	9595	9600	9605	9609	9614	9619	9624	9628	9633
9.2	.9638	9643	9647	9652	9657	9661	9666	9671	9675	9680
9.3	.9685	9689	9694	9699	9703	9708	9713	9717	9722	9727
9.4	.9731	9736	9741	9745	9750	9754	9759	9763	9768	9773
9.5	.9777	9782	9786	9791	9795	9800	9805	9809	9814	9818
9.6	.9823	9827	9832	9836	9841	9845	9850	9854	9859	9863
9.7	.9868	9872	9877	9881	9886	9890	9894	9899	9903	9908
9.8	.9912	9917	9921	9926	9930	9934	9939	9943	9948	9952
9.9	.9956	9961	9965	9969	9974	9978	9983	9987	9991	9996

TABLE E

Powers and
Roots

n	n^2	$\sqrt{n}$	n^3	$\sqrt[3]{n}$	n	n^2	$\sqrt{n}$	n^3	$\sqrt[3]{n}$
1	1	1.000	1	1.000	51	2,601	7.141	132,651	3.708
2	4	1.414	8	1.260	52	2,704	7.211	140,608	3.733
3	9	1.732	27	1.442	53	2,809	7.280	148,877	3.756
4	16	2.000	64	1.587	54	2,916	7.348	157,464	3.780
5	25	2.236	125	1.710	55	3,025	7.416	166,375	3.803
6	36	2.449	216	1.817	56	3,136	7.483	175,616	3.826
7	49	2.646	343	1.913	57	3,249	7.550	185,193	3.849
8	64	2.828	512	2.000	58	3,364	7.616	195,112	3.871
9	81	3.000	729	2.080	59	3,481	7.681	205,379	3.893
10	100	3.162	1,000	2.154	60	3,600	7.746	216,000	3.915
11	121	3.317	1,331	2.224	61	3,721	7.810	226,981	3.936
12	144	3.464	1,728	2.289	62	3,844	7.874	238,328	3.958
13	169	3.606	2,197	2.351	63	3,969	7.937	250,047	3.979
14	196	3.742	2,744	2.410	64	4,096	8.000	262,144	4.000
15	225	3.873	3,375	2.466	65	4,225	8.062	274,625	4.021
16	256	4.000	4,096	2.520	66	4,356	8.124	287,496	4.041
17	289	4.123	4,913	2.571	67	4,489	8.185	300,763	4.062
18	324	4.243	5,832	2.621	68	4,624	8.246	314,432	4.082
19	361	4.359	6,859	2.668	69	4,761	8.307	328,509	4.102
20	400	4.472	8,000	2.714	70	4,900	8.367	343,000	4.121
21	441	4.583	9,261	2.759	71	5,041	8.426	357,911	4.141
22	484	4.690	10,648	2.802	72	5,184	8.485	373,248	4.160
23	529	4.796	12,167	2.844	73	5,329	8.544	389,017	4.179
24	576	4.899	13,824	2.884	74	5,476	8.602	405,224	4.198
25	625	5.000	15,625	2.924	75	5,625	8.660	421,875	4.217
26	676	5.099	17,576	2.962	76	5,776	8.718	438,976	4.236
27	729	5.196	19,683	3.000	77	5,929	8.775	456,533	4.254
28	784	5.292	21,952	3.037	78	6,084	8.832	474,552	4.273
29	841	5.385	24,389	3.072	79	6,241	8.888	493,039	4.291
30	900	5.477	27,000	3.107	80	6,400	8.944	512,000	4.309
31	961	5.568	29,791	3.141	81	6,561	9.000	531,441	4.327
32	1,024	5.657	32,768	3.175	82	6,724	9.055	551,368	4.344
33	1,089	5.745	35,937	3.208	83	6,889	9.110	571,787	4.362
34	1,156	5.831	39,304	3.240	84	7,056	9.165	592,704	4.380
35	1,225	5.916	42,875	3.271	85	7,225	9.220	614,125	4.397
36	1,296	6.000	46,656	3.302	86	7,396	9.274	636,056	4.414
37	1,369	6.083	50,653	3.332	87	7,569	9.327	658,503	4.431
38	1,444	6.164	54,872	3.362	88	7,744	9.381	681,472	4.448
39	1,521	6.245	59,319	3.391	89	7,921	9.434	704,969	4.465
40	1,600	6.325	64,000	3.420	90	8,100	9.487	729,000	4.481
41	1,681	6.403	68,921	3.448	91	8,281	9.539	753,571	4.498
42	1,764	6.481	74,088	3.476	92	8,464	9.592	778,688	4.514
43	1,849	6.557	79,507	3.503	93	8,649	9.644	804,357	4.531
44	1,936	6.633	85,184	3.530	94	8,836	9.695	830,584	4.547
45	2,025	6.708	91,125	3.557	95	9,025	9.747	857,375	4.563
46	2,116	6.782	97,336	3.583	96	9,216	9.798	884,736	4.579
47	2,209	6.856	103,823	3.609	97	9,409	9.849	912,673	4.595
48	2,304	6.928	110,592	3.634	98	9,604	9.899	941,192	4.610
49	2,401	7.000	117,649	3.659	99	9,801	9.950	970,299	4.626
50	2,500	7.071	125,000	3.684	100	10,000	10.000	1,000,000	4.642

Answers to Odd-Numbered Exercises

Section 1.1 (page 7)

1, 3, 5.

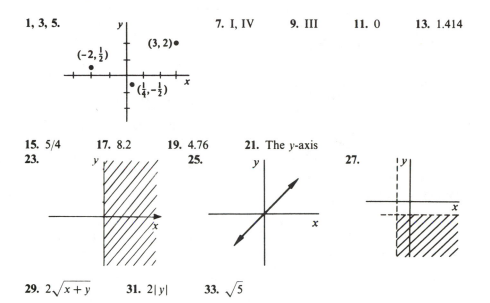

7. I, IV   **9.** III   **11.** 0   **13.** 1.414

15. 5/4   **17.** 8.2   **19.** 4.76   **21.** The y-axis

23.   **25.**   **27.**

29. $2\sqrt{x+y}$ **31.** $2|y|$ **33.** $\sqrt{5}$

Section 1.2 (page 11)

1. $-6, 6$ **3.** $6, -6$ **5.** $44, -4$ **7.** $A = f(r)$ **9.** $P = f(s)$

11. Domain: all reals   **13.** Domain: all reals ≥ 25 **15.** Domain: all reals $\neq 0$
      Range: all reals   Range: all reals ≥ 0 Range: all reals $\neq 0$

17.   **19.**

21.   **23.**

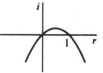

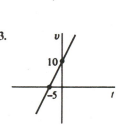

25. yes   **27.** yes   **29.** no   **31.** yes   **33.** no

Section 1.3 (page 15)

1. Three   **3.** Four   **5.** One   **7.** Two   **9.** Four   **11.** 9820   **13.** 54.7
15. 0.0658   **17.** 39.8   **19.** 1.00   **21.** 21.50   **23.** 1.92   **25.** 65   **27.** 3.1
29. 2300   **31.** 139.6

Section 1.4 (page 22)

1. $90°$ **3.** $75°15', 15°5'$ **5.** $109°10', 8°1'20''$ **7.** $130°55'15'', -10°34'45''$
9. $574°10'54'', -92°39'34''$ **11.** $-121°1'10'', 39°35'16''$ **13.** $18.4267°$ **15.** $94.2856°$
17. $283.6083°$ **19.** $183.2444°$ **21.** $48°15'26''$ **23.** $-235°27'0''$ **25.** $45°45'27''$
27. $15°15'27''$ **29.** $-60°$ **31.** $42.5°$ **33.** $135°$ **35.** $-70°$ **37.** $0°$ **39.** $120°$
41. $45°$ **43.** $69°55'$ **45.** $135°$ **47.** $-90°$ **49.** $180°$ **51.** $79°28'23''$ **53.** $0.08°$

Section 1.5 (page 26)

1. A multiple of 2π **3.** $\frac{5}{12}\pi$ **5.** $\frac{8}{3}\pi$ **7.** $-\frac{4}{3}\pi$ **9.** $\frac{19}{36}\pi$ **11.** $\frac{25}{6}\pi$ **13.** 1.61
15. 0.00161   **17.** 4.43   **19.** 0.00785   **21.** $\frac{1}{4}\pi$ **23.** π **25.** $-\frac{3}{4}\pi$ **27.** $-\frac{1}{2}\pi$
29. $-2.28, 229.18°$ **31.** $\pi, 180°$ **33.** $\pi, -540°$ **35.** $-0.53, 5729.58°$ **37.** $0, 18,000°$
39. 2.62 ft $(=\frac{5}{6}\pi)$ **41.** 11.4 in   **43.** 4.2 in   **45.** 1.47 in   **47.** $57.1°$ **49.** 293 in

Section 1.6 (page 33)

1. $74°$ **3.** 127 ft   **5.** 3.7 m   **7.** 8.7 cm   **17.** $a = 8.1, c = 10.3$ **19.** $a = 9.5, b = 3.5$
21. 33 cm   **23.** Yes, because angles are all $60°$. No, angle size may vary.   **25.** 80 m
27. $\sqrt{2}m$ **29.** $\alpha = 7°2', \phi = 58°46'$

Review Exercises for Chapter 1 (page 35)

1. 2, 1, -2 **3.**   **5.** $\sqrt{10}$ **7.** No   **9.** $160°$ **11.** $38.7231°$

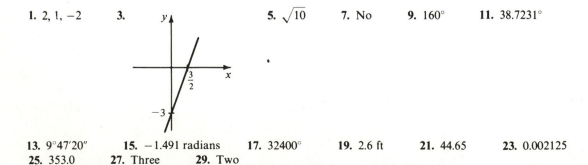

13. $9°47'20''$ **15.** -1.491 radians   **17.** $32400°$ **19.** 2.6 ft   **21.** 44.65   **23.** 0.002125
25. 353.0   **27.** Three   **29.** Two

Section 2.1 (page 43)

(The trigonometric functions are listed in this order: sine, cosine, tangent, cotangent, secant, cosecant.)

1. 4/5, 3/5, 4/3, 3/4, 5/3, 5/4   **3.** 12/13, 5/13, 12/5, 5/12, 13/5, 13/12

5. $1/\sqrt{26}$, $5/\sqrt{26}$, 1/5, 5, $\sqrt{26}/5$, $\sqrt{26}$ **7.** $1/\sqrt{2}$, $1/\sqrt{2}$, 1, 1, $\sqrt{2}$, $\sqrt{2}$

9. $\sqrt{3}/\sqrt{7}$, $2/\sqrt{7}$, $\sqrt{3}/2$, $2/\sqrt{3}$, $\sqrt{7}/2$, $\sqrt{7}/\sqrt{3}$

11. $12/\sqrt{769}$, $25/\sqrt{769}$, 12/25, 25/12, $\sqrt{769}/25$, $\sqrt{769}/12$

13. 4/5, 3/5, 4/3, 3/4, 5/3, 5/4   **15.** 1/3, $\sqrt{8}/3$, $1/\sqrt{8}$, $\sqrt{8}$, $3/\sqrt{8}$, 3

17. $\sqrt{3}/2$, 1/2, $\sqrt{3}$, $1/\sqrt{3}$, 2, $2/\sqrt{3}$ **19.** $\sqrt{5}/\sqrt{6}$, $1/\sqrt{6}$, $\sqrt{5}$, $1/\sqrt{5}$, $\sqrt{6}$, $\sqrt{6}/\sqrt{5}$

21. 10/11, $\sqrt{21}/11$, $10/\sqrt{21}$, $\sqrt{21}/10$, $11/\sqrt{21}$, 11/10

23. u/v, $\sqrt{v^2-u^2}/v$, $u/\sqrt{v^2-u^2}$, $\sqrt{v^2-u^2}/u$, $v/\sqrt{v^2-u^2}$, v/u

25. $\sqrt{1-u^2}$, u, $\sqrt{1-u^2}/u$, $u/\sqrt{1-u^2}$, $1/u$, $1/\sqrt{1-u^2}$ **27.** 1.5   **29.** 0.2925   **31.** 0.9594

Section 2.2 (page 46)

1. 2   **3.** 1/3   **5.** 1/2   **7.** $1/\sqrt{2}$ **9.** $\sqrt{3}/2$ **11.** 5/12   **13.** $1/\sqrt{5}$ **15.** 1/3

17. 2   **19.** $\sqrt{2}/\sqrt{7}$ **21.** 1/3, $\sqrt{8}/3$, $1/\sqrt{8}$, $\sqrt{8}$, $3/\sqrt{8}$, 3   **23.** $1/\sqrt{5}$, $2/\sqrt{5}$, 1/2, 2, $\sqrt{5}/2$, $\sqrt{5}$

25. $\sqrt{2}/\sqrt{3}$, $1/\sqrt{3}$, $\sqrt{2}$, $1/\sqrt{2}$, $\sqrt{3}$, $\sqrt{3}/\sqrt{2}$ **27.** 8.658   **29.** 0.5258

Section 2.3 (page 53)

1. 0.2250   **3.** 0.3153   **5.** 1.0041   **7.** 3.1084   **9.** 0.9681   **11.** 0.1736   **13.** 0.2476

15. 2.3515   **17.** 0.6412   **19.** 2.820   **21.** 0.6421   **23.** 1.011   **25.** 27°30′   **27.** 75°

29. 56°10′   **31.** 72°50′   **33.** 3°20′   **35.** 0.60   **37.** 0.60   **39.** 0.19

Section 2.4 (page 55)

1. 0.7716   **3.** 1.1467   **5.** 21.2285   **7.** 0.9608   **9.** 1.0570   **11.** 0.9968

13. 0.0918   **15.** 3.2539   **17.** 0.9864   **19.** 1.4959   **21.** 2.9029   **23.** 1   **25.** 33.7°

27. 35.3°   **29.** 43.5°   **31.** 83.5°   **33.** 45.5°   **35.** 70°35′   **37.** 38°43′   **39.** 5°0′

41. 41°48′   **51.** 0.9677

Section 2.5 (page 59)

1. 0.6243   **3.** 0.8916   **5.** 1.2460   **7.** 0.9095   **9.** 1.2225   **11.** 0.3830   **13.** 0.6478

15. 0.8543   **17.** 0.9999   **19.** 50°54′   **21.** 54°28′   **23.** 59°17′   **25.** 26°34′

27. 56°15′   **29.** 14°29′   **31.** 0.137   **33.** 0.384   **35.** 1.167   **37.** 0.713   **39.** 0.105

41. 0.795

Section 2.6 (page 65)

1. $A = 29°45'$, $B = 60°15'$, $c = 8.06$ **3.** $A = 36°52'$, $B = 53°8'$, $b = 16$
5. $B = 80°35'$, $b = 30.15$, $c = 30.56$ **7.** 1.37 km   **9.** 8 A.M.   **11.** 5.1 m   **13.** 55°, 35°
15. 67°40', 22°20'   **17.** 20.4 m   **19.** 110.3 ft   **21.** 2 min 25 sec   **23.** 617.2 ft

Section 2.7–2.8 (page 73)

1. $F_x = 17.6$, $F_y = 17.7$ **3.** $F_x = 0.0197$, $F_y = 0.0032$ **5.** $F_x = 14.5$, $F_y = 8.25$ **7.** 25, 36°52'
9. 71.5, 75°50'   **11.** 0.153, 31°36'   **13.** Horiz 73.3, Vert 15.8   **15.** 13.46 lb
17. 216.2 lb and 252.1 lb   **19.** 26.7°
21. Horiz rope: 26.8 lb, Vert rope: $F = 51.7$ lb in each supporting rope   **23.** 28.5 mph, 61°23'
25. 36°52'   **27.** 8.66 mph, 240°   **29.** 353.5°, 1 hr 46 min   **31.** $X = 52$, $\theta = 60°$
33. $Z = 224$, $\theta = 26°34'$ **35.** $R = 43.3$, $Z = 50$

Section 2.9 (page 80)

1. 82°49'   **3.** 0.64 g   **5.** 9947.5 mi   **7.** 19,151 mi   **9.** 84°17'   **11.** 2 km
13. 80°5'   **15.** 33°33'   **17.** 7.97 cm   **19.** 48.5 ft   **21.** 924 meters   **23.** $r = 1200$ miles

Review Exercises for Chapter 2 (page 83)

1. 5/13, 12/13, 5/12, 12/5, 13/12, 13/5   **3.** $\sqrt{2}/\sqrt{3}$, $1/\sqrt{3}$, $\sqrt{2}$, $1/\sqrt{2}$, $\sqrt{3}$, $\sqrt{3}/\sqrt{2}$ **5.** 1.5089
7. 0.6660   **9.** 48°18'   **11.** 8°8'   **13.** 65.28°   **15.** 51.14°   **17.** 1.000   **19.** 0.521
21. $B = 58°$, $b = 4.8$, $c = 5.7$ **23.** $A = 45.0°$, $B = 45.0°$, $b = 29$ **25.** horiz: 40, vertical: 29.9
27. Magnitude: 83.9; angle: 59.2°   **29.** 67.9°

Section 3.1 (page 90)

(The trigonometric functions are listed in order: sine, cosine, tangent, cotangent, secant, cosecant.)
1. $2/\sqrt{5}$, $1/\sqrt{5}$. 2, 1/2, $\sqrt{5}$, $\sqrt{5}/2$ **3.** $16/\sqrt{337}$, $-9/\sqrt{337}$, $-16/9$, $-9/16$, $-\sqrt{337}/9$, $\sqrt{337}/16$
5. $-7/\sqrt{53}$, $2/\sqrt{53}$, $-7/2$, $-2/7$, $\sqrt{53}/2$, $-\sqrt{53}/7$ **7.** $-1/\sqrt{10}$, $3/\sqrt{10}$, $-1/3$, -3, $\sqrt{10}/3$, $-\sqrt{10}$
9. $-1/2$, $-\sqrt{3}/2$, $\sqrt{3}/3$, $\sqrt{3}$, $-2\sqrt{3}/3$, -2 **11.** a. I, II   b. I, IV   c. I, III
13. 0, 1, 0, undefined, 1, undefined   **15.** a. $90° + (360°)m$ or $\frac{1}{2}\pi + 2m\pi$ b. $(360°)m$ or $2m\pi$
19. II   **21.** III   **23.** IV   **25.** $(-)$ **27.** $(+)$ **29.** $(-)$ **31.** $(-)$

Section 3.2–3.3 (page 94)

1. 3/5, 4/5, 3/4, 4/3, 5/4, 5/3   **3.** $-3/5$, $-4/5$, 3/4, 4/3, $-5/4$, $-5/3$
5. 0, -1, 0, undefined, -1, undefined

7. 0, 1, 0, undefined, 1, undefined and 0, −1, 0, undefined, −1, undefined
9. QIII: −1/2, −$\sqrt{3}/2$, $\sqrt{3}/3$, $\sqrt{3}$, −2/$\sqrt{3}$, −2; QIV: −1/2, $\sqrt{3}/2$, −$\sqrt{3}/3$, −$\sqrt{3}$, 2/$\sqrt{3}$, −2
11. 10/$\sqrt{101}$, 1/$\sqrt{101}$, 10, 1/10, $\sqrt{101}$, $\sqrt{101}/10$
13. QI: 5/13, 12/13, 5/12, 12/5, 13/12, 13/5; QIV: −5/13, 12/13, −5/12, −12/5, 13/12, −13/5
15. QII: 0.9529, −0.3033, −π, −1/π, −3.297, 1.049; QIV: −.9529, .3033, −π, −1/π, 3.297, −1.049   **17.** 0.94
19. 0.18   **21.** 0.42   **23.** −0.97   **25.** −0.57   **27.** −0.27   **29.** −2.92   **31.** −0.5
33. −0.71

Section 3.4–3.5 (page 101)

1. −cos 55°   **3.** −sin 45°   **5.** −tan 17°   **7.** −sec 0.58   **9.** cos 1.04   **11.** 0.4384
13. −9.5144   **15.** −1.0439   **17.** 0.2728   **19.** −0.9245   **21.** −0.5063   **23.** −0.8820
25. 1.091   **27.** 0.2403   **29.** −0.9916   **31.** $\sqrt{3}/2$ **33.** undefined   **35.** $\sqrt{3}$
37. 0   **39.** −2/$\sqrt{3}$ **41.** 5.617   **43.** 3.978   **45.** 5.215   **47.** 334°50′   **49.** 198°15′
51. 98°20′

Review Exercises for Chapter 3 (page 102)

1. 5/$\sqrt{29}$, −2/$\sqrt{29}$, −5/2, −2/5, −$\sqrt{29}/2$, $\sqrt{29}/5$ **3.** −3/5, −4/5, 3/4, 4/3, −5/4, −5/3   **5.** 44°42′
7. 61.9°   **9.** 1.53   **11.** 0.4245   **13.** −0.3843   **15.** −6.9273   **17.** 0.0715
19. 234°7′ and 305°53′   **21.** 292.6°   **23.** 5.62 radian

Section 4.1–4.2 (page 109)

1. $c = 39.7$ **3.** $c = 46.9$ **5.** $a = 74.0$ **7.** $C = 80°37′$ **9.** $A = 95°44′$ **11.** $B = 118°4′$
13. $A = 129.7°$, $B = 19.6°$, $c = 2.79$ **15.** $A = 37°48′$, $B = 47°47′$, $c = 195.2$
17. $A = 28°57′$, $B = 46°34′$. $C = 104°29′$ **19.** 3.81 km   **21.** 30°45′   **23.** 13°34′
25. $A = 66°35′$, $C = 53°25′$, $\overline{AC} = 3,775$ m   **27.** 346.4 m

Section 4.3 (page 117)

1. $C = 100°$, $b = 14.02$, $c = 18.58$ **3.** $B = 24.5°$, $C = 110.3°$, $c = 11.7$
5. $B = 23.9°$, $C = 120.5°$, $c = 25.9$ **7.** $B = 41°24′$, $C = 17°46′$, $c = 2.35$
9. $B = 28°57′$, $A = 98°3′$, $a = 37.4$ **11.** $A = 44°58′$, $C = 13°2′$, $c = 7.98$
13. $A = 20°$, $a = 17.9$, $c = 49.1$ **15.** $C = 88.6°$, $a = 42.13$, $b = 45.57$
17. $B = 0.83$, $a = 0.854$, $c = 0.743$ **19.** $A = 1.49$, $b = 8930$, $c = 10,900$ **21.** 5.72 ft
23. 1 hr 31 min   **25.** 4.72 ft   **27.** antenna: 545 ft; hill: 847 ft   **29.** 31°51′

Section 4.4 (page 122)

1. No solution   **3.** Two solutions   **5.** One solution   **7.** No solution
9. $B = 14°32′$, $C = 15°28′$, $b = 7.53$ **11.** No solution

13. $A = 38°45'$, $B = 113°15'$, $b = 29.4$ **15.** $B = 41°49'$, $C = 108°11'$, $c = 570$
     $A = 141°15'$, $B = 10°45'$, $b = 5.96$ $B = 138°11'$, $C = 11°49'$, $c = 122.9$
17. $A = 40°$, $B = 70°$, $a = 0.684$ **19.** $B = 47°48'$, $C = 59°57'$, $c = 0.818$ **21.** $15 \sin 25° < b < 15$

Section 4.5 (page 124)

1. $C = 45°$, $b = 669.2$, $c = 489.9$ **3.** $A = 60°$, $b = 1.20$, $c = 4.46$ **5.** no solution
7. $A = 32°$, $B = 118°$, $c = 283$ **9.** no solution   **11.** no solution
13. $C = 20°$, $a = 1.759$, $c = 0.695$ **15.** $C = 50.5°$, $B = 29.5°$, $c = 1.567$ **17.** no solution
19. 1.93 mi   **21.** Ground Speed = 351 mph
          Heading = $98°42'$

Section 4.6 (page 126)

1. 18.75   **3.** 39.43   **5.** 6   **7.** 0.4479   **9.** 2.239
11. Similar triangles may have unequal areas.   **13.** 85.3 sq in   **15.** 573.2 sq ft
17. 301 m or 499.8 m
19. Other side: 5.55   Other side: 26.3   **21.** 122.3 m^2
     Angles: 16.8°, 35.7°, 127.5°   or   Angles: 163.2°, 9.7°, 7.1°

Review Exercises for Chapter 4 (page 128)

1. $C = 118.9°$, $B = 32.1°$, $a = 15.5$ **3.** $A = 44.3°$, $B = 29.5°$, $C = 106.2°$
5. $A = 148.0°$, $B = 7.8°$, $C = 24.2°$ **7.** $B = 40.2°$, $b = 2175$, $c = 1561$
9. $B = 19.8°$, $C = 135.2°$, $c = 25.0$ **11.** no triangle exists   **13.** Area = 9.88
15. $A = 24.4°$, $C = 17.6°$, $b = 14.7$ **17.** 90.3 m   **19.** 123 ft
     Area = 20.3

Section 5.1 (page 135)

1. (a) $\cos 1 = 0.54$, $\sin 1 = 0.84$ (b) $\cos(-2) = -0.42$, $\sin(-2) = -0.91$
     (c) $\cos 3 = -0.99$, $\sin 3 = 0.14$ (d) $\cos 10 = -0.84$, $\sin 10 = -.54$
     (e) $\cos 3\pi = -1$, $\sin 3\pi = 0$ (f) $\cos(-4) = -0.65$, $\sin(-4) = 0.76$
     (g) $\cos(-4\pi) = 1$, $\sin(-4\pi) = 0$ (h) $\cos(\frac{\pi}{3}) = 0.5$, $\sin(\frac{\pi}{3}) = .87$
     (i) $\cos(\frac{1}{3}) = 0.95$, $\sin(\frac{1}{3}) = 0.33$ (j) $\cos(\frac{1}{2}) = 0.88$, $\sin(\frac{1}{2}) = 0.48$
     (k) $\cos\sqrt{7} = -0.88$, $\sin\sqrt{7} = 0.48$ (l) $\cos 5.15 = 0.42$, $\sin 5.15 = -0.91$
5. 0   **7.** 0.5480   **9.** 0.5   **11.** 1   **13.** -1 **15.** -0.8365 **17.** 0.2867
19. $(2n+1)\pi/2$ **21.** $2n\pi$ **23.** $-\pi/2 < x < \pi/2$, $3\pi/2 < x < 5\pi/2$, etc.   **25.** $x = (4n+1)\pi/2$
27. $x = (2n+1)\pi/2$

Section 5.2 (page 139)

1. Domain: all real numbers
Range: $-1, 1$

3. sine function, period $2\pi/3$

5. cosine function, period 4π

7.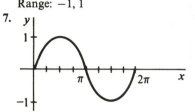

9.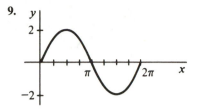

11. period: π
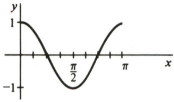

13. period: π
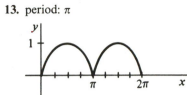

15.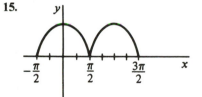

17.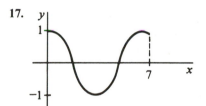

19.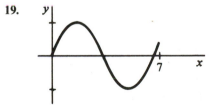

21.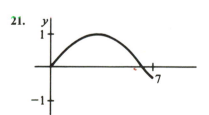

Section 5.3 (page 146)

1. $A = 3$, per. $= 2\pi$ **3.** $A = 6$, per. $= 2\pi$ **5.** $A = 1$, per. $= 3\pi$ **7.** $A = 1$, per. $= 2$

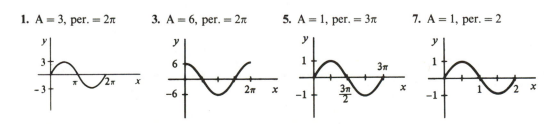

9. $A = \frac{1}{2}$, per. $= \pi$

11. $A = 8.2$, per. $= 5\pi$

13. $A = 1$, per. $= 5\pi$

15. $A = \pi$, per. $= \pi/50$

17. $A = 12$, per. $= 10\pi$

19. $A = 1/50$, per. $= 1/500$

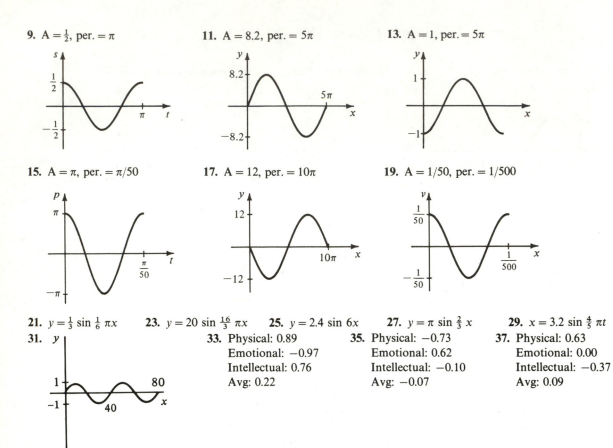

21. $y = \frac{1}{3} \sin \frac{1}{6} \pi x$ **23.** $y = 20 \sin \frac{16}{3} \pi x$ **25.** $y = 2.4 \sin 6x$ **27.** $y = \pi \sin \frac{2}{3} x$ **29.** $x = 3.2 \sin \frac{4}{5} \pi t$

31. y

33. Physical: 0.89
Emotional: -0.97
Intellectual: 0.76
Avg: 0.22

35. Physical: -0.73
Emotional: 0.62
Intellectual: -0.10
Avg: -0.07

37. Physical: 0.63
Emotional: 0.00
Intellectual: -0.37
Avg: 0.09

Section 5.4 (page 152)

1. $A = 1$, avg. $= 3$, per. $= \pi$

3. $A = 8$, avg. $= 6$, per. $= 2\pi$

5. $A = 1$, avg. $= -2$, per. $= 4\pi$

7. $A = 3$, avg. $= 3$, per. $= 2\pi/3$

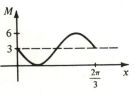

9. $A = 1$, ps $= -\frac{\pi}{3}$, per. $= 2\pi$

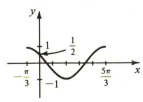

11. $A = 2$, ps $= \pi$, per. $= 4\pi$

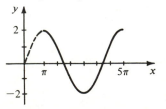

13. $A = 1$, ps $= -\frac{\pi}{2}$, per. $= \pi$

15. $A = 4$, ps $= -\pi$, per. $= 6\pi$

17. $A = 1$, ps $= \frac{1}{4}$, per. $= 2$

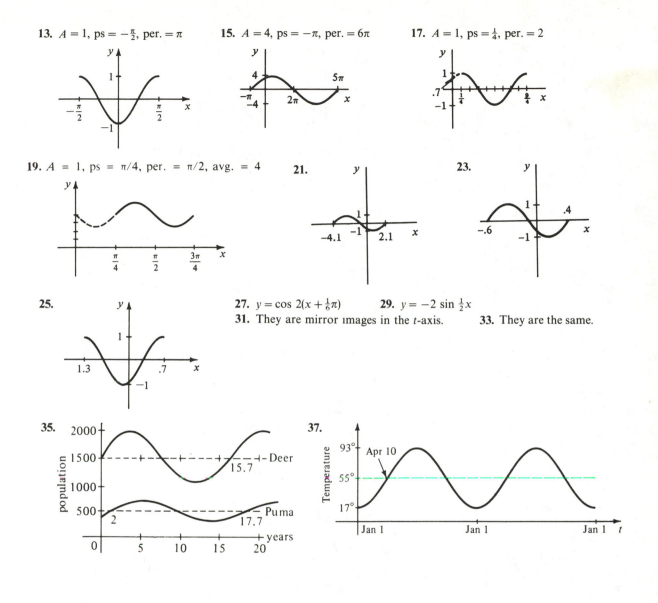

19. $A = 1$, ps $= \pi/4$, per. $= \pi/2$, avg. $= 4$

21.

23.

25.

27. $y = \cos 2(x + \frac{1}{6}\pi)$ **29.** $y = -2 \sin \frac{1}{2}x$

31. They are mirror images in the t-axis.   **33.** They are the same.

35.

37.

Section 5.5 (page 156)

1. 3/4   **3.** 2   **5.** 0.72

7.

9.

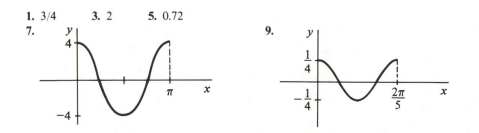

11.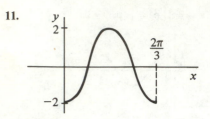

13.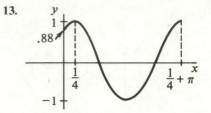

15. a. no effect   b. increase in mass causes increase in period   **17.** 1.96 sec, 1.9 m above equilibrium

19. 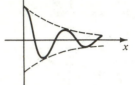   **21.** $x = 2 \cos 2\pi t$, $y = 2 \sin 2\pi t$, $(-2, 0)$, $(0, -2)$, $(2, 0)$

Section 5.6 (page 159)

1.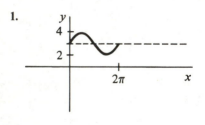

3.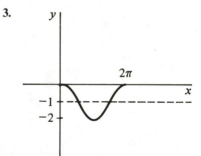

5.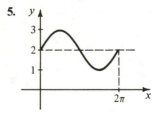

7.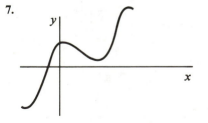

9.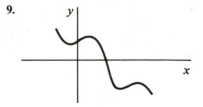

11.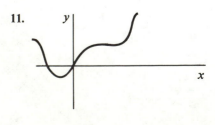

13.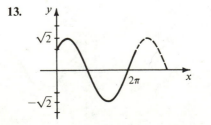

15.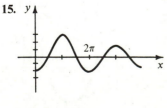

17.

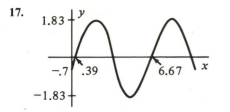

19.

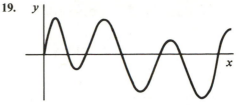

21. They are the same.

Section 5.7–5.8 (page 163)

1. 2π **3.** 2   **5.** 2   **7.** $12\pi/5$

9. Period $\pi/2$,
phase shift 0

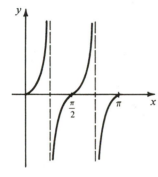

11. Period π,
phase shift $\pi/4$

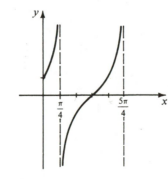

13. Period $\pi/2$,
phase shift $-\pi/6$

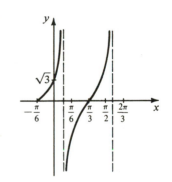

15. Period π, phase shift $3\pi/2$

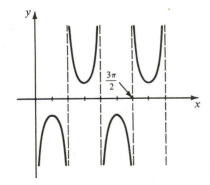

17. Period π, phase shift $\pi/4$

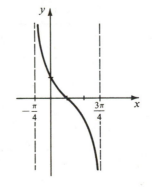

19. $\tan x = -\cot (x + \pi/2)$ **21.** They are the same.   **23.** They are mirror reflections in the y-axis.

Section 5.9 (page 165)

3.

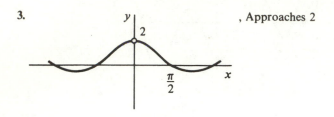

, Approaches 2

Review Exercises for Chapter 5 (page 166)

1. a. 1.683   b. 0   c. 1.683   **3.** $\frac{1}{18}\pi + \frac{1}{3} + \frac{2}{3}n\pi$

5. $A = 1$, p.s. $= \pi/3$, period $= 2\pi$, average $= 2$ **7.** p.s. $= -\frac{1}{4}\pi$, period $= 2\pi$

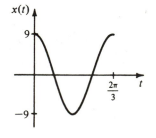

9. p.s. $= \pi$, period $= 6\pi$, average $= 2$ **11.** $A = 9$, period $= 2\pi/3$

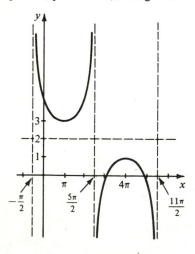

13. $x = -3 + (2n + 1)\pi/2$

Section 6.1 (page 173)

1. $\sec\theta$ **3.** $\sec x$ **5.** $\cot x$ **7.** $\cot x$ **9.** $-\tan^2 x$ **11.** $\sin x$ **13.** 1
15. 1   **17.** $\sec x$ **19.** $\cos x$ **21.** $a\sec\theta,\ x/\sqrt{a^2+x^2}$ **23.** $\sin\theta$ **25.** $\sqrt{3}\cos\theta$

Section 6.2 (page 178)

83. Try $t=\pi$ **85.** Try $x=0$ **87.** Try $x=0$ **89.** Try $x=\frac{1}{2}\pi$ **91.** Try $x=\frac{1}{2}\pi$
93. Try $x=1$ **95.** Identity   **97.** $\pi/2, 3\pi/2$ **99.** Identity   **101.** Identity   **103.** Identity

Section 6.3 (page 184)

1. $\pi/6, 5\pi/6$

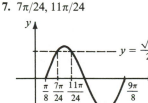

3. $\pi/3$

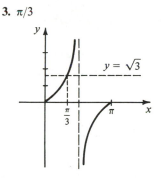

5. $\pi/3, 2\pi/3$

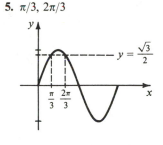

7. $7\pi/24, 11\pi/24$

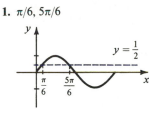

9. $15\pi/4, 23\pi/4$

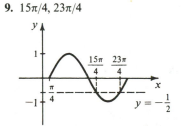

11. $7\pi/6, 11\pi/6$

13. 0   **15.** π **17.** $0, \pi, \pi/6, 5\pi/6$ **19.** no solution   **21.** $\pi/4, 5\pi/4$ **23.** no solution
25. $\pi/6, 5\pi/6, 3\pi/2$ **27.** $\pi/2$ **29.** $0, \pi/4, \pi, 5\pi/4$ **31.** $0, \pi, \pi/4, 3\pi/4, 5\pi/4, 7\pi/4$ **33.** 0
35. $\pi/2, \pi$ **37.** $\pi/4, 3\pi/4, 5\pi/4, 7\pi/4$ **39.** $45°, 135°, 225°, 315°$ **41.** $90°, 270°$
43. $90°, 150°, 210°$ **45.** $0°, 180°, 240°, 300°$ **47.** $0°, 45°, 180°, 225°$ **49.** $30°, 150°$
51. $120°, 300°$ **53.** $0.58, 2.56, 3.72, 5.70$ **55.** $0, 3.14, 3.48, 5.94$ **57.** $0.875, 2.265, 3.59, 5.83$
59. $0.67, 2.47$

Section 6.4–6.6 (page 190)

1. $x^2+y^2=1$ **3.** $x^2+y^2=1$ **5.** $x^2+ay=a^2$ **7.** $y=x,\ -1\le x\le 1$ **9.** $y=x,\ 0\le x\le 1$

11. 0

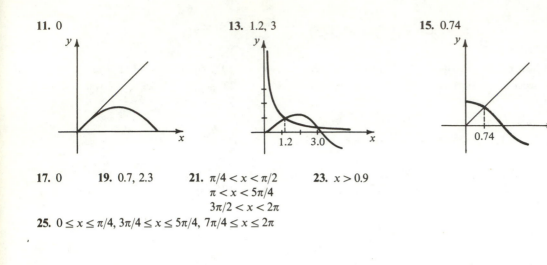

13. 1.2, 3

15. 0.74

17. 0   **19.** 0.7, 2.3   **21.** $\pi/4 < x < \pi/2$ **23.** $x > 0.9$
$\pi < x < 5\pi/4$
$3\pi/2 < x < 2\pi$

25. $0 \le x \le \pi/4,\ 3\pi/4 \le x \le 5\pi/4,\ 7\pi/4 \le x \le 2\pi$

Review Exercises for Chapter 6 (page 190)

11. $4\pi/3,\ 5\pi/3$ **13.** $7\pi/6,\ 11\pi/6,\ \pi/2$ **15.** $\pi/4,\ 3\pi/4,\ 5\pi/4,\ 7\pi/4$

17.   **19.**

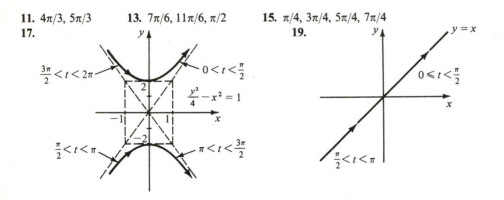

21. 0.88   **23.** $0 \le x < \pi/4,\ 5\pi/4 < x \le 2\pi$

Section 7.1–7.2 (page 198)

7. $\dfrac{\sqrt{2}}{4}(1 + \sqrt{3})$ **9.** $\dfrac{-\sqrt{2}}{4}(1 + \sqrt{3})$ **21.** $\cos 8x$ **23.** $(\sqrt{3} + \sqrt{8})/6$ **25.** 3/5

27. $(5\sqrt{8} + \sqrt{11})/18$ **29.** Amplitude $= \sqrt{2}$, phase shift $= -\pi/4$ **31.** Amplitude $= 2$, phase shift $= \pi/6$

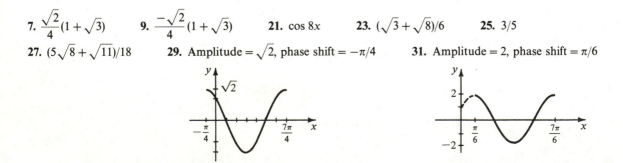

Section 7.3 (page 202)

1. $\dfrac{\sqrt{2}}{4}(1+\sqrt{3})$ **3.** $\dfrac{\sqrt{2}}{4}(\sqrt{3}+1)$ **5.** $\dfrac{\sqrt{3}-1}{\sqrt{3}+1}$ **15.** $\tan x$ **17.** $\tan(x+y+z)$

19. 36/325, 36/323 **21.** $\sqrt{2}\sin(2x+\tfrac{1}{4}\pi)$ **23.** $2\sin(\pi x+\tfrac{1}{6}\pi)$

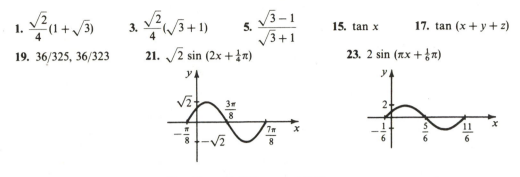

Section 7.4 (page 207)

1. $\sin 6x$ **3.** $-\cos 8x$ **5.** $\tan\tfrac{1}{3}x$

7. max $=1/2$ at $\pi/8$ **9.** max $=-2$ at $3\pi/4$, undefined at $0, \pi/2, \pi$

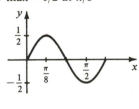

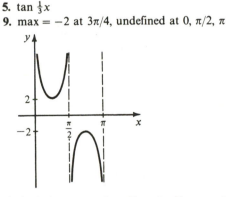

11. Undefined at $x=0, \pi/2, \pi$. Period $=\pi$ (same graph as Exercise 9). **13.** 24/25, 7/25, 24/7

15. 336/625, 527/625, 336/527 **17.** $\sqrt{2-\sqrt{2}}/2$ **19.** $-\sqrt{2}/(2+\sqrt{2})$ **21.** $\sqrt{2+\sqrt{3}}/2$

43. 1/5 **45.** $\dfrac{7\sqrt{2}}{10}$ **47.** $\pi/4, 5\pi/4$ **49.** $0, \pi, 2\pi$ **51.** $3\pi/8, 7\pi/8, 11\pi/8, 15\pi/8$

53. $\pi/6, \pi/3, 2\pi/3, 5\pi/6, 4\pi/3, 7\pi/6, 5\pi/3, 11\pi/6$

Section 7.5–7.6 (page 212)

1. $2\sin 2\theta\cos\theta$ **3.** $2\sin 5x\cos 3x$ **5.** $-2\sin 40°\sin 10°$ **7.** $2\cos 1/2\sin 1/4$

9. $1/2[\sin(\tfrac{3}{2}x)-\sin(\tfrac{1}{2}x)]$ **11.** $1/2(\cos 8x+\cos 4x)$ **19.** $\dfrac{-2\sin\frac{1}{2}(2x+\Delta x)\cos\frac{1}{2}\Delta x}{\Delta x}$

21. $0, \pi/6, \pi/2, 5\pi/6, \pi$ **23.** $0, 2\pi/5, 4\pi/5$ **25.** Amplitude $=2\cos 1$

Period $=2\pi/3$

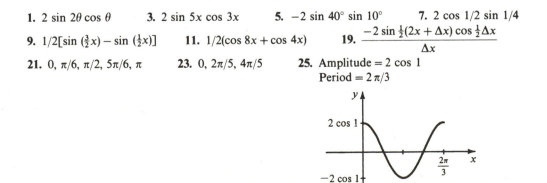

Review Exercises for Chapter 7 (page 213)

1. $(6 - 4\sqrt{5})/15$ **3.** $-24/7$ **21.** $A = 13$, per $= \pi$, p.s. $= \frac{1}{2}$ arctan $(-12/5)$
23. $A = \sqrt{2}$, per $= 2\pi/3$, p.s. $= -\pi/12$ **25.** $A = 1$, per $= 2\pi/3$, p.s. $= \pi/3$

Section 8.1 (page 219)

1. $3/2$ **3.** $5/2$ **5.** 3; None   **7.** $\{(7, 3), (9, 5), (3, 7), (5, 9)\}$ **9.** No inverse function
11. $\{(3, -2), (4, -1), (0, 0)\}$ **13.** No inverse function   **15.** $y = x + 3$ **17.** $y = (1 - x)/x$
19. No inverse function   **21.** $y = (1 + x)/(1 - x)$ **23.** a, c, and d have inverses.
25. Inverse functions   **27.** Not inverse functions   **29.** Inverse functions   **31.** Inverse functions

Section 8.2 (page 225)

1. $\pi/6$ **3.** $\pi/4$ **5.** $5\pi/6$ **7.** $2\pi/3$ **9.** 0   **11.** $-\pi/3$ **13.** $12/13$ **15.** $2/\sqrt{5}$
17. $\sqrt{15}/4$ **19.** $\sqrt{15}/8$ **21.** $1/\sqrt{82}$ **23.** $(2\sqrt{3} - 1)/(2 + \sqrt{3})$ **25.** $-4/3$
27. $x/(1 - x^2)^{1/2}$ **29.** $y\sqrt{1 - x^2} + x\sqrt{1 - y^2}$ **31.** 0.9711   **33.** No solution   **35.** 2.5722
37. -4.2556 **39.** 0.7800   **49.**   **53.** 1.211   **55.** 1.397
57. 1.407   **59.** 0.840

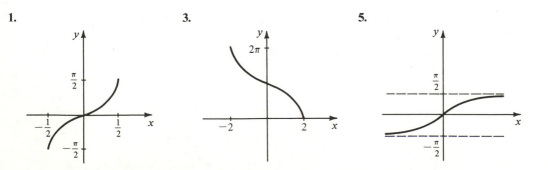

Section 8.3 (page 231)

1.   **3.**   **5.**

7.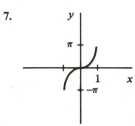

9.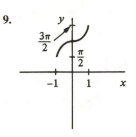

11.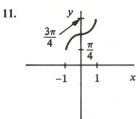

13.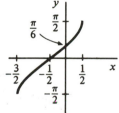

15.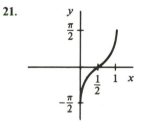

17.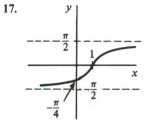

19.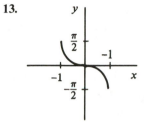

21.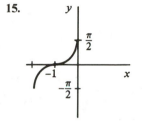

Review Exercises for Chapter 8 (page 231)

1. 4   **3.** $y = (x - 5)/2$ **5.** $y = \frac{1}{2}(5 + \arcsin(x - 3))$ **7.** $\pi/3$ **9.** $-2/\sqrt{21}$

11. $-24/25$ **13.** $-(1 + \sqrt{120})/12$ **15.** 1   **17.**   **19.**

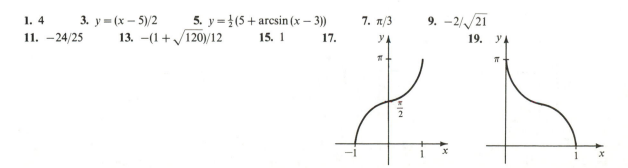

21.

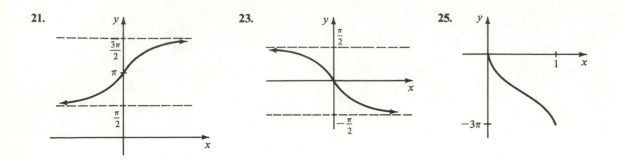

23.

25.

Section 9.1 (page 236)

1. $7 + 5i$ **3.** $-2 + 3i$ **5.** 4 **7.** $3 + i$ **9.** $-7 + 22i$ **11.** 26 **13.** $13 + 8\sqrt{3}i$

15. $18 + 24i$ **17.** $(5 - i)/2$ **19.** $(6 + 9i)/13$ **21.** $-i/5$ **23.** $\dfrac{(-4 + 3\sqrt{2}) - (12 + \sqrt{2})i}{18}$

Section 9.2 (page 238)

1. $7 + 6i$ **3.** $2 + 4i$ **5.** $-1 - 3i$ **7.** $-1 + 3i$ **9.** 10 **11.** $-1 + 8i$

13. $3 + (\sqrt{3} + 1)i$ **15.** $4i$

Section 9.3 (page 243)

1.

3.

5.

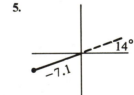

7.

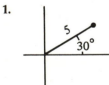

9.

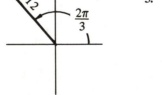

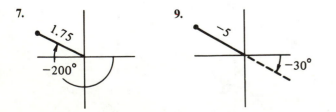

11. $r(2\cos\theta + 3\sin\theta) = 6$ **13.** $r = 4\cos\theta$ **15.** $r^2(\cos^2\theta + 4\sin^2\theta) = 4$ **17.** $r\cos^2\theta = 4\sin\theta$

19. $x^2 + y^2 = 25$ **21.** $x^2 + y^2 = 10y$ **23.** $x^2 + y^2 = (x^2 + y^2)^{1/2} + 2y$ **25.** $y^2 = 25 - 10x$

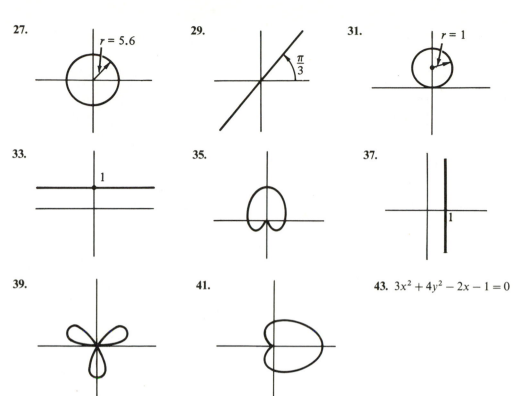

27. $r = 5.6$

29. $\frac{\pi}{3}$

31. $r = 1$

33. 1

35.

37. 1

39.

41.

43. $3x^2 + 4y^2 - 2x - 1 = 0$

Section 9.4 (page 247)

1. $2 \operatorname{cis}(-60°)$ **3.** $3 \operatorname{cis} 41°49'$ **5.** $9 \operatorname{cis} 0°$ **7.** $5 \operatorname{cis}(-53°8')$ **9.** $\sqrt{61} \operatorname{cis}(-50°12')$

11. $\sqrt{3} + i$ **13.** $\dfrac{5\sqrt{2}}{2}(-1 + i)$ **15.** $\dfrac{-3 - \sqrt{3}i}{2}$ **17.** $\dfrac{3 - 3\sqrt{3}i}{2}$ **19.** $9.397 + 3.420i$

21. $12 \operatorname{cis} 90°$ **23.** $2 \operatorname{cis} 330°$ **25.** $20 \operatorname{cis} 135°$ **27.** $10 \operatorname{cis} 23°8'$ **29.** $5 \operatorname{cis}(-60°)$

31. $2 \operatorname{cis} 7°30'$ **33.** $\dfrac{\sqrt{2}}{2} \operatorname{cis}(-75°)$ **35.** $\dfrac{4}{\sqrt{2}} \operatorname{cis}(-45°)$

Section 9.5 (page 251)

1. $8 \operatorname{cis} 0°$ **3.** $9 \operatorname{cis} 240°$ **5.** $128\sqrt{2} \operatorname{cis} 315°$ **7.** $128 \operatorname{cis} 330°$ **9.** $841 \operatorname{cis} 272°48'$

11. $1 \operatorname{cis} 0° = 1$
  $1 \operatorname{cis} 72° = .3090 + .9511i$
  $1 \operatorname{cis} 144° = -.8090 + .5878i$
  $1 \operatorname{cis} 216° = -.8090 - .5878i$
  $1 \operatorname{cis} 288° = .3090 - .9511i$

13. $1 \operatorname{cis} 22°30' = .9239 + .3827i$
  $1 \operatorname{cis} 112°30' = -.3827 + .9239i$
  $1 \operatorname{cis} 202°30' = -.9239 - .3827i$
  $1 \operatorname{cis} 292°30' = .3827 - .9239i$

15. $\sqrt[4]{2}$ cis 22°30′
     $\sqrt[4]{2}$ cis 202°30′

17. $\sqrt[6]{2}$ cis (25° + M · 60°), M = 0, 1, 2, 3, 4, 5

19. $\sqrt[4]{2}(\sqrt{3}/2 + i/2)$
     $\sqrt[4]{2}(-1/2 + i\sqrt{3}/2)$
     $\sqrt[4]{2}(-\sqrt{3}/2 - i/2)$
     $\sqrt[4]{2}(1/2 - i\sqrt{3}/2)$

23. 4 cis 60° $= 2 + 2\sqrt{3}i$
     4 cis 180° $= -4$
     4 cis 300° $= 2 - 2\sqrt{3}i$

Review Exercises for Chapter 9 (page 252)

1. $9 - 3i$ **3.** 5   **5.** $8 - 6i$ **7.** 0   **9.** 16   **11.** $(1 - 3i)/-2$ **13.** i

15. $\frac{19}{2} - \frac{9}{2}i$ **17.** 1 cis 90°   **19.** $\sqrt{2}(\text{cis } 45°)$ **21.** $\sqrt{2}(1 + i)$

23. $4(\cos 20° - i \sin 20°) = 3.8 - 1.4i$ **25.** $\frac{27}{2}(\sqrt{3} + i)$

27. $\tan \theta = 2$ **29.** $r^2 = 1/(4 \cos^2 \theta + \sin^2 \theta)$ **31.** $x^2 + y^2 = 4$

33. $x^2 + 4y = 4$ **35.** $x^2 + y^2 - 3x = 0$

37. $(x^2 + y^2)^{3/2} = 2xy$ **39.** $x^2 + y^2 = 1$

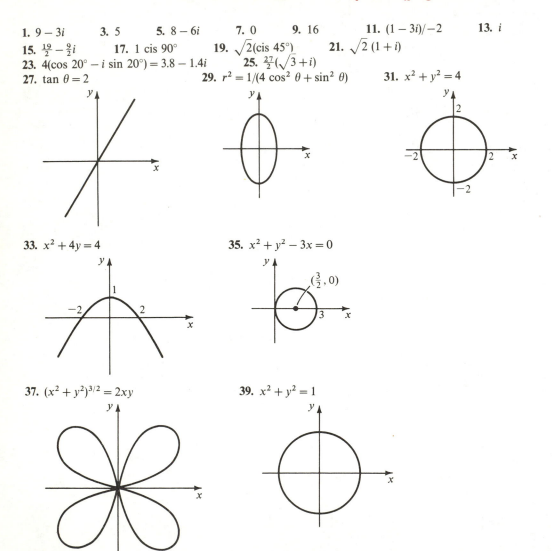

41. $2^{3/2}$ cis $3\pi/4$ **43.** 32 cis $(\pi/6)$ **45.** 10^5 cis 3.06

47. 1 cis $(\pi/5)$, 1 cis $(3\pi/5)$, 1 cis (π), 1 cis $(9\pi/5)$, 1 cis $(9\pi/5)$ **49.** $2^{1/6}$cis $(\pi/12)$, $2^{1/6}$ cis $(3\pi/4)$, $2^{1/6}$ cis $(17\pi/12)$

Section 10.1 (page 255)

1. $\log_2 x = 3$ **3.** $\log_5 M = -3$ **5.** $\log_7 L = 2$ **7.** $b = 2$ **9.** $b = 10$ **11.** $x = 10{,}000$
13. $x = 10$ **15.** $x = 4$ **17.** $x = 9$ **19.** $x = 2$ **21.** $x = a$ **23.** $x = 8$ **25.** $x = 36$
27. $x = 6$ **29.** a. 2   **31.** 2 days
            b. -3
            c. 4

Section 10.2 (page 258)

1. 9   **3.** $\frac{1}{2}$ **5.** 6   **7.** 24   **9.** 0.1761   **11.** 1.0791   **13.** 1.9542   **15.** 0.3495
17. 3.3801   **19.** -2.8539 **21.** $\log_2 x$ **23.** 0   **25.** $\log \dfrac{5t(t^2 - 4)^2}{\sqrt{t + 3}}$ **27.** $\log \dfrac{u^3}{(u + 1)^2 (u - 1)^5}$
31. The same for $x > 0$. For $x < 0$ $\log x^2$ is defined, $2 \log x$ is not.   **33.** Both are -2.
35. (a) Multiply each ordinate value by p.   (b) Move curve up $\log p$ units.   (c) Move curve p units to the left.
(d) Move curve down $\log p$ units.

Section 10.3 (page 261)

1. a. 0.7332   b. 3.7332   c. $7.7332 - 10$ **3.** a. 0.9557   b. $6.9557 - 10$ c. 5.9557
5. a. 1.2825   b. 0.1578   **7.** a. 1.3390   b. $7.4740 - 10$ **9.** a. 5.9763   b. $8.8469 - 10$
11. a. 1.820   b. 1.82   c. 0.0182   **13.** a. 15.2   b. 0.000000152   c. 1,520,000
15. a. 4.03   b. 0.00403   c. 0.403   **17.** $x = 1.3979$ **19.** $x = 0.6505$ **21.** 1.7466
23. 2.1271   **25.** 0.3243   **27.** -2.4078 or $-3 + 0.5922$ **29.** 1306.17   **31.** 5.785
33. 0.13848   **35.** 0.00000135

Section 10.4 (page 263)

1. 0.3731   **3.** $7.9811 - 10$ **5.** 0.4343   **7.** 5.8077   **9.** 3.0022   **11.** 129.1
13. 65,170,000   **15.** 0.005282   **17.** 1.778   **19.** 1,385   **21.** $x = 0.4343$ **23.** $x = 0.8686$
25. $x = 0.1681$ **27.** $N = 1{,}385$ **29.** $N = 3.162$ **31.** 0.5396   **33.** 2, Exact

Section 10.5 (page 266)

1. 0.2700   **3.** 13.68   **5.** 1.514   **7.** 33.37   **9.** 0.4034   **11.** -5740 13. 0.3709
15. 1.632   **17.** 1.652   **19.** -0.4062 **21.** 1.9 sec

Review Exercises for Chapter 10 (page 267)

1. $x = 1000$ **3.** $x = 4$ **5.** $x = 2$ **7.** $x = 10$ **9.** 4   **11.** 1.5   **13.** 21
15. $3 \log x$ **17.** $\log_2 (2x^4)$ **19.** $\log_5 (x^3/(2x - 3))$ **21.** 11800   **23.** 19.7   **25.** 24.66
27. 0.9924   **29.** 16.7

Index

319

Trigonometric Identities

(1) $\quad \sin A = \dfrac{1}{\csc A}$

(2) $\quad \cos A = \dfrac{1}{\sec A}$

(3) $\quad \tan A = \dfrac{1}{\cot A}$

(4) $\quad \tan A = \dfrac{\sin A}{\cos A}$

(5) $\quad \cot A = \dfrac{\cos A}{\sin A}$

(6) $\quad \sin^2 A + \cos^2 A = 1$

(7) $\quad 1 + \tan^2 A = \sec^2 A$

(8) $\quad 1 + \cos^2 A = \csc^2 A$

(9) $\quad \sin(A + B) = \sin A \cos B + \cos A \sin B$

(10) $\quad \sin(A - B) = \sin A \cos B - \cos A \sin B$

(11) $\quad \cos(A + B) = \cos A \cos B - \sin A \sin B$

(12) $\quad \cos(A - B) = \cos A \cos B + \sin A \sin B$

(13) $\quad \tan(A + B) = \dfrac{\tan A + \tan B}{1 - \tan A \tan B}$

(14) $\quad \tan(A - B) = \dfrac{\tan A - \tan B}{1 + \tan A \tan B}$

(15) $\quad \sin 2A = 2 \sin A \cos A$

(16) $\quad \cos 2A = \cos^2 A - \sin^2 A = 2 \cos^2 A - 1 = 1 - 2 \sin^2 A$

(17) $\quad \tan 2A = \dfrac{2 \tan A}{1 - \tan^2 A}$

(18) $\quad \sin \tfrac{1}{2}x = \pm \sqrt{(1 - \cos x)/2}$

(19) $\quad \cos \tfrac{1}{2}x = \pm \sqrt{(1 + \cos x)/2}$

(20) $\quad \tan \tfrac{1}{2}x = \dfrac{\sin x}{1 + \cos x}$

(21) $\quad \sin A + \sin B = 2 \sin \tfrac{1}{2}(A + B) \cos \tfrac{1}{2}(A - B)$

(22) $\quad \sin A - \sin B = 2 \cos \tfrac{1}{2}(A + B) \sin \tfrac{1}{2}(A - B)$

(23) $\quad \cos A + \cos B = 2 \cos \tfrac{1}{2}(A + B) \cos \tfrac{1}{2}(A - B)$

(24) $\quad \cos A - \cos B = -2 \sin \tfrac{1}{2}(A + B) \sin \tfrac{1}{2}(A - B)$

(25) $\quad \sin A \cos B = \tfrac{1}{2} \sin\{(A + B) + \sin (A - B)\}$

(26) $\quad \cos A \sin B = \tfrac{1}{2} \sin\{(A + B) - \sin (A - B)\}$

(27) $\quad \cos A \cos B = \tfrac{1}{2} \cos\{(A + B) + \cos (A - B)\}$

(28) $\quad \sin A \sin B = \tfrac{1}{2} \cos\{(A - B) - \cos (A + B)\}$